Membrane recycling

The Ciba Foundation is an international scientific and educational charity. It was established in 1947 by the Swiss chemical and pharmaceutical company of CIBA Limited—now CIBA-GEIGY Limited. The Foundation operates independently in London under English trust law.

The Ciba Foundation exists to promote international cooperation in medical and chemical research. It organizes international multidisciplinary meetings on topics that seem ready for discussion by a small group of research workers. The papers and discussions are published in the Ciba Foundation series.

The Foundation organizes many other meetings, maintains a library which is open to all graduates in science or medicine who are visiting or working in London, and provides an information service for scientists. The Ciba Foundation also functions as a centre where scientists from any part of the world may stay during working visits to London.

Membrane recycling

Ciba Foundation symposium 92

1982

Pitman

© Ciba Foundation 1982

ISBN 0 272 79656 5

Published in October 1982 by Pitman Books Ltd, London.
Distributed in North America By CIBA Pharmaceutical Company (Medical Education Administration), Summit, NJ 07006, USA

Suggested series entry for library catalogues:
Ciba Foundation symposia

Ciba Foundation symposium 92
x + 310 pages, 58 figures, 12 tables

British Library cataloguing in publication data:
Membrane recycling.—(Ciba Foundation
symposium; 92)
1. Cell membranes—Congresses
I. Evered, David II. Collins, Geralyn M.
III. Series
574.87′5 QH601

Text set in 10/12 pt Linotron 202 Times, printed and bound
in Great Britain at The Pitman Press, Bath

QH
601
.M45
1982

Contents

Symposium on: Membrane recycling, held at the Ciba Foundation, London, 19–21 January 1982
Editors: David Evered (Organizer) and Geralyn M. Collins

G. E. PALADE (*Chairman*) Problems in intracellular membrane traffic 1

Z. A. COHN and R. M. STEINMAN Phagocytosis and fluid-phase pinocytosis 15
Discussion 28

I. S. MELLMAN Endocytosis, membrane recycling and Fc receptor function 35
Discussion 51

A. HELENIUS and M. MARSH Endocytosis of enveloped animal viruses 59
Discussion 69

J. L. GOLDSTEIN, R. G. W. ANDERSON and M. S. BROWN Receptor-mediated endocytosis and the cellular uptake of low density lipoprotein 77
Discussion 89

P. CUATRECASAS Epidermal growth factor: uptake and fate 96
Discussion 104

GENERAL DISCUSSION I Receptor-mediated endocytosis of asialoglycoproteins in the hepatocyte 109, Recycling of the transferrin receptor in HeLa cells 115

J. E. ROTHMAN The Golgi apparatus: roles for distinct 'cis' and 'trans' compartments 120
Discussion 130

S. KORNFELD, M. L. REITMAN, A. VARKI, D. GOLDBERG and C. A. GABEL Steps in the phosphorylation of the high-mannose oligosaccharides of lysosomal enzymes 138
Discussion 148

M. G. FARQUHAR Membrane recycling in secretory cells: pathway to the Golgi complex 157
Discussion 174

M. J. RINDLER, I. E. IVANOV, E. J. RODRIGUEZ-BOULAN and D. D. SABATINI Biogenesis of epithelial cell plasma membranes 184
Discussion 202

R. RODEWALD and D. R. ABRAHAMSON Receptor-mediated transport of IgG across the intestinal epithelium of the neonatal rat 209
Discussion 226

GENERAL DISCUSSION II Development of a cell line with a marked proliferation of crystalloid endoplasmic reticulum 233, Early events in the receptor-modulated endocytosis of epidermal growth factor and α_2-macroglobulin–protease complexes 239 Coated vesicles—variations in morphology and function 242

B. M. F. PEARSE Structure of coated pits and vesicles 246
Discussion 257

M. S. BRETSCHER Endocytosis, the sorting problem, and cell locomotion in fibroblasts 266
Discussion 272

FINAL GENERAL DISCUSSION Membrane recycling under experimental control 282, Crude export from ER to Golgi, and the role of coated vesicles 287

G. E. PALADE Chairman's closing remarks 293

Index of contributors 298

Subject index 299

Participants

P. F. BAKER F.R.S.   Department of Physiology, King's College London, Strand, London WC2R 2LS, UK

A. D. BLEST   Department of Neurobiology, Research School of Biological Sciences, Box 475 PO, Canberra City, ACT 2601, Australia

D. BRANTON   Cell and Developmental Biology, The Biological Laboratories, Harvard University, Cambridge, Massachusetts 02138, USA

M. S. BRETSCHER   MRC Laboratory of Molecular Biology, Medical Research Council Centre, University Medical School, Hills Road, Cambridge CB2 2QH, UK

Z. A. COHN   Cellular Physiology and Immunology Laboratory, The Rockefeller University, 1230 York Avenue, New York, NY 10021, USA

P. CUATRECASAS   Department of Molecular Biology, Wellcome Research Laboratories, Burroughs Wellcome Co., 3030 Cornwallis Road, Research Triangle Park, North Carolina 27709, USA

R. T. DEAN   Department of Applied Biology, School of Biological Sciences, Brunel University, Uxbridge, Middlesex UB8 3PH, UK

M. G. FARQUHAR*   Section of Cell Biology, Yale University School of Medicine, 333 Cedar Street, P.O. Box 3333, New Haven, Connecticut 06510, USA

M. J. GEISOW   Biophysics Division, National Institute for Medical Research, The Ridgeway, Mill Hill, London NW7 1AA, UK

J. L. GOLDSTEIN   Department of Molecular Genetics, University of Texas Health Sciences Center at Dallas, 5323 Harry Hines Boulevard, Dallas, Texas 75235, USA

* Unable to attend the symposium

A. HELENIUS Section of Cell Biology, Yale University School of Medicine, 333 Cedar Street, P.O. Box 3333, New Haven, Connecticut 06510, USA

C. R. HOPKINS Department of Histology and Cell Biology, The Medical School, University of Liverpool, P.O. Box 147, Liverpool L69 3BX, UK

A. L. HUBBARD Department of Cell Biology and Anatomy, Johns Hopkins Medical School, 725 North Wolfe Street, Baltimore, Maryland 21205, USA

G. W. JOURDIAN Rackham Arthritis Research Unit, School of Medicine, University of Michigan, Ann Arbor, Michigan 48109, USA

D. E. KNIGHT Department of Physiology, King's College London, Strand, London WC2R 2LS, UK

S. KORNFELD Division of Hematology-Oncology, Washington University School of Medicine, Box 8125, 660 South Euclid, St Louis, Missouri 63110, USA

J. MELDOLESI Facolta di Medicina e Chirurgia, Istituto di Farmacologia, Universita Degli Studi, Via Vanvitelli 32, 20129 Milano, Italy

I. S. MELLMAN Section of Cell Biology, Yale University School of Medicine, 333 Cedar Street, P.O. Box 3333, New Haven, Connecticut 06510, USA

G. E. PALADE Section of Cell Biology, Yale University School of Medicine, 333 Cedar Street, P.O. Box 3333, New Haven, Connecticut 06510, USA

B. M. F. PEARSE MRC Laboratory of Molecular Biology, Medical Research Council Centre, University Medical School, Hills Road, Cambridge CB2 2QH, UK

M. RAFF Department of Zoology, University College London, Gower Street, London WC1E 6BT, UK

R. RODEWALD Institut de Biochimie, Université de Lausanne, CH-1066 Epalinges, Switzerland. *Permanent address*: Department of Biology, Gilmer Hall, University of Virginia, Charlottesville, Virginia 22901, USA

J. E. ROTHMAN Department of Biochemistry, Stanford University Medical Center, Stanford, California 94305, USA

D. D. SABATINI Department of Cell Biology, New York University Medical Center, New York, NY 10016, USA

Y.-J. SCHNEIDER International Institute of Molecular and Cellular Pathology, Université Catholique de Louvain 7539, 75 avenue Hippocrate, B-1200 Bruxelles, Belgium

W. S. SLY Department of Biochemistry, Stanford University Medical Center, Stanford, California 94305, USA. *Permanent address*: Division of Medical Genetics, Department of Pediatrics, Washington University School of Medicine, St Louis, Missouri 63110, USA

C. C. WIDNELL Department of Anatomy and Cell Biology, School of Medicine, University of Pittsburgh, Pittsburgh, Pennsylvania 15261, USA

Problems in intracellular membrane traffic

GEORGE E. PALADE

Section of Cell Biology, Yale University School of Medicine, 333 Cedar Street, P.O. Box 3333, New Haven, Connecticut 06510, USA

Abstract Eukaryotic cells operate an extensive and well regulated traffic of membrane-bound vesicles to: (a) transport intracellularly, and eventually discharge by exocytosis, macromolecular products; (b) take up by endocytosis molecules and particles from the environment; (c) transport macromolecules across epithelial barriers; and (d) move membranes from their site of assembly to their final locations. Vesicular transport appears to be the equivalent of a discontinuous circulatory system in which vesicles recycle between the termini of each transport pathway, so that balanced membrane distribution is maintained among cell compartments and the cell's surface. Although the general outline of the process is reasonably clear, much remains to be learned about the number and types of pathways, the types and quantities of membranes, and the rates of vesicular movement. Since each vesicular carrier finds its specific terminus (and fuses with it), vesicular traffic is strictly controlled. By analogy with the control of intracellular protein traffic, it may be assumed that vesicular traffic is regulated by the mutual recognition of protein signals and receptors affixed, in this case, with appropriate asymmetry to the surface of the interacting membranes. Since vesicular transport operates without loss of specific chemistry and function of various cellular membranes, cells can counteract effectively the randomization of membrane proteins and lipids, which becomes possible whenever two membranes establish continuity of their fluid bilayers, and when membrane is removed from one or both termini of a recycling pathway. Specific selection of termini and prevention of randomization among membrane components are major unsolved problems in vesicular transport. Their solution in terms of molecular interactions requires further work.

Membrane specificity

A eukaryotic cell is an extensively compartmented system. Within its boundaries, which are set by the cell's membrane or plasmalemma, it accommodates a relatively large number of intracellular compartments, each of them delimited by its own membrane. At present, we can distinguish in an animal eukaryote at least 10 different types of membrane, each type being

---

1982 Membrane recycling. Pitman Books Ltd, London (Ciba Foundation symposium 92) p 1–14

identified by its specific structure, chemistry and function, as well as by the characteristic morphology of the compartment it outlines.

Chemical specificity at the subcellular level is clearly expressed by the concept of marker component (or marker enzyme) proposed by de Duve and his collaborators more than 25 years ago (see de Duve 1963). It applied originally to subcellular particles, and it implied that certain molecular components (or activities) are restricted in their distribution to specific types of subcellular structure. The concept can be extended to cellular membranes since many subcellular particles are membrane-bounded compartments, and many markers are intrinsic membrane proteins with well defined enzymic activities. At a general and not particularly stringent level, the marker concept is still valid and useful, but in detail and at critical levels it should be redefined to take into account more recent findings and developments in cell biology, including membrane recycling.

Cellular membranes are diverse in their functions—primarily as a result of specific differences among their protein components—but all have a common structural and biochemical denominator: a bimolecular film of polar lipids—the lipid bilayer (for short)—that functions as a diffusion barrier for hydrophilic solutes between the cell and its environment, as well as between intracellular compartments. The repertoire of membrane lipids is considerably more limited than that of membrane proteins; moreover, differences from one type of membrane to another are essentially quantitative: the same lipid classes are found—at different relative concentrations—in all cellular membranes; (the only exception is diphosphatidylglycerol, restricted in its distribution to the inner mitochondrial membrane).

Membrane fluidity

By now it has been convincingly established that the lipid bilayers of all cellular membranes are fluid at the temperature of the cell's immediate environment (Singer & Nicolson 1972), and that cells in general have the ability to modify the chemistry of their complex lipids so as to maintain the average transition temperature of the bilayers well below ambient temperature (Cronan & Gelmann 1975), whether the ambience is a polar sea, a hot spring or a homeothermic organism.

We now begin to understand why cells insist on maintaining membrane fluidity: fluid bilayers are prerequisites for membrane growth—by expansion of pre-existing membranes—in anticipation of cell division (Palade 1978); and bilayer fluidity is also required for effective membrane fusion–fission without lysis. This type of membrane interaction occurs at the last step in cell division and at the first step in zygote formation. Membrane fusion–fission is also required in other, 'every-minute' cell activities, such as vesicular transport of macromolecules and particles from one compartment to another, including endocytosis, exocytosis and compensatory membrane recycling. A point that deserves to be stressed is that membrane fusion–fission leads to continuity being established not only between compartments, but also between the fluid bilayers of interacting membranes.

The corollary of membrane fluidity is a rapid diffusion in the plane of the lipid bilayer of both lipid and integral membrane proteins (Schlessinger et al 1977)*, and hence a potential randomization of the chemistry of any membrane as well as of any pair of interacting membranes. And here we encounter a set of opposite requirements, a contradiction in principles, at the basic level of cellular organization. Cells insist on having membranes that are generally fluid and yet chemically specific, irrespective of the frequency and extent of their interactions. Eukaryotic cells have clearly solved—a few billion years ago—the problem posed by such contradictory demands. It is now our turn to solve it again (so to speak), by deciphering the elements of the solution.

Differentiated domains and stabilizing infrastructures

Eukaryotic cells can circumvent randomization and create differentiated domains within the continuity of their plasmalemma and—most probably—of other membranes. The domains are of varied but specific chemistry, transient or durable existence, varied origin and varied sizes (from < 80 nm to $> 10\ \mu$m diam). Most of those identified at present appear to arise from interactions between integral membrane proteins and an infrastructure that is generated by peripheral membrane proteins which act as selectors and stabilizing agents for the corresponding domain. This stabilizing infrastructure may involve (1) geodetic cages formed by clathrin and associated proteins, as for coated pits (Pearse 1975, 1978); or (2) fibrillar proteins of still unidentified nature (perhaps actin), as for certain discharging secretory granules (Palade 1975); or (3) more extensive structures which, in extreme cases, immobilize practically all integral membrane proteins (Elgsaeter & Branton 1974), as do spectrins, ankyrin and actin in erythrocytes. A specific stabilizing infrastructure is also the formula used by an enveloped virion in selecting out its own glycoproteins (from the plasmalemmal proteins of the

---

* Coefficients for lateral diffusion (within the plasmalemma) measured by these workers are about 9×10^{-9} cm^2 s^{-1} for lipids and about 2×10^{-10} cm^2 s^{-1} for proteins. Linear approximations of these values are 1 μm s^{-1} for lipids and 1 μm min^{-1} for proteins.

host cell), and in stabilizing the patch of cell membrane that, upon budding, will become the envelope of the virion (Compans & Klenk 1979).

Other stabilizing interactions

In principle, differentiated microdomains could also be created by interactions among integral membrane proteins within the membrane or within the lipid bilayer itself. Many such interactions have been suggested but none has been well documented. Moreover, in multicellular organisms, differentiated microdomains within the plasmalemma of a given cell can be generated by 'in phase' interactions with an adjacent partner cell. In such cases, stability is achieved by interactions within each membrane as well as between membranes, as in 'gap' (or communicating) junctions, or by multiple interactions with stabilizing infrastructures, with homologous molecules within one membrane, and with complementary molecules in the plasmalemma of the partner cells, as in the elements of junctional complexes. Some junctional elements, such as the occluding zonules of epithelial cells, create apparently uninterrupted barriers within the continuity of the plasmalemma and thereby they separate from one another relatively large plasmalemmal domains that face different extracellular compartments and differ significantly in structure, chemistry and function.

Thus, eukaryotic cells have evolved procedures to counteract randomization of membrane proteins and to generate differentiated domains of stabilized chemistry by a variety of interactions. This ability plays an important part in vesicular transport and membrane recycling, since it allows the cell to perform such operations without losing the chemical specificity of the interacting membranes. In fact, some of the differentiated microdomains already mentioned—the coated pits and the stabilized membranes of discharging secretion vacuoles (or granules)—can be recognized at present as termini of certain pathways of vesicular transport.

Vesicular transport

Eukaryotic cells make extensive use of vesicular containers to export—by *exocytosis*—macromolecules and macromolecular complexes into their environment, and to import—by *endocytosis*—macromolecules and particles, together with a variable amount of fluid, from their external medium.

Endocytosis

In endocytosis, vesicles of various sizes and at least four different types (i.e. phagocytic vacuoles, pinocytic vacuoles, plasmalemmal vesicles, or micropinocytic vesicles, and coated vesicles) transport imported molecules and particles from the cell's surface to secondary lysosomes (Silverstein et al 1977), either directly or through one or two intermediary compartments. The membrane of these vesicular carriers is eventually returned to the cell surface so that, in steady state conditions, each compartment or interacting membrane retains a relatively constant surface area. The cells therefore move vesicles along a circuit (the endocytic circuit) in which the inward arc is directly involved in importing matter, whereas the outward arc ensures that membrane is retrieved from either a lysosomal or a pre-lysosomal compartment. The extent to which the membranes of all these carriers are similar to, or different from, one another—and from the membranes of their terminal or intermediary stations—is discussed in this volume primarily by Cohn & Steinman, Mellman, Pearse and Bretscher. Membrane recycling usually refers to the second arc of the circuit and implies that the membranes of the corresponding vesicular carriers are repeatedly reutilized.

Exocytosis and intracellular transport

In *overt* secretion the export of macromolecules involves a series of intracellular compartments, namely the endoplasmic reticulum (ER), the Golgi complex, and secretion vacuoles (or granules) that eventually discharge their contents into the external medium (Palade 1975). Part of the same pathway is used to transport lysosomal hydrolases from the ER to the Golgi complex and eventually to primary and secondary lysosomes in *covert* secretion. The same pathway appears to be involved in the transport of membrane proteins (already assembled into a lipid bilayer) from the ER to the plasmalemma, and the same probably applies for membrane proteins with other destinations, such as Golgi membranes or lysosomal membranes. In such cases, the species of primary interest are the membrane proteins (rather than the content proteins) of the vesicular carriers.

Along the secretory pathway, vesicular transport operates at two junctions, namely between the ER and the Golgi complex (Jamieson & Palade 1967a,b), and between the Golgi complex and the plasmalemma or (for exocrine cells) the specialized luminal domain of the plasmalemma (Palade 1975). Small vesicles of about 50 nm diameter appear to operate at the first junction, and larger vesicles of up to 500 nm diameter, usually designated as secretion vacuoles or granules, are the carriers at the second junction. The sites at which vesicular transport operates along the secretory pathway may, however, be more numerous, since the Golgi complex apparently consists of a series of functionally distinct compartments (as discussed in this volume by Kornfeld et al and Rothman), which may also be connected to one another by vesicular carriers.

In secretory cells, transporting vesicles continuously move macromolecular products from one compartment to another along the secretory pathway. Yet, under steady state conditions, each compartment and each interacting membrane retains a relatively constant surface area, because membrane appears to be continuously recovered by the donor compartment from its receiving partner. In such cells, two circuits can be recognized at present: one between the ER and the Golgi complex, and the other between the Golgi complex and the plasmalemma. At each site, the centrifugal (outward) arc of each circuit carries secretory products, whereas the centripetal (inward) arc is involved in the recovery of container membranes (see Farquhar, this volume).

It can be assumed that vesicular carriers are also involved in the transport of lysosomal hydrolases from the Golgi complex to lysosomes (Friend & Farquhar 1967) and in the transport of pre-assembled membrane proteins from some subcompartment of the Golgi complex to different plasmalemmal domains, and perhaps to other cellular membranes. In certain epithelial cells, vesicular carriers function in the transport of macromolecules from one extracellular compartment to another, across the epithelium, in an operation described as *diacytosis* or *transcytosis*, which appears to be differentially amplified in the vascular endothelium (Palade et al 1979). Diacytosis in the intestinal epithelium is discussed in this volume by Rodewald & Abrahamson (see also p 109-115). Recycling of vesicular containers has also been postulated in diacytosis (Palade et al 1979, Simionescu et al 1981).

A eukaryotic animal cell therefore appears to operate a number of discontinuous circulatory systems, present in all cell types, but individually amplified in some as a result of cell differentiation. What is still uncertain is the extent to which these multiple pathways are interconnected, or use common vesicular carriers or different carriers that move along common arcs. A connection between the endocytic and exocytic pathway has been detected in secretory cells (see Farquhar, this volume) and a link between the transcytotic, endocytic and exocytic pathways apparently exists in intestinal epithelia (see Rodewald & Abrahamson, this volume).

An attempt to rationalize vesicular transport

The existence of distinct compartments connected by shuttling vesicular carriers can be rationalized in terms of a series of well established findings and reasonable postulates. Protein synthesis (for whatever final destination) is essentially centralized in a single cell compartment, i.e. the cytoplasmic matrix or cytosol*; all proteins produced for overt or covert secretion are

---

* (Only a small fraction ($\leq 2\%$) of the cell's proteins is produced in the mitochondrial matrix by mitochondrial ribosomes.)

translocated across the ER membrane into the cisternal space of the ER, the first compartment of the secretory pathway. A still unknown, but presumably large number of integral membrane proteins, destined for the plasmalemma (and presumably for other cellular membranes), is also inserted into the ER membrane (Blobel et al 1979, Sabatini et al 1982). All these proteins are modified co- or post-translocationally by complex biochemical reactions (e.g. partial proteolysis, glycosylation of a variety of types, sulphation, phosphorylation and amino acid residue modifications) specific for each compartment of the secretory pathway (i.e. ER, Golgi complex and secretion vacuoles). From an initially common duct system, i.e. the ER, content proteins are sorted out into secretory and lysosomal proteins and are eventually separated into distinct and different compartments (secretion vacuoles versus lysosomes). The sorting-out operation is probably more elaborate for membrane proteins, since each of them eventually reaches its proper destination via vesicular carriers that presumably depart from the Golgi complex. In addition to modifications incurred by individual macromolecules, the macromolecular solutes may be extensively concentrated (as in condensing vacuoles and secretion granules), or the properties of their solvent may be changed (as in lysosomes). Each of these numerous operations probably requires a specific environment, and hence a distinct compartment, to which access can be controlled. Vesicular transport may be the way of supplying macromolecular substrates to each of these operations with a minimum of disturbance to the conditions prevailing within each receiving compartment. A continuous conduit with several functionally differentiated segments (in the style of a nephron) may not provide adequate environmental separation for the series of operations mentioned above. Vesicular transport of secretory proteins is known to be vectorial (in the few cases so far studied), the receiving compartment acting as an efficient sink so that backflow or diffusion up the pathway is prevented. A similar performance may be difficult to achieve in a continuous conduit in the absence of rapid flow.

The secretory pathway resembles a modern industrial assembly line, since it is built to perform a series of operations in which each step depends on the completion of the previous one. The cellular 'assembly line' requires, however, a different environment for most—if not each—of these operations, and it therefore takes the form of a series of connectable compartments rather than of a continuous, moving band.

Problems generated by vesicular transport

Many direct observations show that eukaryotic cells have evolved efficient procedures for the control of vesicular transport. Notwithstanding the large number of diverse carriers, the multiplicity and complexity of circuits, the crowding and the complex geometry of the cell's interior, each carrier appears to recognize its proper destination and to interact with its proper partner, to which it eventually delivers either its content or its membrane. So far no obvious 'mistakes' have been recorded under normal operational conditions.

Traffic control

Vesicular traffic is obviously carefully controlled but the means by which control is achieved is still completely unknown. We can draw an analogy with the extensively studied mechanisms that control the traffic of newly synthesized proteins from the cytosol to a multiplicity of final destinations (Walter et al 1981, Walter & Blobel 1981). Thus, we may assume that the mutual recognition of a signal (affixed to a vesicular carrier) by a signal receptor (borne by the membrane of its compartmental partner) leads to a series of interactions that eventually result in binding, followed—in this case—by membrane fusion–fission between the vesicular carrier and its appropriate receiving compartment. The assumption implies that signals as well as receptors are membrane-bound, that their recognition sites are exposed on the cytoplasmic aspect of the corresponding membranes, and that each circuit is provided with its specific set of pilots (or signals) and ports of entry (or signal receptors).

Because many homologous compartments, such as mitochondria, lysosomes and secretion granules, often undergo fusion–fission (restricted specifically to members of the same set of compartments), it can be further postulated that in vesicular transport each interacting membrane carries both the signal (**S**) and its receptor (**R**) as reciprocal doublets, so that **S–R** recognizes and is recognized by **R–S**.

Control of membrane specificity

The second major problem connected with vesicular transport concerns the retention of chemical specificity by all the membranes involved in a given circuit, i.e. usually the membranes of the two compartments that function as donor and receiving termini and the membrane of the shuttling vesicular connector (Palade 1976).

Unless prevented by appropriate means, loss of specificity becomes unavoidable at two different steps for two different reasons. (1) In the process of fusion–fission of the membrane of a vesicular carrier with that of a receiving compartment, continuity between the two corresponding bilayers is established and rapid lateral diffusion (in the plane of the fused membranes) is to be expected. It should lead to intermixing and eventual randomization of the components of the interacting membranes, unless one or the other membrane (or both) has its chemistry stabilized for the duration of bilayer continuity. (2) When excess membrane is removed from the receiving compartment to be returned to the donor compartment, so as to maintain the balanced membrane distribution that characterizes the steady-state operation of the circuit, randomization again becomes possible (this time for a different reason), unless prevented. If the cell removes from the receiving terminus, and returns to the donor terminus, a membrane patch equivalent in surface area, but not in chemistry, to the membrane of the vesicular carrier already received, balanced distribution of membranes can be maintained but randomization of chemistry and function will unavoidably ensue throughout the circuit. To prevent randomization, removal of excess membrane must follow a non-random procedure which ensures that the membrane removed is the same in surface area and chemistry as the membrane received (Palade 1976). But for a removal of this type to be possible, the membrane of the incoming vesicle must be stabilized throughout its residence in the membrane of the receiving terminus. Stabilization of the incoming vesicle membrane could be achieved by some of the means already discussed, in conjunction with differentiated microdomains within the plasmalemma. In fact the same means of stabilization (e.g. interactions with a stabilizing infrastructure) could prevent randomization at the time of bilayer fusion and make possible the subsequent non-random removal of the carrier membrane.

A priori, stabilization would be required at both the donor and receiving termini if the vesicular carrier had a membrane of specific chemistry distinct from that of each of the termini. But stabilization would be needed at one terminus only if the membrane of the carrier were identical to that of the other terminus. Some evidence already suggests that cells use both alternatives. The membranes of some endocytic vacuoles or vesicles appear to be qualitatively similar to the plasmalemma (see Mellman, this volume), and vesicles that transport secretory proteins and lipoproteins from the ER to the Golgi complex in hepatocytes appear to have a number of antigens and enzymic activities in common with the ER membrane (Ito & Palade 1978). But, in receptor-mediated endocytosis a special (coated) vesicle, distinct from the rest of the plasmalemma is involved. (Pearse & Bretscher 1981, Goldstein et al, this volume). Secretion granule membranes, isolated from the pancreas, parotid (Castle et al 1975) or adrenal (Winkler 1971), appear to be much simpler membranes than the plasmalemma, and possibly the Golgi membranes, but in all these examples the evidence is incomplete, in that a reliable comparison of the membrane chemistry of the carrier and the relevant domains at both termini is not yet on record.

In principle, cells have the option of stabilizing either the membrane of the vesicular carriers (as discussed above) or the membrane of one (or both) termini; and the available evidence suggests that they use both options. Vascular endothelial cells are remarkable for the diversity and surface density of their differentiated plasmalemmal microdomains (Simionescu et al 1981). Plasmalemmal vesicles involved in diacytosis (or transcytosis) are part of the local repertoire of differentiated microdomains. Studies indicate so far that this vesicle membrane is different from the plasmalemma proper and that the latter (rather than the vesicles themselves) is backed by a fibrillar infrastructure that presumably has a stabilizing function. A similar situation is encountered in neuromuscular junctions (Ceccarelli et al 1979) and presumably in synapses. Such an infrastructure may also define docking sites for incoming and departing vesicles, in addition to stabilizing the chemistry of the plasmalemma proper. The same alternative may apply for the membranes of Golgi cisternae which are kept in situ in characteristic alignment*. Interactions of this type may explain why Golgi cisternae can be isolated as intact stacks by certain fractionation procedures (see Farquhar & Palade 1981).

A second look at stabilizing infrastructures

At present the outstanding example of stabilizing infrastructure is provided by the budding of enveloped virions. A structure that consists of matrix or nucleocapsid proteins is found immediately under the plasmalemma in phase with a patch of membrane within which viral glycoproteins are concentrated, whereas host membrane proteins are usually excluded (Compans & Klenk 1979). Viral glycoproteins are transmembrane proteins with a relatively small endodomain. It is assumed, but not proven, that the budding process is controlled (in part) by interactions between these endodomains and elements of the stabilizing infrastructure. These interactions probably ensure selection of viral glycoproteins, and exclusion of host proteins; in addition, they apparently control the direction of budding and the size and shape of viral particles (see Rindler et al, this volume).

Except for an inverted geometry—'budding' proceeds towards the cytosol rather than towards the external medium—the geodetic cages formed by clathrin (and associated proteins) have many properties in common with viral matrix and nucleocapsid proteins. The cage may act as both selector and stabilizer, and may also control the direction of 'budding' as well as the size and shape of the coated vesicles finally produced. Selection of membrane

* (Fibrillar structures have been detected [see Farquhar & Palade 1981] in between Golgi cisternae in a few cases.)

proteins undoubtedly occurs but the degree of selection—and conversely of exclusion—is still unknown for proteins as well as for lipids (Pearse & Bretscher 1981; Pearse, this volume; Bretscher, this volume). The stability of the interactions between the clathrin infrastructure and the integral proteins of the membrane appears to vary. In the rather well documented case of coated pits involved in receptor-mediated endocytosis, for instance, some receptors (such as LDL receptors) appear to be permanently located in the pits, while others (such as the receptors for epidermal growth factor, insulin, transferrin, asialoglycoproteins and different types of enveloped virus) are collected into coated pits only after interacting with their respective ligands before being endocytosed (see contributions by Goldstein et al, Cuatrecasas, Helenius and Hubbard in this volume).

The stabilizing infrastructures of animal eukaryotes thus seem to be more versatile than their viral counterparts. They may allow the membrane components of an incoming vesicle to disperse into the membrane matrix of a terminus, and they may depend on changes in affinity (induced by ligand binding or clustering) in order to recapture these components before non-random removal from that terminus is effected, as seems to happen for many plasmalemmal receptors.

Although coated pits and coated vesicles have attracted attention mostly as agents involved in the early steps of receptor-mediated endocytosis, it should be stressed that similar structures are found to be associated with many (but not all) intracellular compartments. Equivalents of coated pits are detected on endosomes (see Helenius, this volume), Golgi cisternae, Golgi condensing vacuoles (Jamieson & Palade 1967a,b, Farquhar & Palade 1981) and, in a simplified version, in the transitional elements of the ER (Palade 1975, Palade & Fletcher 1977). Hence, clathrin and associated proteins may function as selectors and stabilizers of vesicular carrier membranes at the termini of a number of vesicular transport circuits.

A certain amount of information is currently available about clathrin and its associated proteins (Pearse & Bretscher 1981, Pearse 1982) as well as about conditions involved in their co-polymerization into geodetic cages (Unanue et al 1981). But, except for the presence of the receptors already mentioned, we know much less about the membrane components of coated pits and coated vesicles, and we depend exclusively on assumptions and speculations when considering the nature of the transmembrane interactions involved in the assembly of these structures.

As already mentioned, other types of stabilizing infrastructure and interaction should be considered in relation to the control of membrane specificity in vesicular transport.

A wide-open question concerns the mechanisms operating in situ to assemble and dismantle these stabilizing infrastructures and the possibility that elements of the locomotor apparatus of the cell play some part in the movement of coated vesicles to their appropriate destinations.

Perspectives

Membrane recycling, viewed as an integral part of vesicular transport, will substantially affect our thinking and our experimental approaches over the next few years. We can expect to see different types of cytoplasmic free vesicle and coated vesicle, similar perhaps in appearance but different in membrane chemistry, each type representing a specific carrier for one of the five or six circuits that we have identified so far in animal cells. Lumping together all smooth vesicles under the term smooth ER (and, hence, smooth microsomes) and all coated vesicles under another single heading will become a thing of the past. Transient residence of stabilized vesicular carriers in one or both termini of a given circuit may lead to a non-coincidence of structural and chemical boundaries among cellular compartments. We assume, for instance, that a Golgi cisterna is a typical, unitary, structural element of the Golgi complex; chemically, it may turn out to be a mosaic. The same may apply to the plasmalemma: a coated pit may be a carrier vesicle in transient residence in the cell membrane; chemically that pit may belong to a different intracellular membrane system*. In other words, we may be obliged to reorder our traditional categories of cellular compartments and membranes.

For the same reasons, the marker concept—used in guiding present-day fractionation procedures and in interpreting their results—will have to be redefined, because the transient residence of a vesicular carrier of specific chemistry in the membrane of a terminal compartment of different chemistry may introduce an apparently 'extraneous' marker in the cell fraction that represents the terminal compartment. That 'extraneous' marker will be found there on account of function, not necessarily on account of particulate contamination, as we usually assume at present. In addition, we shall need more refined separation procedures for subcellular components, based whenever possible on specific chemical interactions (see Ito & Palade 1978, Merisko et al 1982) rather than on general non-specific physical parameters, to be able to sort out the new categories of vesicle involved in vesicular transport and recycling.

An awareness of the diversity and extent of vesicular transport should prompt us to work towards a better understanding of the molecular mechan-

---

* A different interpretation is proposed, however, by Pastan and his collaborators (see Willingham & Pastan 1980); they assume that coated pits are stable plasmalemmal structures which generate uncoated vesicles that transport ligands to 'receptosomes'.

isms that control vesicular traffic and membrane specificity, not to mention those involved in membrane fusion–fission.

In lieu of conclusion

This paper is primarily a logical construct based on premises provided by established facts and currently recognized limitations within which eukaryotic animal cells must function. In the ultimate analysis, these facts and limitations indicate that membrane traffic and membrane specificity must be controlled. Yet this construct extends far beyond established facts: it hangs in the unknown, supported only by untested logical extrapolations. Therefore, some of its aspects may prove to be wrong, but others may have the merit of being reasonable working hypotheses that, in due time, could be put to experimental test.

REFERENCES

Blobel G, Walter P, Chang CH, Goldman BM, Erickson AH, Lingappa VR 1979 Translocation of proteins across membranes: the signal hypothesis and beyond. Symp Soc Exp Biol 33:9-36

Castle JD, Jamieson JD, Palade GE 1975 Secretion granules of the rabbit parotid gland. Isolation, subfractionation and characterization of the membrane and content subfractions. J Cell Biol 64:182-210

Ceccarelli B, Grohoraz F, Hurlbut WP 1979 Freeze fracture studies of frog neuromuscular junctions during intense release of neurotransmitter. J Cell Biol 81:163-177

Compans RW, Klenk H-D 1979 Viral membranes. In: Fraenkelconrat H, Wagner RR (eds) Comprehensive Virology. Plenum Press, New York, vol 13:293-407

Cronan JE, Gelmann EP 1975 Physical properties of membrane lipids: biological relevance and regulation. Bacteriol Rev 39:232-256

de Duve C 1963 The separation and characterization of subcellular particles. Harvey Lect 59:49-87

Elgsaeter A, Branton D 1974 Intramembrane particle aggregation in erythrocyte ghosts. I: The effects of protein removal. J Cell Biol 63:1018-1030

Farquhar MG, Palade GE 1981 The Golgi apparatus (complex)–(1954–1981)–from artifact to center stage. J Cell Biol 91:77s-103s

Friend DS, Farquhar MG 1967 Functions of coated vesicles during protein absorption in the rat vas deferens. J Cell Biol 35:357-376

Ito A, Palade GE 1978 Presence of NADPH-cytochrome P450 reductase in rat liver Golgi membranes: evidence obtained by an immunoadsorption method. J Cell Biol 79:590-597

Jamieson JD, Palade GE 1967a Intracellular transport of secretory proteins in the pancreatic exocrine cell. I: Role of the peripheral elements of the Golgi complex. J Cell Biol 34:577-596

Jamieson JD, Palade GE 1967b Intracellular transport of secretory proteins in the pancreatic exocrine cell. II: Transport to condensing vacuoles and zymogen granules. J Cell Biol 34:597-615

Merisko EM, Farquhar MG, Palade GE 1982 Coated vesicle isolation by immunoadsorption on *Staphylococcus aureus* cells. J Cell Biol 92:846-857

Palade GE 1975 Intracellular aspects of protein secretion. Science (Wash DC) 189:347-358

Palade GE 1976 Interactions among cellular membranes; problems and perspectives. In: Passino R (ed) Biological and artificial membranes and desalination of water: proceedings. Elsevier, Amsterdam (Pontif Acad Sci Scr Varia) vol 40:85-97

Palade GE 1978 Membrane biogenesis. In: Solomon AK, Karnovsky M (eds) Molecular specialization and symmetry in membrane function. Harvard University Press, Cambridge, p 3-30

Palade GE, Fletcher M 1977 Reversible changes in the morphology of the Golgi complex induced by the arrest of secretory transport. J Cell Biol 75:371a

Palade GE, Simionescu M, Simionescu N 1979 Structural aspects of the permeability of the microvascular endothelium. Act Physiol Scand Suppl 463:11-32

Pearse BM 1975 Coated vesicles from pig brain: purification and chemical characterization. J Mol Biol 97:93-98

Pearse BM 1978 On the structural and functional components of coated vesicles. J Mol Biol 126:803-812

Pearse BMF 1982 Coated vesicles from human placenta carry ferritin, transferrin and immunoglobulin G. Proc Natl Acad Sci USA 79:451-455

Pearse BMF, Bretscher MS 1981 Membrane recycling by coated vesicles. Annu Rev Biochem 50:85-101

Sabatini DD, Kreibich G, Morimoto T, Adesnik M 1982 Mechanisms for the incorporation of proteins in membranes and organelles. J Cell Biol 92:1-22

Schlessinger J, Axelrod D, Koppel DE, Webb WW, Elson EL 1977 Lateral transport of a lipid probe and labeled proteins on a cell membrane. Science (Wash DC) 195:307-309

Silverstein SC, Steinman RM, Cohn ZA 1977 Endocytosis. Annu Rev Biochem 46:669-721

Simionescu N, Simionescu M, Palade GE 1981 Differentiated microdomains on the luminal side of the capillary endothelium. I: Preferential distribution of anionic sites. J Cell Biol 90:605-613

Singer SJ, Nicolson GL 1972 The fluid mosaic model of the structure of cell membranes. Science (Wash DC) 175:720-731

Unanue ER, Ungewickell E, Branton D 1981 The binding of clathrin triskelions to membranes from coated vesicles. Cell 26:439-446

Walter P, Blobel G 1981 Translocation of proteins across the endoplasmic reticulum. II: Signal recognition protein (SRP) mediates the selective binding to microsomal membranes of in vitro assembled polysomes synthesizing secretory proteins. J Cell Biol 91:557-561

Walter P, Ibrahimi I, Blobel G 1981 Translocation of proteins across the endoplasmic reticulum. I: Signal recognition protein (SRP) binds to in vitro assembled polysomes synthesizing secretory proteins. J Cell Biol 91:545-556

Willingham MC, Pastan I 1980 The receptosome: an intermediate organelle of receptor-mediated endocytosis in cultured fibroblasts. Cell 21:67-77

Winkler H 1976 The composition of adrenal chromaffin granules: an assessment of controversial results. Neuroscience 1:65-80

Phagocytosis and fluid-phase pinocytosis

ZANVIL A. COHN and RALPH M. STEINMAN

Cellular Physiology and Immunology Laboratory, The Rockefeller University, 1230 York Avenue, New York, NY 10021, USA

Abstract The generation, flow, directionality and fusion of phagocytic and fluid-phase pinocytic vesicles in cultured macrophages and fibroblasts are reviewed. Specific plasma membrane (PM) receptors, receptor mobility, contractile cytoplasmic elements and lipid composition of the PM serve to regulate the flow of large phagosomes into the perinuclear zone. Fluid-phase vesicles are constitutively generated and carry large quantities of PM, fluid and solutes into the cytoplasm. Quantitative information is cited on the rates of vesicular generation, fusion with other members of the vacuolar system, fluid and solute uptake, and digestion and solute release. The nature and composition of fluid-phase vesicles, phagocytic vacuoles and PM are compared. Once interiorized, PM and its component polypeptides rapidly cycle back to the cell surface. The flow rates of both the centrifugal and the centripetal compartments as well as the fate of a minor degradation pool are illustrated and compared to the turnover of individual membrane polypeptides. Implications of membrane flow for cell shape, motility and new PM insertion are discussed.

We wish to discuss the bi-directional flow of solutes and membrane into and out of cells in culture. For this purpose we must first delineate the cellular compartments involved, the nature of the solutes, the composition of membranes and the motive forces used by the macrophage and fibroblast to interiorize membrane, to move the newly formed endocytic vesicle (endosome) into the cytoplasm, to fuse it with other members of the vacuolar apparatus and, finally, to dispose of solutes and to recycle membrane constituents. For this purpose we shall use results from structural, cytochemical and biochemical techniques to arrive at quantitative estimates.

Vacuolar apparatus

Most of the information that we have accrued over the past decade concerns

---

1982 Membrane recycling. Pitman Books Ltd, London (Ciba Foundation symposium 92) p 15–34

the macrophage, which will serve as the focus of our paper, but we shall also present corroborative and comparative information relating to the fibroblast.

The macrophage is a rather plastic cell which, through its endocytic activity, can modify the relative sizes of its vacuolar compartments. In steady-state conditions of culture in 10–20% fetal bovine serum, the mouse peritoneal cell maintains a stable surface area as it remains adherent to substrates of plastic or glass, where there is an intermediate layer of serum and cellular-derived proteins on which the macrophage resides. In addition to the plasma or limiting membrane, the vacuolar apparatus consists of the variety of vesicles and vacuoles in transit through the cytoplasm, many of which are of endocytic origin. About 20% of these are visible by phase-contrast optics in the living cell (Cohn 1966). Apparently, not all vesicles seen in the living cell flow inwards, but follow a centrifugal path towards the plasma membrane.

We must also consider a third form of vesicle in our discussion of the vacuolar apparatus. This is the primary lysosome derived from the Golgi–endoplasmic reticulum complex (GERL). These vesicles are usually indistinguishable from small pinocytic vesicles except for their more central distribution and their content of acid hydrolases (Cohn et al 1966). To our knowledge the number and flow rates of primary lysosomes have not been quantitated in any cell type. Although the major flow of vesicles bounded by membranes and originating from GERL is directed toward the lysosomal compartment, small amounts of cationized ferritin, as in other cell types, have been reported to accumulate in presumed Golgi saccules.

Table 1 gives the major dimension of the components of the macrophage vacuolar apparatus. The plasma membrane area, 825 μm^2, of macrophages as

TABLE 1 Steady-state dimensions of pinocytic vesicles and secondary lysosomes in macrophages

	Pinocytic vesicles	*Secondary lysosomes*
Number per cell	240	1100
Diameter (μm)	0.202	0.136
Area (μm^2)	0.196	0.109
Volume (μm^3)	0.0143	0.0074
% Total volume	2.5	2.5
% Total surface area	12.5	18.0
Total MØ volume = 395 (μm^3)		
Total plasma membrane surface area = 825 (μm^2)		

derived by stereological techniques (Steinman et al 1976), is about 3.3-fold higher than one would have estimated from calculations based on a simple sphere. The difference in surface area is the result of many folds of redundant plasma membrane (PM), particularly in circulating macrophages or in cells washed from the peritoneal cavity. As the cell adheres to substrates, this preformed membrane is utilized for the spreading process (Bianco et al 1976). Similarly, if large numbers of non-digestible polystyrene latex particles are ingested, thus trapping PM around them, the cell surface becomes progressively smoother as the number of ingested particles increases, and before the overall cell shape changes markedly. Eventually, if enough particles are ingested, cell processes are retracted and rounding of the cell takes place.

The two other significant members of the vacuolar system are largely derived from the PM: the populations of lucent pinocytic vesicles and dense secondary lysosomes. Table 1 gives surface area and volume measurements of macrophages based on stereological techniques and on the use of horseradish peroxidase (HRP) as the fluid-phase solute (Steinman et al 1976). At any given time about 250 vesicles are present in the macrophage and are in transit to the centrosphere region. Fusion occurs frequently, resulting in larger vesicles. The mean pinocytic vesicle diameter of 0.202 μm was measured by using short HRP pulses before extensive fusions. Pinocytic vesicles are present throughout the cytoplasm and originate in an apparently random fashion from all segments of the membrane. The average secondary lysosome is somewhat smaller in size although this compartment may show a wide range of sizes and composition. In these steady-state conditions the combined volumes of the pinocytic vesicle and secondary lysosome compartments is 5.0% of the total macrophage volume whereas their combined fractional surface area is 30% of macrophage surface area. The size of the secondary lysosome compartment can easily be altered, and the storage of a wide variety of poorly or non-digestible products leads to more than a two-fold increase in total lysosomal volume (Cohn & Ehrenreich 1969).

The stereological information obtained by grid intersections and by point counting from sectioned material is in general agreement with results from compartmental analysis by other techniques. For example, Werb & Cohn (1971a), in studies of membrane exchange of cholesterol noted that 70% of the cholesterol was in a rapidly exchanging compartment and 30% was in a more slowly exchanging pool. These calculations, based on a two-compartment model, were extended to show that the larger pool represented PM and the smaller one secondary lysosomes (Werb & Cohn 1971b). In contrast, alveolar macrophages, with their large number of lysosomes, had a slow pool which was enlarged to 54% of the total.

Analogous information was obtained by Edelson & Cohn (1976a) when they examined the cellular compartments of 5′-nucleotidase (EC 3.1.3.5) activity in peritoneal macrophages. This is an ectoenzyme expressed on the cell surface and inhibited with the diazonium salt of sulphanilic acid (DASA)—an impermeant reagent. When peritoneal macrophages were exposed to appropriate concentrations of DASA, about 70% of the 5′-nucleotidase activity was lost, leaving the remaining 30% in an intracellular pool (Edelson & Cohn 1976b). The majority of the intracellular enzyme was considered to be in an endocytic compartment and was probably distributed in both pinocytic vesicles and secondary lysosomes. Immediately after the uptake of a load of latex particles this pool expanded by as much as three-fold. Thereafter, the intracellular pool of enzyme surrounding the latex particles was inactivated within a few hours. This is in keeping with the rapid inactivation of 5′-nucleotidase when PM is interiorized about latex particles (Werb & Cohn 1972).

Membrane interiorization—quantitative aspects

We shall review and compare here the uptake of soluble and particulate molecules by macrophage, although a discussion of the many factors that influence these two forms of endocytosis is beyond our present scope. Table 2

TABLE 2 Comparison of fluid-phase pinocytosis with phagocytosis in macrophages

	Pinocytosis	*Phagocytosis*
Vesicle diameter (μm)	0.1–1.0 (mean 0.2)	approx 1–>10
Continuous process	Yes	No
Activation energy		
(kcal mol^{-1})	18	54
19:0 enriched cell[a]	17	90
Temperature range (°C)	2–>38	17–20
Inhibited by cytochalasin	No	Yes
Vesicle fusion inhibited by concanavalin A	Yes	No
Vesicle fusion inhibited by macroanions	No	Yes

[a] 19 carbons: 0 unsaturated bonds

lists some of the more obvious differences between fluid-phase pinocytosis and phagocytosis. Perhaps the most significant is the continuous and constitutive nature of pinocytosis. Pinocytic vesicle formation is set at a fairly constant rate, and we are not aware of any natural inducers in the extracellular environment. Nevertheless, the addition of certain lectins, (e.g. Concanavalin A) will lead to a short-lived increase in solute uptake (Edelson & Cohn 1974a,b). In contrast, the uptake of particulates is a discontinuous process in which the solid is surrounded by a tightly fitting sleeve of membrane, the size of which is governed by particle size. According to the temperature coefficients and calculated activation energies, the Fc receptor-mediated uptake of erythrocytes requires more energy than does the pinocytosis of HRP. Interestingly, manipulations that decrease the fluidity of the PM, such as its enrichment in saturated fatty acids (19 carbons : 0 unsaturated bonds) (Mahoney et al 1977, 1980), decrease the extent of both pinocytosis and phagocytosis. However, only for phagocytosis is the activation energy increased. It is unclear whether this implies that membrane domains are more fluid, and thus that they allow residual pinocytic activity to continue.

Another rather striking difference is the ability of the cytochalasins to inhibit phagocytosis without influencing pinocytosis. This is in keeping with the structural studies showing the aggregation of sub-plasmalemmal actin filaments beneath attached particles. Presumably actin and perhaps other proteins of the contractile network are necessary for large-particle ingestion and are not required for pinocytosis. Clathrin baskets have been associated with both latex phagosomes and coated vesicles of macrophages (personal communication, J. Aggelar and Z. Werb).

Several other membrane perturbants influence the two forms of endocytosis in dissimilar ways. For example, tetravalent Concanavalin A, which cross-links exteriorly disposed PM glycoproteins, leads to large endosomes which fail to fuse with pinocytic vesicles but fuse readily with latex particles (Edelson & Cohn 1974a). The reverse happens after the storage of a variety of large and highly sulphated or carboxylated anions in secondary lysosomes. The mechanisms underlying these distinctions are not clear.

Phagocytosis

Depending on the nature, size and digestibility of the particle, fairly large amounts of PM may be interiorized around the particle. One example will illustrate this point. Macrophages with the vacuolar dimensions of those shown in Table 1 are allowed to ingest 100, 1.1 μm diameter polystyrene latex beads which each have a surface area of 3.8 μm^2. With tight apposition between membrane and particle, approximately 380 μm^2, or 46%, of the total PM surface area (825 μm^2) is interiorized. One would expect some rounding and changes in cell outline, but very little appears to happen, perhaps because we have failed to consider the remainder of the vacuolar system, i.e. pinocytic vesicles and secondary lysosomes, which add some 251 μm^2 of surface area, making a total surface area of 1076 μm^2 for the whole vacuolar apparatus. If phagosomes and lysosomes fuse extensively and vacuolar membrane is redistributed to the cell surface, a quite different answer results. An irreducible minimum of 380 μm^2 is trapped around the beads, leaving as much as 700 μm^2 to cover the cell surface, i.e. there is a decrease in PM as small as 16% ($\sim$ 129 μm^2). We shall deal later with the reverse flow of vacuolar membrane to the cell surface.

Pinocytosis

The influx of pinocytic vesicle membrane in steady-state conditions has been analysed by stereological techniques and reported by Steinman et al (1976). Table 3 gives values for peritoneal macrophages. During each minute, 0.43%

TABLE 3 Influx of macrophage pinocytic vesicles

	per Minute	*per Hour*
Number	125	7500
Volume (μm^3)	1.7	102
Surface area (μm^2)	26	1560
Fractional volume (% total)	0.43	26
Fractional surface area (% total)	3.1	186

of the total volume and 3.1% of the surface area is interiorized as pinocytic vesicles that are labelled with HRP. During each hour, 186% of the total surface is internalized; or the entire surface area of the macrophage is internalized every 33 minutes. The constancy of the high pinocytic rate, the cell surface area and the cell volume in steady-state conditions *in vitro* place some constraints on the choice of possible mechanisms. This finding was intuitively recognized by Lewis in his initial description of pinocytosis (Lewis 1931). Comparative information from L cell fibroblasts is given in Table 4.

TABLE 4 L Cell dimensions and fluxes

Complete L cell		*Vacuolar compartments* *Pinocytic vesicle*	*Secondary lysosome*
Volume (μm^3)	1765	5.7	15.7
Surface area (μm^2)	2100	89	136
Total fractional influx			
% surface area min^{-1}		0.9	
% volume min^{-1}		0.05	

Vesicle generation

Pinocytic vesicles are generated from the PM with restricted diameter, whereas phagocytic vesicles have diameters that are governed by particle size and they may approach a large fraction of the total cell volume. Fluid-phase pinocytosis is not associated with known ligands but is often associated with undulating or ruffled segments of the membrane. In most cases vesicles are generated on the cell surface although in others deep channels invaginate into the cytoplasm and liberate vesicles at their distal ends. These channels which are common on certain cell types, e.g. Kupffer cells, may trap solutes and thus lead to spurious uptake values.

More information is available about phagocytic vesicle (phagosome) formation, which is a ligand–receptor interaction. Griffin & Silverstein (1974) have defined the selective nature of the macrophage membrane and its disciminatory ability, which make the phagocytic stimulus a highly local phenomenon. Further work has led to the 'zipper' model of particle ingestion (Griffin et al 1975, 1976). This indicates that receptors on the macrophage membrane bind ligands on the particle in a sequential fashion. Such step-wise interactions are necessary for the movement of membrane around the particle and for the polar fusion of the membrane. The model predicted that particles coated with ligand on only one hemisphere would bind but would fail to be ingested. This was demonstrated in a number of experimental conditions. More recently the mobility of the Fc receptor in the plane of the membrane has been examined by observations of its modulation from the upper cell surface when cells are plated on immobilized immune complexes (Michl et al 1979).

Nature of internalized endocytic vesicle membrane

Our previous information on the nature of PM that was internalized around phagocytic particles came from work on the activity of 5′-nucleotidase. A more exacting technique was necessary to compare the PM with the endosome or with the phagolysosome (fused phagosome and lysosome). Cell fractionation by the usual velocity or isopycnic techniques consistently failed to give clean enough fractions for this analysis. A major advance in this area occurred when Hubbard developed the enzymic iodination technique and applied it to the mouse erythrocyte and mouse L cell (Hubbard & Cohn 1972, 1975a,b). Careful analysis revealed that through the use of soluble lactoperoxidase, and glucose oxidase(EC 1.1.3.4)-generated fluxes of peroxide and ^{125}I, preferential labelling of exteriorly disposed polypeptides of the PM could be achieved. Since iodination took place at 4 °C the membrane compartments did not move significantly. It was soon apparent that the pattern of 20 labelled polypeptides from the PM and that from the phagolysosomes generated from labelled PM were indistinguishable when displayed on continuous gradient, sodium dodecyl sulphate (SDS) slab gels. This suggested that representative samples of PM were interiorized and were then present immediately in the membrane of latex phagolysosomes. Since these labelled peptides were selected by virtue of their available tyrosine groups it was possible that other qualitative or quantitative differences existed between these segments of the vacuolar apparatus.

Two other approaches have used lactoperoxidase-mediated iodination in either a soluble or a particulate form to label pinocytic vesicle (pinosome) membrane or phagolysosome membrane respectively. Mellman et al (1980) used lactoperoxidase as a fluid-phase marker to label the luminal surface of pinocytic vesicles. Several lines of evidence suggested that selective labelling of pinocytic vesicles and not PM took place. These included electron microscopy (EM) autoradiography, low-temperature inhibition, latency of the labelled organelle and inaccessibility to monoclonal antibodies. When Nonidet P-40 lysates were displayed on gels the 15–20 labelled polypeptides of both pinocytic vesicles and PM were essentially indistinguishable. The major exception was the presence of labelled lactoperoxidase in the pinocytic vesicle fraction. The pinocytic vesicle membrane, as for the L cell phagosome, contained a representative sample of PM polypeptides. Further experiments demonstrated that eight PM antigens, recognized by distinctive monoclonal antibodies, were interiorized to the same relative extent as they were expressed on the PM. These antigens were labelled with ^{125}I, but they represented minor components that were not recognized on SDS gels as discrete bands.

Further evidence for the similarity of polypeptides between the phagosome–lysosome and PM of macrophage will be presented in a later section.

Vesicle flow and directionality

Pinocytic and phagocytic vesicles arising from the PM follow a course that leads them to the centrosphere region of the cell in the peri-nuclear 'hof'. In cells attached to substrates, phagosomes arise from the superior PM surface with which particles come in contact and move short distances before they fuse with lysosomes to form phagolysosomes. Pinocytic vesicles follow a more heterogeneous route from their origins. When visualized in living cells the vesicles demonstrate a linear directionality through pseudopods as they home into and fuse in the peri-Golgi zone. Their forward movement is a saltatory one—stop and go. Although the skeletal framework for this movement is not completely understood a few observations are noteworthy. First, on transmission EM sections of macrophage projections, pinocytic vesicles are found in a central location surrounded by a rete-like network of microtubules (Cohn 1966). These originate from the two centrioles and radiate to all poles of the cell, acting as a framework along which pinocytic vesicles migrate. Both the propulsive force moving vesicles and the saltatory motion are unexplained. Current speculation suggests that this involves reversible bridges between vesicles and structural proteins, or a dynamic polymerization–depolymerization process in microtubules, or both.

Vesicle fusion

We have recently adapted a technique described by D'Arcy Hart and his colleagues to quantitate the rate and extent of phagosome–lysosome fusion in macrophage (see Kielian & Cohn 1980). These studies have described a variety of membrane and cytoskeletal perturbants that do or do not modify the fusion process. It is of interest that fusion is an energetic event with a Q_{10} of 2.5 and an E_{act} of 16.4 kcal mol^{-1}—a value similar to that for the generation of pinocytic vesicles (18 kcal mol^{-1}) and considerably less than that for phagocytosis (54 kcal mol^{-1}). However, fusion is completely but reversibly blocked below about 15 °C, a cut-off value similar to that for receptor-mediated phagocytosis and dissimilar from the linear pinocytic relationship between 2–27 °C. As noted in Table 2, different factors modify the fusion of pinocytic and phagocytic vacuoles. We have recently studied the influence of large anions which are stored within lysosomes and block subsequent fusion between phagosomes and lysosomes without influencing the input of solutes via pinocytosis. Anions such as highly sulphated dextran have slight effects on intralysosomal pH. They interact strongly with plasma lipoproteins and the resulting complex is rapidly taken up into lysosomes where it inhibits fusion within a few hours. The lipoprotein portion of the complex serves as the carrier and the dextran sulphate inhibits fusion. Studies on the fluorescence polarization of β-parinaric acid in isolated lysosomes containing dextran (which does not inhibit fusion) or dextran sulphate revealed that dextran sulphate decreases the fluidity of the limiting phagolysosome membrane. Presumably direct ionic interactions occur between membrane proteins or lipids, or both (Kielian & Cohn 1982).

Bidirectional flow of membrane polypeptides and the process of recycling

The stereological information on membrane influx suggested an extremely rapid rate of interiorization in the absence of total cell surface area and volume modifications. PM components were either re-utilized or degraded and resynthesized extensively and (seemingly) wastefully. Initial studies on the turnover of individual PM glycoproteins indicated relatively long half-lives, varying between 24–80 h depending on the labelling procedures (Hubbard & Cohn 1975a, Kaplan et al 1979, Steinman et al 1980). These findings as well as earlier observations on the relatively slow net synthetic rate of PM components made it unlikely that a synthetic event was involved.

The more reasonable assumption was for membrane that was interiorized as a vesicle to deposit its solutes in the vacuolar apparatus and then to return to the cell surface for re-use. To accumulate more direct and definitive evidence for this process novel procedures were required; for example, one in which we could preferentially tag the membrane of lysosomes and then follow the efflux of membrane components. The procedure which met these requirements is outlined in Table 5. It depended on the rapid delivery of a

TABLE 5 Lactoperoxidase–latex iodination of phagolysosomes

1. Rapid internalization, fusion and residence in lysosomes
2. Labelling after addition of ^{125}I and H_2O_2 (GO + Glucose) at 0–4 °C
3. A non-reutilizable label on tyrosyl residues of polypeptides (MIT); no lipid labelling
4. $>$ 98% of label associated with phagolysosomal membranes on their luminal surfaces
5. Grains on electron microscopy autoradiographs over lactoperoxidase–latex beads; 97% of cell labelled
6. Label distributed on ~24 polypeptides as displayed on single-dimension gels

particle-bound iodinating system to the phagolysosome and on the subsequent labelling of membrane polypeptides (Muller et al 1980a). This was accomplished via the enzyme lactoperoxidase, covalently coupled to carboxylated latex beads, and resulted in the labelling of the luminal surface of secondary lysosomes without significant PM labelling. The entire procedure was done at low temperature (0–4 °C) to minimize membrane movement and to lead to the localization of 75% of the autoradiographic grains over phagolysosomes containing latex beads (Muller et al 1980b). When the temperature was increased to 37 °C for 30 min a redistribution of grains occurred and 78% were randomly associated with PM, with a concomitant reduction in the phagolysosomes containing latex. Subsequent experiments demonstrated that redistribution occurred more rapidly (± 5 min) and then reached an equilibrium maintained for at least two hours.

We next observed that the polypeptides which were iodinated on the inner surface of the phagolysosome membrane were indistinguishable from those labelled on the cell surface via the soluble lactoperoxidase system. This finding was consistent with the information discussed previously for L cell and J774 phagolysosomes (see Kaplan et al 1979) and emphasized that the cell membrane was related to the phagolysosome membrane. To carry the analysis further it was necessary to know which polypeptides were moving from the phagolysosome to the cell surface—were the autoradiographic grains represented by a restricted sample or by a representative sample of membrane glycoproteins?

This question was answered by trapping PM around latex beads and analysing the spectrum of labelled polypeptides. Phagolysosomes were labelled with lactoperoxidase—latex at low temperature (0–4 °C). Immediately thereafter and at subsequent intervals the cells ingested a much larger load of

0.527 μm-diameter styrene butadiene latex beads of low buoyant density which could be separated from lactoperoxidase–latex beads on sucrose gradients. Within 30 min all the polypeptides labelled in the phagolysosome were distributed on the PM. This suggested that the movement of phagolysosomal membrane polypeptides occurred as a unit in a relatively synchronous and perhaps quantal fashion. Labelled membrane peptides presumably reached the cell surface via vesicles, fused with the PM and were then laterally distributed in the plane of the membrane—a suggestion consistent with the uniform grain distribution over all segments of PM.

The rapid centrifugal flow of phagolysosomal membrane polypeptides was also accompanied by the simultaneous influx of pinocytic vesicles and exogenous solute. For example, HRP was entering the cell continuously and was accumulating in the lactoperoxidase–latex-filled phagolysosomal compartment at the same rate as it did in other secondary lysosomes. This rapid bi-directional flow of relatively long-lived vacuolar membrane polypeptides provides evidence for the recycling of membrane proteins. It implies a continuous and balanced shuttle in which representative samples of membrane glycoproteins are moving between the PM and the secondary lysosome.

Implications and queries

The flux of vesicles between the PM and the lysosomal compartment appears random and leads to a continuum of membrane in the vacuolar apparatus. Selective enrichment of specific receptors may well occur during adsorptive endocytosis of either solutes or particles, yet this clustering presumably occurs on a rather uniform polypeptide base. This constant flow in the macrophage may be associated with discriminatory mechanisms that allow for the selective degradation of vesicles as units or individual polypeptides in the membrane. Vesicles flowing to the cell surface could be focused into specific areas of the cytoplasm leading to changes in cell shape and motility. This possibility emphasizes our ignorance about the factors controlling the direction and flow of particles.

The lack of distinctive polypeptides in the secondary lysosome membrane suggests that the primary lysosome may carry 'PM' units in addition to newly synthesized acid hydrolases. These organelles arising from the Golgi region may therefore insert quanta of new membrane directly into the secondary lysosome and thence into the cell surface. Another route would lead to direct PM fusion thus by-passing the secondary lysosomes and liberating the contents into the extracellular milieu. Some of these possibilities are presented in Fig. 1.

Another route for the pinocytic vesicle is for it to liberate its contents

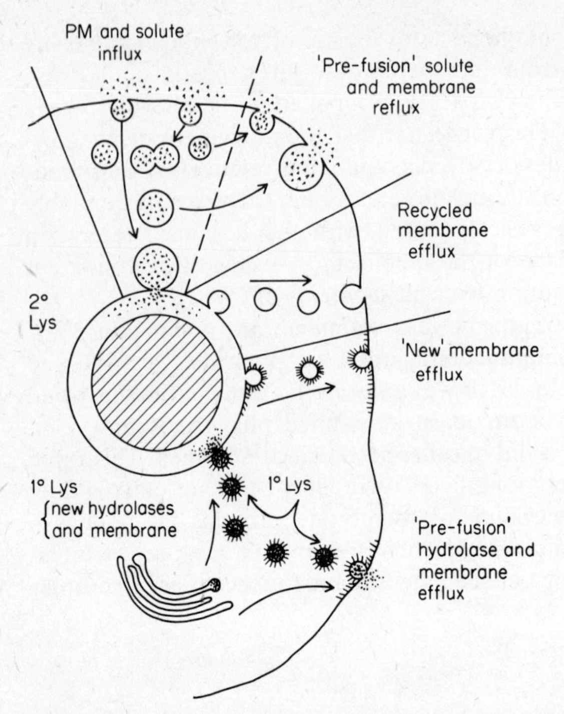

FIG. 1. Pathways that may be involved in the recycling of membrane and solute. The secondary lysosome (2° Lys) serves as the intermediate member of the vacuolar apparatus, accepting membrane and solutes from the extracellular environment and from endogenous synthetic sources. Fluid-phase vesicles flow, fuse and occasionally reflux their contents back into the medium. After the vesicles discharge their contents into the secondary lysosome, an equivalent amount of membrane may form a vesicle, in the absence of appreciable lysosomal content, and flow centrifugally to the plasma membrane (PM). It is possible that 'new' membrane enters the system via the primary lysosome (1° Lys). This may occur through direct fusion with secondary lysosome and subsequent recycling to the cell surface or through immediate fusion with the plasma membrane.

shortly after interiorization and before fusion with secondary lysosomes. Although we have no direct evidence for this mechanism Besterman et al (1981) suggest that a portion of interiorized sucrose follows this path. Finally, it should be pointed out that in lower forms such as *Entamoeba histolytica* we have evidence for a large intracellular compartment—non-lysosomal—through which certain solutes flow into and out of the cell. Entry into the lysosomal compartment requires long periods of incubation and then leads to either storage or digestion of the solute (S. Aley, W. A. Scott and Z. A. Cohn, unpublished observations).

Acknowledgements

Zanvil A. Cohn is supported by U.S.P.H.S. grant AI-07012, and Ralph M. Steinman by U.S.P.H.S. grant AI-13013.

REFERENCES

Besterman JM, Airhart JA, Woodworth RC, Low RB 1981 Exocytosis of pinocytosed fluid in cultured cells: kinetic evidence for rapid turnover and compartmentation. J Cell Biol 91:716-727

Bianco C, Eden A, Cohn ZA 1976 The induction of macrophage spreading. J Exp Med 144:1531-1544

Cohn ZA 1966 The regulation of pinocytosis in mouse macrophages: I. J Exp Med 124:557-571

Cohn ZA, Ehrenreich BA 1969 The uptake, storage and intracellular hydrolysis of carbohydrates by macrophages. J Exp Med 129:201-225

Cohn ZA, Fedorko ME, Hirsch JG 1966 The *in vitro* differentiation of mononuclear phagocytes. V: The formation of macrophage lysosomes. J Exp Med 123:757-766

Edelson PJ, Cohn ZA 1974a Effects of Concanavalin A on mouse peritoneal macrophages. I: Stimulation of endocytic activity and inhibition of lysosome formation. J Exp Med 140:1364-1386

Edelson PJ, Cohn ZA 1974b Effects of Concanavalin A on mouse peritoneal macrophages. II: Metabolism of endocytized proteins and reversibility of the effects by mannose. J Exp Med 140:1387-1403

Edelson PJ, Cohn ZA 1976a 5°-Nucleotidase activity of mouse peritoneal macrophages. I: Synthesis and degradation in resident and inflammatory populations. J Exp Med 144:1581-1595

Edelson PJ, Cohn ZA 1976b 5°-Nucleotidase activity of mouse peritoneal macrophages. II: Cellular distribution and the effects of endocytosis. J Exp Med 144:1596-1608

Griffin FM Jr, Silverstein SC 1974 Segmental response of the macrophage plasma membrane to a phagocytic stimulus. J Exp Med 139:323-336

Griffin FM Jr, Griffin JA, Leider JE, Silverstein SC 1975 Studies on the mechanism of phagocytosis. I: Requirements for circumferential attachment of particle-bound ligands to specific receptors on the macrophage plasma membrane. J Exp Med 142:1263-1282

Griffin FM Jr, Griffin JA, Silverstein SC 1976 Studies on the mechanism of phagocytosis. II: The interaction of macrophages with anti-immunoglobulin IgG-coated bone marrow-derived lymphocytes. J Exp Med 144:788-809

Hubbard AL, Cohn ZA 1972 The enzymatic iodination of the red cell membrane. J Cell Biol 55:390-405

Hubbard AL, Cohn ZA 1975a Externally disposed plasma membrane proteins. I: Enzymatic iodination of mouse L cells. J Cell Biol 64:438-460

Hubbard AL, Cohn ZA 1975b Externally disposed plasma membrane proteins. II: Metabolic fate of iodinated polypeptides of mouse L cells. J Cell Biol 64:461-479

Kaplan G, Unkeless JC, Cohn ZA 1979 Insertion and turnover of macrophage plasma membrane proteins. Proc Natl Acad Sci USA 76:3824-3828

Kielian MC, Cohn ZA 1980 Phagosome–lysosome fusion. Characterization of intracellular membrane fusion in mouse macrophages. J Cell Biol 85:754-765

Kielian MC, Cohn ZA 1982 The intralysosomal accumulation of polyanions. II: Polyanion internalization and its influence on lysosomal pH and membrane fluidity. J Cell Biol 93:875-882
Lewis WH 1931 Pinocytosis. Bull Johns Hopkins Hosp 49:17-27
Mahoney EM, Hamill AL, Scott WA, Cohn ZA 1977 Response of endocytosis to altered fatty acyl composition of macrophage phospholipids. Proc Natl Acad Sci USA 74:4895-4899
Mahoney EM, Scott WA, Landsberger FR, Hamill AL, Cohn ZA 1980 Influence of fatty acyl substitution on the composition and function of macrophage membranes. J Biol Chem 255:4910-4917
Mellman IS, Steinman RM, Unkeless JC, Cohn ZA 1980 Selective iodination and polypeptide composition of pinocytic vesicles. J Cell Biol 86:712-722
Michl J, Pieczonka MM, Unkeless JC, Silverstein SC 1979 Effects of immobilized immune complexes on Fc- and complement-receptor function in resident and thioglycollate-elicited mouse peritoneal macrophages. J Exp Med 150:607-621
Muller WA, Steinman RM, Cohn ZA 1980a The membrane proteins of the vacuolar system. I: Analysis by a novel method of intralysosomal iodination. J Cell Biol 86:292-303
Muller WA, Steinman RM, Cohn ZA 1980b The membrane proteins of the vacuolar system. II: Bidirectional flow between secondary lysosomes and plasma membrane. J Cell Biol 86:304-314
Steinman RM, Brodie SE, Cohn ZA 1976 Membrane flow during pinocytosis. A stereologic analysis. J Cell Biol 68:665-687
Steinman RM, Nogueira N, Witmer M, Tydings JD, Mellman IS 1980 Lymphokine enhances the expression and synthesis of Ia antigens on cultured mouse peritoneal macrophages. J Exp Med 152:1248-1261
Werb Z, Cohn ZA 1971a Cholesterol metabolism in the macrophage. I: The regulation of cholesterol exchange. J Exp Med 134:1545-1569
Werb Z, Cohn ZA 1971b Cholesterol metabolism in the macrophage. II: Alteration of subcellular exchangeable cholesterol compartments and exchange in other cell types. J Exp Med 134:1570-1585
Werb Z, Cohn ZA 1972 Plasma membrane synthesis in the macrophage following phagocytosis of polystyrene latex particles. J Biol Chem 247:2439-2446

DISCUSSION

Bretscher: I would like to ask about the effect of temperature on fluid-phase pinocytosis. Professor Cohn mentioned the linear increase in rate of pinocytosis between 2–27 °C. Is that distinctly different from the uptake for low density lipoprotein (LDL)?

Goldstein: LDL uptake begins at about 16 °C. The rate of uptake at 37 °C is seven to eight-fold higher than at 16 °C.

Bretscher: Does this therefore provide a clear distinction between coated vesicle-mediated uptake and fluid-phase uptake?

Cohn: Yes. I would distinguish between the nature of the solutes rather than the nature of the vesicles that are carrying the solute. I am uncertain whether fluid-phase vesicles in the macrophage are coated or not; we have not done the requisite studies. Perhaps all generated vesicles have coats at some time in their origin. However, there is a clear distinction, in terms of the nature of the solute, between the temperature coefficient for LDL uptake and that for uptake of horseradish peroxidase (HRP) or sucrose.

Bretscher: I don't understand. To measure fluid-phase pinocytosis in fibroblasts one can use sucrose. Does this uptake stop below 16 °C?

Cohn: It depends on the sensitivity of the method used to measure the exact point between 16 and 4 °C. Obviously very little solute is transferred in those conditions but, nevertheless, transfer is going on. If incubations are prolonged sufficiently, solutes are obviously interiorized.

Palade: But one should consider both coated and uncoated small pinocytic vesicles. The remarks about continued uptake at low temperature (< 16 °C) may apply primarily to regular (uncoated) plasmalemmal vesicles, of the type found at high volume density in vascular endothelia. The temperature limitation that you are speaking about may apply only to *coated* vesicles; so this point deserves to be clarified.

Bretscher: I am trying to draw a distinction between uptake by coated and uncoated vesicles. In your endocytic vesicles you find an unselected sample of plasma membrane proteins, whereas one gets the impression that coated vesicles are highly selective in what they take in. By that criterion, also, your vesicles might have been uncoated.

Cohn: We did not have the appropriate fixation procedures to ensure whether the vesicles were coated or not, at the time we did these studies. We have not repeated the experiments so, as I mentioned, I can't tell you what type of vesicle we were working with.

Dean: Do you find a comparable temperature dependence between adsorptive endocytosis and fluid-phase endocytosis for sucrose?

Cohn: We have not measured adsorptive endocytosis by using the ligands. Ira Mellman's work (see p 35-58) on the use of a ligand that is an immune complex may prove helpful here.

Baker: It is striking that the quantity of membrane in the different compartments is roughly in balance. Is it possible, for instance by changing temperature, to put the mechanisms out of balance so that membrane accumulates in one or other compartment?

Cohn: I can't answer the specific example concerned with temperature, but the macrophage will sometimes interiorize very large amounts of non-digestible particles such as latex and it will take in considerable amounts (roughly 50%) of its plasma membrane. It will 'round up' in those circumstances, with its endocytic activity very reduced for a period of many hours. Over the next 24 h it will spread out on the surface again, maintain the same irreducible minimum amount of membrane around those polystyrene latex particles, and it will presumably have synthesized new plasma membrane to allow the spreading to occur. Its surface area is then greater than it was initially. This is associated with net increments in phospholipid composition and sterol content. So it can adapt slowly, over 12–24-h periods.

Helenius: By monitoring the expression of Semliki Forest virus infection we have a sensitive way of detecting endocytic uptake. We find a cut-off temperature similar to the one you mentioned—below 16 °C cells are not infected.

Cohn: That work probably involves more of a phagocytic activity.

Helenius: By electron microscopy, the viruses appear to enter via coated vesicles, and the uptake is probably pinocytic (Marsh & Helenius 1980). There is no detectable endocytosis below 15–16 °C.

Hubbard: Is there really no internalization of LDL below 15–16 °C? In the asialoglycoprotein system there is internalization at temperatures down to about 10 °C but fusion with liposomes does not follow (Dunn et al 1980).

Goldstein: The rate of LDL uptake is so low at temperatures below 15 °C that it is difficult to obtain reliable measurements. Professor Cohn, if you radiolabel the surface of the macrophage with [^{125}I]iodine and then use your monoclonal antibody to immunoprecipitate the Fc receptor, is the amount of ^{125}I-precipitable Fc receptor enriched in the membrane that is taken up with latex particles?

Cohn: Muller et al (see 1980a,b) studied latex-containing phagosomes and did not find an enrichment of Fc receptor in those. Nor would one expect latex to interact with Fc receptors in those conditions.

Raff: Can one extrapolate, from the findings on the phagosome membranes and the secondary lysosomes to which they go, to fluid-phase endocytic membranes? Does one know the composition of the latter membranes?

Cohn: Dr Mellman's work (see p 35-58) shows that by specific ^{125}I-Iabelling of the fluid-phase vacuoles and by comparison of their polypeptide composition with that of the plasma membrane, it is impossible to distinguish between the membranes.

Raff: This suggests then that in macrophages, at least, fluid-phase endocytosis does not involve coated vesicles.

Sly: Did you label any *content markers?* You call these vesicles secondary lysosomes but what is the evidence that they are secondary lysosomes as opposed to pre-fusion vesicles?

Cohn: Cytochemistry and fractionation procedures show that the vesicles contain acid hydrolases, but very little of the content is labelled; if we disrupt the phagolysosomes and separate the membrane from the content, about 1% of the label is found in the content, and the rest is in the membrane. The membranes that surround the latex particles are very tightly opposed, which presumably explains the distribution of the label.

Meldolesi: Is fluid-phase endocytosis influenced by changes in the ionic composition of the medium bathing the cells?

Cohn: Hardly at all. We know of no exogenous agent that modifies the steady-state level of fluid-phase endocytosis except for short-lived effects which can be induced with agents such as concanavalin A, leading to increased vesicle flow for a temporary period of 3–4 h.

Meldolesi: Are any extracellular ions essential for the process?

Cohn: We are not sure about the monovalent cations, but divalent cations do not appear to be required. The process is obviously different from endocytosis in some of the free-living amoebae, in which monovalent cations specially stimulate channel formation.

Goldstein: In relation to whether these vesicles are coated or not, the recent work of Aggeler et al (1981) indicates that clathrin molecules are present underneath the plasma membranes of macrophages.

Cohn: Those studies indicate that coated vesicles are being generated in the macrophage. But the proportion of such vesicles is unclear from that work. Interestingly, latex phagolysosomes are also coated with clathrin baskets as they are interiorizing (Z. A. Cohn et al, unpublished results).

Rothman: If a representative sample of plasma membrane proteins is taken in, and if the flux is such that the bulk of the plasma membrane is endocytosed repeatedly, then why is it that these proteins are metabolically stable; are they resistant to protease? If they are cycling into a secondary lysosomal compartment where hydrolases are present, why are they so long-lived?

Cohn: This question was raised initially in 1963 at the Ciba Foundation meeting on lysosomes, and it really hasn't yet been answered! However, this type of recycling suggests that the rate at which the membrane is captured from the secondary lysosome is influential here. In other words, the membrane may not stay very long in the secondary lysosomal compartment and may move out rapidly. Alternatively, as many people have suggested, these membrane polypeptides may be inherently resistant to acid hydrolases.

Rothman: If one has labelled the proteins with iodine from within the secondary lysosomes and then allowed most of them to return to the plasma membrane, can one then chill the cells again and add protease to see if they are susceptible to the protease?

Cohn: A rather small proportion of them is susceptible.

Rothman: So they may well be inherently resistant.

Baker: Is that done at a sufficiently acid pH to mimic the interior of the lysosome?

Cohn: No. This was done at neutral pH with living cells.

Baker: Therefore the question of stability within the lysosome can't strictly be answered.

Rothman: Could one use a crude lysosomal extract as a source of proteolytic enzymes?

Cohn: Yes. A good test has not yet been done.

Goldstein: What is the time-course of recycling of the plasma membrane that is internalized with the latex particle?

Cohn: We have to look at the process in two cycles—internalization and centrifugal flow. We know that internalization can take place as rapidly as within three minutes into a lysosomal compartment. New solute enters these lactoperoxidase–latex phagolysosomes at the same rate as it enters other secondary lysosomes. Reflux back to the cell surface appears to occur within five minutes.

Sly: Is there any evidence for involvement of an intermediate vesicle in particle uptake?

Cohn: Yes. One sees intermediate vesicles easily under phase-contrast and time-lapse cinephotography. There are many fusion products of endocytic vesicles—large, electron-lucent structures which move into the centrosphere region and then increase in density as they become lysosomes. This is one of the more obvious endocytic vacuoles that is formed in the cell. Only about 10% of the vesicles are large enough to be visible under phase contrast. *Amoeba histolytica*, the pathogen, has a large intermediate compartment; about 50% of the intracellular vesicles in this organism are vacuoles that have the same pH as the outside medium. When we use a solute such as fluoresceinated dextran, it is all taken up rapidly into this compartment and reaches a plateau after about an hour. But as soon as one removes the fluoresceinated dextran from the medium, it also comes right out from the cells. So it enters and leaves the cell quickly. It does not enter a lysosomal compartment unless one incubates the cell for periods of up to eight hours, in which case one sees another series of vacuoles which have the pH and acid hydrolase content of typical secondary lysosomes.

Sly: Would this be analogous to the reflux of material taken up by receptors that you mentioned?

Cohn: No; this is much more extensive quantitatively and involves all the solute that has been taken in.

Palade: What is known about interactions between polystyrene particles and the cell surface? What kind of interactions are involved at that stage? What cell-surface groups bind to polystyrene particles?

Cohn: We don't know.

Palade: Is there any special interaction that could explain the results of the experiment?

Cohn: There may be.

Palade: Are the results of such experiments likely to be affected by the fact that the ingested objects are large and cannot be reduced in size by digestion? In such cases, membranes may not be recovered from the endosomes or secondary lysosomes in which the particles are located, and therefore membrane components may not escape degradation.

Cohn: In some circumstances when L cells take in polystyrene latex particles many of the labelled polypeptides that surround the particle are rapidly degraded. We don't know what reconstitutes that membrane, but new synthesis could be the explanation. Probably such particles perturb the system but we cannot yet assess the extent of perturbation.

Geisow: The nature of the attraction of the latex to the macrophage surface is an important point. Have you tried using opsonized latex, in which case you would presumably bring Fc receptors into the secondary lysosome? You might see a difference in intracellular events between unopsonized and opsonized latex, because in the last case you are introducing a specialized receptor.

Mellman: A difficulty in the use of opsonized latex beads is that the interaction between latex and the plasma membrane is stronger than the interaction between the opsonin and the Fc receptor. For example, our anti-Fc receptor antibodies do not inhibit or even decrease appreciably the uptake of IgG-coated latex beads under conditions where we can completely abolish the binding and/or uptake of IgG-coated erythrocytes (I. S. Mellman, unpublished). In any event, the presence of Fc receptor as one of the many labelled membrane components of phagolysosome membrane has been documented by immune precipitation with our anti-receptor antibodies (W. Muller, unpublished). In my paper (p 35-58) I shall be discussing the selective inclusion of these receptors in internalized membrane during the receptor-mediated phagocytosis of opsonized erythrocytes.

Cohn: Coating latex with a variety of proteins does not change their fate or rate of internalization appreciably.

Geisow: If one coats latex beads with gelatine, their uptake by macrophages is reduced. By then modifying this surface one might have the opportunity to achieve specific opsonization.

Rothman: I believe that latex beads are made by emulsion polymerization in the presence of SDS and are coated with a tightly packed monolayer of SDS. It is therefore conceivable that when the latex bead interacts with the membrane, the SDS may cause some randomization.

REFERENCES

Aggeler J, Heuser J, Werb Z 1981 The distribution of clathrin in phagocytosing macrophages. J Cell Biol 91:264a (abstr)

Ciba Foundation 1963 Lysosomes. (Ciba Found Symp), Churchill, London

Dunn WA, Hubbard AL, Aronson NN Jr 1980 Low temperature selectively inhibits fusion between pinocytic vesicles and lysosomes during heterophagy of ^{125}I-asialofetuin by the perfused rat liver. J Biol Chem 255:5971-5978

Marsh M, Helenius A 1980 Adsorptive endocytosis of Semliki Forest Virus. J Mol Biol 142:439-454
Muller WA, Steinman RM, Cohn ZA 1980a The membrane proteins of the vacuolar system. I: Analysis by a novel method of intralysosomal iodination. J Cell Biol 86:292-303
Muller WA, Steinman RM, Cohn ZA 1980b The membrane proteins of the vacuolar system. II: Bidirectional flow between secondary lysosomes and plasma membrane. J Cell Biol 86:304-314

Endocytosis, membrane recycling and Fc receptor function

IRA S. MELLMAN

Section of Cell Biology, Yale University School of Medicine, 333 Cedar Street, P.O. Box 3333, New Haven, Connecticut 06510, USA

Abstract We have studied the composition and fate of plasma membrane internalized during both fluid-phase and receptor-mediated endocytosis in mouse macrophages. Particular attention has been paid to the macrophage Fc receptor, an intrinsic membrane glycoprotein that we have isolated and characterized biochemically and immunologically. Monoclonal and polyclonal antibodies directed against the receptor and against a series of other unrelated plasma membrane proteins have been used. In addition, we have used radioiodination techniques to label selectively the polypeptides of pinocytic vesicle membrane from within intact cells. Our results indicate that fluid pinocytosis in macrophages involves the internalization of a largely representative sample of plasma membrane polypeptides. Significantly, the Fc receptor seems to be internalized at a rate similar to that of most other membrane proteins. However, selective internalization of the receptor is induced during the endocytosis of certain ligands. The phagocytosis of immunoglobulin G (IgG)-coated erythrocyte ghosts results in the selective and largely irreversible removal of Fc receptors from the macrophage surface. Selectively internalized receptors do not recycle but are rapidly degraded. Fc receptors also appear to be preferentially interiorized during the rapid pinocytosis of IgG-containing soluble immune complexes. Uptake is accompanied by a sharp decrease in the number of surface receptors, which is partially reversed after the removal of ligand.

It is now generally accepted that endocytosis brings about the bi-directional flow of plasma membrane (PM) into and out from the cytoplasm. The magnitude and rapidity of this process was first demonstrated by R. M. Steinman, Z. A. Cohn and their associates (see Cohn & Steinman 1982, Steinman & Cohn 1972, Steinman et al 1974, 1976) who found that cultured cells continuously internalize more than 50% of their cell surface as pinocytic vesicles each hour. Moreover, pinocytic vesicles form, fuse repeatedly with

1982 Membrane recycling. Pitman Books Ltd, London (Ciba Foundation symposium 92) p 35-58

each other, or with lysosomes, or both, and apparently return to the cell surface, all in a matter of minutes.

As impressive as the extent of this membrane flow is its specificity. Endocytic vesicles fuse only with a highly restricted subset of intracellular membranes, namely elements of the vacuolar apparatus (PM, endocytic vesicles, lysosomes and Golgi). In addition, various mechanisms allow even these organelles to maintain their diversity and integrity in spite of extensive communication with the PM. Central to understanding these events is a detailed knowledge of the composition and dynamics of the membrane that is internalized during endocytosis. Little is known about this, since most work on endocytosis and membrane recycling has concentrated on analysing the contents of endocytic vesicles—that is, markers of endocytosis which either bind to specific PM receptors or are internalized in the fluid phase. Until quite recently it has been difficult to think in terms of identifying individual components of the plasma membrane and following their behaviour directly.

We have concentrated particularly on the membrane proteins that are internalized during both fluid-phase and receptor-mediated endocytosis in mouse macrophages. We have used a series of well characterized monoclonal antibodies directed against discrete PM polypeptides to examine the composition and dynamics of the membrane involved in the endocytic event. We have also used radiolabelling techniques which permit the selective iodination of endocytic vesicle membrane from within intact cells. Particular attention has been paid to the macrophage receptor for the Fc region of immunoglobulin G(IgG). This receptor is an intrinsic membrane glycoprotein that we have recently isolated and characterized biochemically and immunologically. We have thus been able to study quantitatively the internalization and fate of a well known cell-surface receptor, relative to other unrelated PM proteins, in both the presence and the absence of specific ligand. Moreover, since the Fc receptor mediates the endocytosis of both large particles (via phagocytosis) and small immune complexes (via pinocytosis), we have been able to examine the effects of a ligand's valency and of its pathway of internalization on receptor dynamics.

Experimental procedures

Cells

The cells used in these studies were either thioglycollate-elicited mouse peritoneal macrophages or the mouse macrophage-like cell line J774, which was grown either in suspension or monolayer culture. Both types of cell were maintained in 10% fetal bovine serum.

Monoclonal antibodies

Rat monoclonal antibodies directed against mouse macrophage PM proteins were produced according to a modification of the procedure of Gefter et al (1977), using animals immunized with J774 cells (Unkeless 1979, Mellman et al 1980).

Radioiodination

Cells were radioiodinated using modifications of the lactoperoxidase–glucose oxidase procedure of Hubbard & Cohn (1975a) as described previously (Mellman et al 1980). Neither the cell surface nor intracellular labelling procedures resulted in significant loss of viability or iodination of cell lipid. Isolated proteins were labelled using Iodogen (Pierce Chemical Co., Chicago) and unincorporated iodine was removed by passing the reaction mixture over a 0.2 ml column of Dowex 1-X8.

Immune precipitation

Antibody–antigen complexes were isolated at 4 °C by precipitation either with $F(ab')_2$ fragments of affinity-purified rabbit anti-rat IgG coupled to Sepharose 4B or with protein A-Sepharose (Mellman et al 1980). Cell lysates were prepared using 0.5% Nonidet P-40 containing the protease inhibitors, aprotinin and phenylmethane sulphonyl fluoride (PMSF), and were centrifuged for 15 min at 40 000 ***g*** just before use. ^{125}I-labelled antigens were displayed on 4–11% sodium dodecyl sulphate (SDS)–polyacrylamide gels, autoradiographed and excised from the gel from quantitation.

Preparation of soluble immune complexes

Immune complexes were formed using standard procedures (e.g. Leslie 1980) by combining dinitrophenyl (DNP)–bovine serum albumin (> 10 DNP/mole) with affinity-purified rabbit or mouse anti-DNP IgG for 30 min at 37 °C in medium with or without fetal bovine serum. Various molar ratios of IgG:albumin were used, but most experiments used 2.5:1 complexes. The complexes remained soluble for at least 24 h, and IgG concentrations of 1–80 μg ml^{-1} yielded quantitatively similar results with respect to binding to Fc receptors.

Quantitative binding and endocytosis assays

Surface antigens were measured at 0 °C, using cells plated in 16 mm wells, and ^{125}I-labelled monoclonal antibodies (Steinman et al 1980, I. S. Mellman, unpublished results). Non-specific binding was measured using low specific activity controls. Cells were harvested with a cotton-tipped swab and counted in a gamma counter. Binding and internalization of immune complexes was determined at 0 °C and 37 °C respectively, as above, except that non-specific binding and uptake were measured by a simultaneous incubation of cells with 10 μg ml^{-1} of the anti-Fc receptor monoclonal antibody 2.4G2; see below (Unkeless 1979, Mellman & Unkeless 1980). On J774 cells, 'non-specific' binding was usually less than 15% of total. Surface-bound complexes could be specifically eluted at 0 °C using 50 μg ml^{-1} 2.4G2 IgG (2 h) or 1 mg ml^{-1} of the bacterial protease subtilisin BPN (Sigma) which degrades bound IgG (2 h; 10 mM-dithiothreitol, 5 mM-EDTA).

Results and discussion

Fc receptor structure

Macrophages and many lymphocytes bear receptors for the Fc portion of the IgG molecule. Moreover, at least on mouse macrophages it is clear that there is considerable receptor heterogeneity, with three discrete receptors having been identified by biochemical, genetic and (most recently) immunological criteria (Unkeless et al 1981). The most prevalent of these is specific for IgG1/IgG2b immune complexes and is trypsin-resistant; a second is trypsin-sensitive and binds monomeric IgG2a, and a third binds only IgG3. Fc receptors are of considerable interest not only because of their ability to mediate the pinocytosis or phagocytosis of bound ligands but also because of the various other effects on cell physiology that can be triggered by ligand binding (Unkeless et al 1981).

The many attempts to isolate Fc receptors via ligand-mediated affinity chromatography have yielded variable or contradictory results (Unkeless et al 1981). A new approach to the isolation of this molecule was suggested by the recent production of a high-affinity monoclonal antibody, designated 2.4G2, which was found to inhibit the activity of the IgG1/IgG2b immune complex receptor on intact cells (Unkeless 1979). Monovalent Fab fragments of the 2.4G2 antibody blocked the rosetting of IgG1/IgG2b-coated but not of IgG2a-coated erythrocytes to J774 cells; it bound only to cell types bearing the Fc receptor and it did not bind to Fc receptor-negative mutants of J774 cells.

We have been able to use Fab fragments of monoclonal antibody 2.4G2 coupled to Sepharose to immunoprecipitate Fc receptors from labelled cells and to isolate more than 300 μg of this receptor from J774 tumours (Mellman & Unkeless 1980). Depending on the cell type, the Fc receptor was found to consist of one or two glycosylated peptides with relative molecular masses (M_r) of 45 000–70 000. J774 cells exhibit two bands, M_r 60 000 and 47 000, while thioglycollate-elicited macrophages have only the smaller peptide. However, two-dimensional peptide mapping has suggested that these peptides are closely related, perhaps by post-translational proteolysis (I. S. Mellman & J. C. Unkeless, unpublished results).

Significantly, we were able to show that the isolated material exhibited Fc receptor activity. Fc receptor from J774 cells was iodinated *in vitro* and found to bind specifically to IgG-coated Sephadex beads or erythrocytes (Mellman & Unkeless 1980). Both the 60 000 and the 47 000 M_r peptides bound, and this binding was inhibited by 2.4G2.

We have now produced additional monoclonal and polyclonal antibodies raised against the isolated Fc receptor (Mellman et al 1981, and unpublished results). The latter, which has been used in some of the work to be discussed below, precipitates the same molecule(s) as 2.4G2 and inhibits the rosetting of opsonized erythrocytes as dilutions (of the original serum) of 1:1600. It also recognizes domains of the Fc receptor distinct from its active site since it will quantitatively immunoprecipitate receptor bound to either 2.4G2 or IgG.

Internalization of Fc receptor and other membrane during fluid endocytosis

The macrophage is a particularly good cell type in which to study endocytosis and PM recycling, given the rapid rate at which these events take place. Pinocytosis of extracellular fluid by macrophages continues for days in culture and brings about the interiorization of almost two cell-surface equivalents of PM every hour (Steinman et al 1976). PM internalized in this way is presumably recycled since estimates of PM protein turnover are usually given with half-lives of more than 20 h (Hubbard & Cohn 1975b, Kaplan et al 1979) (see below). Alternatively, if some or all of these PM proteins are excluded from nascent pinocytic vesicle membrane, their recycling need not be invoked (Pearse & Bretscher 1981). Thus, it was of interest to investigate the polypeptide composition of internalized membrane relative to its cell-surface progenitor.

Two additional questions relate specifically to Fc receptor function: (1) is the receptor internalized in the absence of ligand; and (2) if so, is it internalized selectively (as has been suggested for other receptors) or at a rate similar to that for other PM polypeptides?

Since pinocytic vesicles have proved to be difficult to isolate by subcellular fractionation, we developed an alternative approach which involved modifying the classical technique of lactoperoxidase–glucose oxidase (LPO–GO)-catalysed iodination to permit the selective labelling of pinocytic vesicles from within intact cells (Mellman et al 1980). Since LPO, like horseradish peroxidase (HRP) (Steinman & Cohn 1972, Steinman et al 1974), was found not to bind to the PM, suspension cultures of J774 cells could be incubated in LPO (0.5–1.5 mg ml^{-1}) and GO (< 20 μg ml^{-1}) at 37 °C for 3–5 min then quickly chilled and washed to remove extracellular enzyme. Addition of [^{125}I]sodium iodide at 0 °C resulted in the incorporation of radiolabel into pinocytic vesicle membrane protein and contents. The localization of incorporated ^{125}I was confirmed in several ways: (1) iodination was greatly reduced (> 95%) if pinocytic uptake of LPO or GO or both was inhibited (by low temperature, or by omitting the enzyme); (2) quantitative electron microscopy and autoradiography indicated that most radiolabel was present in the peripheral cytoplasm and not on the PM; (3) iodinated membrane antigens were largely inaccessible to specific monoclonal antibodies added to intact cells; and (4) the pinocytic vesicle content (e.g. LPO, IgG) was iodinated and could not be eluted from the cell surface (see below).

The spectrum of iodinated polypeptides from pinocytic vesicle membrane was found by SDS gel electrophoresis to be quite similar to the PM proteins labelled after a standard cell-surface iodination, i.e. extracellular LPO, GO and [^{125}I]sodium iodide at 0 °C (Mellman et al 1980) (Fig. 1). Consistent differences between the two patterns were noted in three or four bands of low M_r (< 40 000); also, a species of 330 000 M_r was often more heavily labelled after iodination of the pinocytic vesicles. Most of the other differences were either inconsistent between experiments or were differences in band intensity (e.g. the 50 000 M_r species in Fig. 1). A major 80 000 M_r protein present only in the pinocytic vesicle profile was found to co-migrate with authentic LPO, and probably represents self-iodinated enzyme sequestered intracellularly. Using this method, similar results have subsequently been obtained for Chinese hamster ovary cells (Storrie et al 1981).

Iodinated pinocytic vesicle and PM proteins were also compared quantitatively by immune precipitation using anti-Fc receptor antibody and other anti-macrophage PM monoclonal antibodies. Each of the eight antigens studied, including the Fc receptor, was detected on pinocytic vesicle membrane and was labelled to the same relative extent as it was on the PM (Table 1). This finding suggests that most PM proteins, including the Fc receptor in the absence of ligand, are equally susceptible to being included in pinocytic vesicle membrane. One exception was the 22 000 M_r antigen, 2.6, which was always more heavily labelled intracellularly. Preliminary results indicate that this polypeptide may also be enriched in the membrane of phagocytic vesicles.

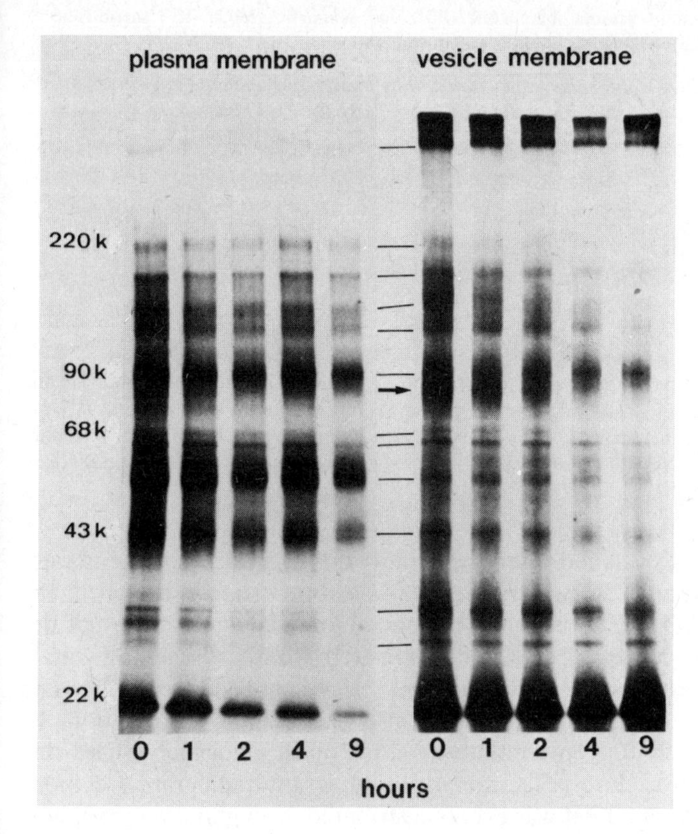

FIG. 1. Iodination and turnover of pinocytic vesicle membrane (right) and plasma membrane (left) proteins in J774 cells. Cells were radioiodinated as described in the text and recultured on 100 mm plates in medium containing 10% fetal bovine serum. At the indicated time intervals, cells were harvested, lysed in Nonidet P-40, and analysed by SDS-polyacrylamide gel electrophoresis under reducing conditions using 4–11% gradient gels. Relative molecular masses ($\times 10^3$) are given in left-hand column. Arrow marks the radiolabelled band that co-migrates with authentic lactoperoxidase.

While these findings support the concept that endocytosis and recycling involve most components of PM protein (Muller et al 1980, Cohn & Steinman 1982) they do not give any information about the protein:lipid ratio of internalized membrane. Thus, we do not yet know whether PM proteins are uniformly concentrated or depleted in pinocytic vesicle membranes. It is clear, however, that these proteins do not represent a distinct subpopulation destined for rapid degradation. As shown in Fig. 1, iodinated polypeptides of both the PM and the pinocytic vesicle membrane turn over at similarly slow rates ($t_{\frac{1}{2}} > 26$ h, determined by quantifying trichloroacetic acid-precipitable ^{125}I).

TABLE 1 Composition of plasma membrane (PM) and pinocytic vesicle (PV) membrane as determined by the relative labelling of eight membrane antigens

Monoclonal antibody[b]	*Relative ^{125}I labelling after isolation by immunoprecipitation*[a] *PM iodination*[c]	*PV iodination*[d]
2D2C (90k)	1.000	1.000
1.21J (94k, 180k)	0.552	0.452
2E2A (82k)	0.150	0.166
F480 (150k)	0.170	0.092
2F44 (42k)	0.035	0.024
25–1 (44k, 12k)	0.065	0.041
2.4G2 (60k, 47k)	0.022	0.026
2.6 (20k)	0.008	0.017

[a] Representative data from four paired experiments. [b] Numbers in parentheses indicate the relative molecular masses ($\times 10^{-3}$) of the polypeptide(s) precipitated by each antibody. [c] 7 744 counts per minute (c.p.m.) (expt 1); 127 781 c.p.m. (expt 2); 3 691 c.p.m. (expt 3); 55 045 c.p.m. (expt 4). [d] 11 341 c.p.m. (expt 1); 84 900 c.p.m. (expt 2); 4 712 c.p.m. (expt 3); 59 848 c.p.m. (expt 4). From Mellman et al 1980.

Similar results were obtained when the turnover of two individual membrane proteins was compared by immunoprecipitation. Interestingly, the presumptive content marker, LPO, is degraded much more rapidly than any of the major membrane proteins. ($t_{\frac{1}{2}} < 4$ h, estimated from gel-scanning data). Together, these results suggest that labelled vesicles can mediate the transfer of their contents to lysosomes while themselves escaping degradation. In addition, it seems that a representative sample of all pinocytic vesicles was labelled by our procedure. Electron microscopic autoradiography demonstrated that incorporated ^{125}I was evenly distributed throughout the peripheral cytoplasm, most of it apparently being associated with small vesicles (Mellman et al 1980).

We are now adapting intracellular iodination to facilitate study of receptor-mediated pinocytosis by preparing covalent conjugates of LPO and ligands such as IgG. In early experiments, however, ligand and peroxidase have been added separately. J774 cells were incubated at 0 °C in soluble immune complexes (monoclonal mouse anti-DNP IgG:DNP-albumin), washed and warmed for 5 min in phosphate-buffered saline with or without LPO–GO as above. After either intracellular or cell-surface iodination, labelled membrane proteins were compared and found to be again largely similar. Interestingly, considerable iodination of IgG was detected (by immunoprecipitation with anti-mouse IgG) in the pinocytic vesicle compartment. Very little of this IgG could be eluted from intact cells at 0 °C using high concentrations of 2.4G2, thus confirming its intracellular localization; in contrast, this treatment removed most of the labelled IgG after cell-surface iodination (Fig. 2). This assay provides a convenient method for confirming

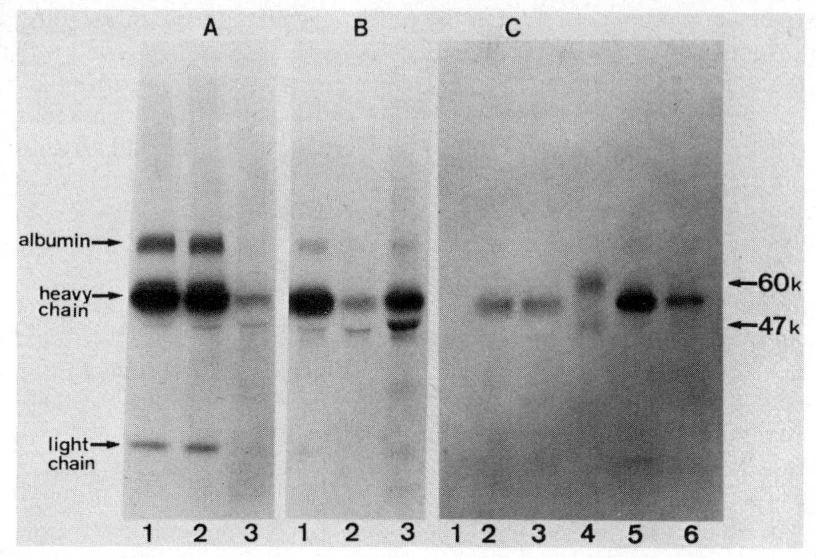

FIG. 2. Iodination and elution of Fc receptor-bound mouse IgG-immune complexes. *Panels A and B:* J774 cells were incubated at 0 °C with complexes, washed and warmed to 37 °C in phosphate-buffered saline for 5 min in the presence (A) or absence (B) of lactoperoxidase–glucose oxidase. Cells were then subjected to intracellular (A) or cell-surface (B) iodination. Each sample was divided in half, with the first aliquot lysed immediately in 0.5% Nonidet P-40 (lane 1) and the second incubated at 0 °C with 20 μg ml^{-1} 2.4G2 IgG for 2 h before lysis (lane 2). Lysates and the 2.4G2 eluate (lane 3) were incubated for 1 h at 0 °C with rabbit anti-mouse IgG followed by adsorption with protein A-Sepharose. Precipitates were analysed on 4–11% gels. *Panel C:* Cells were incubated with mouse immune complexes as above and subjected to cell-surface (lanes 1, 2 and 5) or intracellular (3 and 6) iodination. Cells in lane 3 were surface-labelled without prior incubation with complexes. Nonidet P-40-lysates were incubated with pre-immune IgG (lane 1) rabbit anti-mouse IgG (lanes 2 and 3) or rabbit anti-Fc receptor sera (lanes 4–6) followed by immunoprecipitation with protein A-Sepharose.

whether intracellular iodination has been achieved. Moreover, the fact that internalized IgG was labelled at all suggests that at least some Fc receptor-bound ligand entered the cells in the same vesicles as did the fluid-phase marker, LPO.

Coated vesicles and pinocytic vesicles

An important unknown in these studies is the primary route of entry of LPO into the cells. The traditional view, based on the cytochemical observations of Steinman & Cohn (1972) and Steinman et al (1974, 1976) is that fluid markers of pinocytosis are internalized in small (< 0.2 μm diameter) smooth-surfaced vesicles which may then fuse with each other, with lysosomes, or possibly with the PM (Besterman et al 1981). On the other hand, the uptake of receptor-bound or non-specifically adsorbed molecules has long been associated with the formation of clathrin-coated vesicles which are defrocked soon after their interiorization (Pearse & Bretscher 1981). However, recent quantitative studies of the uptake of Semliki Forest virus by baby hamster kidney cells (Marsh & Helenius 1980, Helenius & Marsh 1982) have suggested that most fluid pinocytosis in these cells can be accounted for by coated vesicles. In other words, fluid and adsorptive uptake may be reflections of the same continuous pinocytic process.

These suggestions are not necessarily in conflict with the findings of Steinman et al (1974, 1976) since there are numerous reasons why it would have been difficult to detect the HRP cytochemical reaction product in an evanescent population of coated vesicles. However, the function of these structures in fluid pinocytosis remains to be directly demonstrated either in fibroblasts or macrophages. It is interesting that a role for coat formation in macrophage phagocytosis has recently been implied (Aggeler et al 1982). Phagocytic vesicles are well known to contain a largely representative sample of PM proteins (Hubbard & Cohn 1975b, Muller et al 1980). The composition of coated pinocytic vesicle membrane is less well understood. Bretscher et al (1980) obtained evidence that two membrane antigens and their bound antibodies may be excluded from coated pits on fibroblasts. However, the increasing number of specific ligands (and presumably their receptors) as well as 'opportunistic' ligands (e.g. viruses and cationized ferritin, which presumably do not bind to specific receptors) that are found to be taken up in coated vesicles would imply that a wide range of PM proteins might be present in coated vesicle membrane.

Internalization and fate of Fc receptors during phagocytosis of IgG-coated erythrocyte ghosts

The above results indicate that, relative to other PM proteins, the Fc receptor is neither selectively included in nor excluded from pinocytic vesicle membrane. However, as has been indirectly suggested for other receptors, such selective internalization can be induced by the binding of specific ligand. Unfortunately, we have not been able to study this phenomenon using intracellular iodination since the binding of IgG to Fc receptor effectively prevents the LPO-mediated iodination of the receptor (Fig. 2). We have therefore used other approaches.

We first studied the internalization and fate of the Fc receptor and other PM components during the endocytosis of a large, multivalent ligand, the IgG-coated erythrocyte ghost (Mellman et al 1981, and unpublished results).

For these experiments, thioglycollate-elicited macrophages (2×10^5 Fc receptors per cell) were plated and allowed to bind opsonized sheep red cells (5×10^9 molecules of IgG per red cell) at 4 °C to form rosettes of 10–15 erythrocytes per macrophage. The red cells were lysed hypotonically, and quantitative binding studies were done with ^{125}I-labelled anti-Fc receptor antibody (2.4G2 Fab) and anti-PM antibodies. Rosetted macrophages bound less than 50% of the amount of 2.4G2 Fab bound by control macrophages; antibody binding to three other PM antigens (H-2D, 1.21J, and 2D2C) was unchanged. Thus, the binding of IgG-coated ghosts involved at least half of the surface pool of Fc receptors.

The cultures were warmed to 37 °C and the ghosts were rapidly internalized (< 15 min). Antibody-binding studies done at various times thereafter indicated that internalization was accompanied by a rapid but transient decrease (approximately 20%) in the expression of each of the PM antigens studied, including an additional 20% decrease in the amount of 2.4G2 Fab bound. While the amounts of surface 1.21J, H-2D, and 2D2C returned to control levels within 30 min, the number of Fc receptors remained at only 50% of control, increasing only slightly during the next 24 h. Thus, the binding and internalization of the IgG-coated ghosts resulted in the selective removal of Fc receptors from the PM; that this removal was largely irreversible indicated that these receptors, unlike other PM proteins, were not free to recycle. This selective interiorization of Fc receptors did not occur after the phagocytosis of 'non-specifically' ingested particles such as zymosan or latex beads.

The failure of these internalized receptors to return to the PM was a result of their rapid and apparently selective degradation in phagolysosomes. This was demonstrated by a series of turnover studies in which the macrophage PM was first iodinated and then the cells were returned to culture to bind and ingest IgG-coated red cell ghosts. At various time intervals, cells were harvested and lysed, and iodinated Fc receptors and other PM antigens were quantified by immunoprecipitation. To detect Fc receptors, a rabbit anti-Fc receptor serum was used since (unlike 2.4G2) it precipitated the Fc receptor completely, irrespective of whether receptor-bound ligand was present. In fact, the antiserum could specifically precipitate ^{125}I-labelled 2.4G2 Fab when the latter was bound to unlabelled Fc receptors. In control cells, Fc receptor was found to turn over with a $t_{\frac{1}{2}}$ of 10 h. On the other hand, in cells that had ingested IgG-coated ghosts, more than 50% of the labelled receptor was rapidly degraded, with $t_{\frac{1}{2}}$ equal to 2 h, while the remainder was turned over at the control rate. The other membrane antigens that we studied turned over at similarly slow rates ($t_{\frac{1}{2}}$ = 18–22 h) in both control and ingesting cells.

Because the Fc receptor is internalized during fluid pinocytosis and normally has a relatively long (10 h) half-life the unbound receptor is probably capable of recycling. Why it should be removed from the recycling pathway when it has ingested IgG-coated ghosts is not yet clear. Ligand binding may cause some conformational change in the receptor which makes it more susceptible to proteolysis. Alternatively, the internalization of a large segment of PM selectively enriched in Fc receptor–ligand complexes may in effect deliver receptor to the lysosomal compartment faster than it can be recycled.

Internalization of Fc receptors during pinocytosis of soluble immune complexes

We have recently begun to investigate the dynamics of the Fc receptor during receptor-mediated pinocytosis. These studies were partly intended to determine whether the inability of the receptor to recycle after phagocytosis of Ig-coated ghosts was a consequence of the large size of the PM segment that was interiorized during this process. The ligand we have used is a soluble, IgG-containing immune complex, formed from rabbit anti-DNP IgG and DNP-bovine serum albumin at a molar ratio of 2.5:1. These complexes bind avidly and specifically to Fc receptors on J774 cells and on thioglycollate-elicited macrophages. At 0 °C, saturation occurs at 1.3×10^{-7} M–IgG (approximately 20 μm ml^{-1}) and about 80% of this binding is inhibited by 2.4G2. Moreover, the number of complexes bound per cell agrees well with the number of surface Fc receptors measured by using ^{125}I-labelled 2.4G2 Fab (500 000 receptors per J774 cell) (Mellman & Unkeless 1980). It seems likely that bound complexes are internalized at 37 °C via coated pits and coated vesicles, and our preliminary results using colloidal gold conjugates support this concept (see below).

The behaviour of the Fc receptors during pinocytosis of soluble complexes seems to involve both a rapid redistribution of receptors from the PM to some intracellular compartment(s) and the possible re-utilization or recycling of at least some of these receptors. The uptake of ^{125}I-labelled IgG by J774 cells at 37 °C is rapid, reaching a maximum (by 1 h) which is maintained for as long as 24 h. On the other hand, degradation of the ligand, which is detectable within 30 min, is relatively slow, with only about 25% of the total cell-associated ^{125}I released as trichloroacetic acid-soluble material in the medium every hour. Interestingly, however, most of the cell-associated IgG is intracellular. This can be demonstrated by immunofluorescence (using rhodamine-labelled goat anti-rabbit IgG) on intact or Triton-treated cells after a 2-h continuous exposure to saturating concentrations of complexes (Fig. 3). These results suggested that not only were bound complexes internalized, but also that the number of cell-surface Fc receptors (and accordingly the rate of uptake) had been reduced.

The kinetics of these processes were examined quantitatively by studying

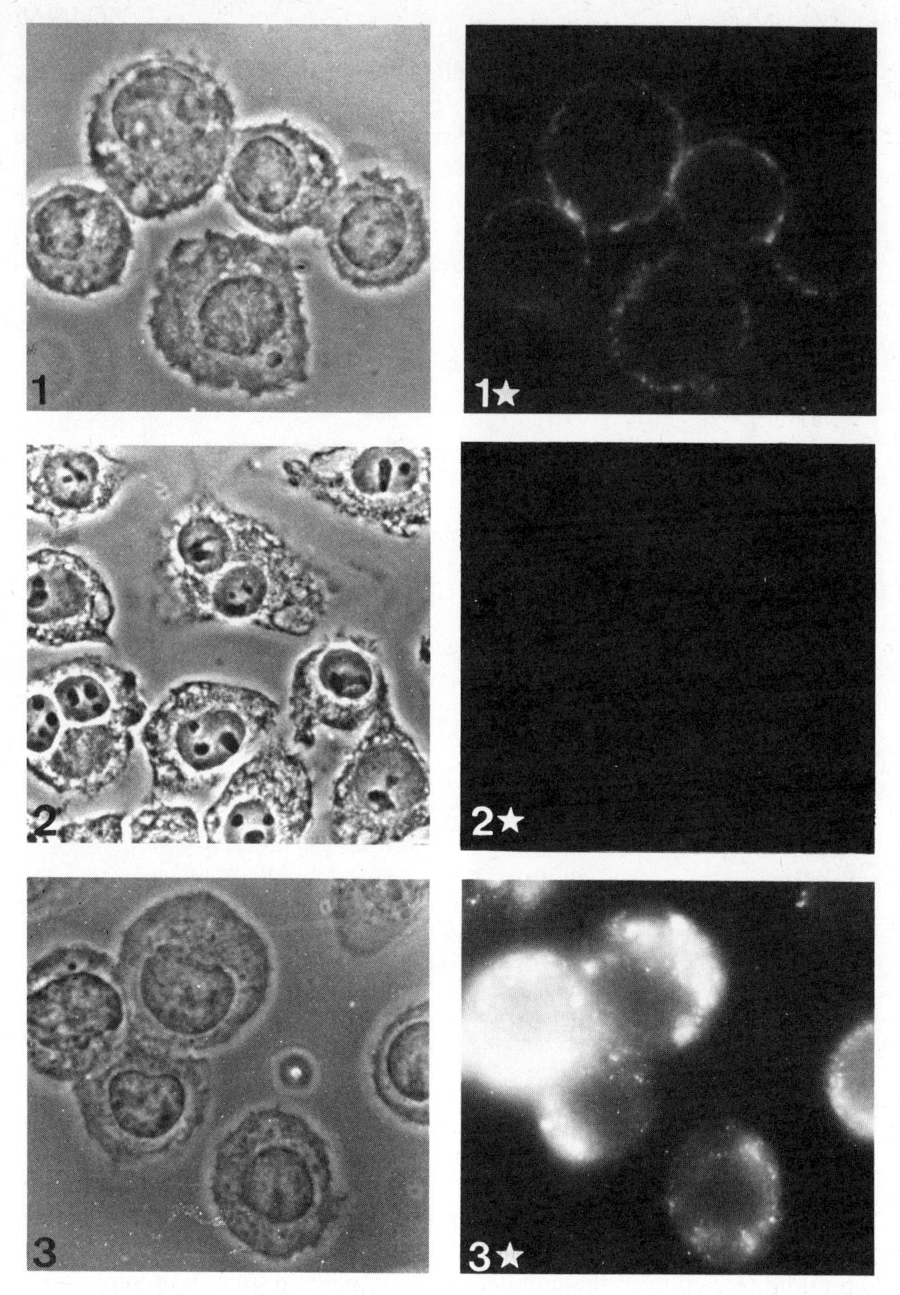

FIG. 3. Visualization of immune-complex binding and endocytosis by immunofluorescence. (1) J774 cells incubated with immune complexes at 0 °C for 2 h, fixed with 1% paraformaldehyde in phosphate-buffered saline, and stained with rhodamine-goat anti-rabbit IgG. (2) Cells incubated for 2 h at 37 °C with complexes and stained as above. (3) Cells incubated as in panel 2 and treated with 0.1% Triton X-100 before staining.

the time course of uptake of ^{125}I-labelled complexes and by determining the amount of intracellular (subtilisin-resistant) and PM-associated (subtilisin-releasable) IgG at each time point (Fig. 4). These experiments demonstrated that internalization of bound complexes is very rapid, perhaps as fast or faster that the $t_{\frac{1}{2}}$ for the initial binding of ligand to receptor. Within 15 min, more than half of the cell-associated IgG was intracellular. This amount increased to a plateau of about 80% by 2 h. Moreover, the amount of ^{125}I-labelled IgG

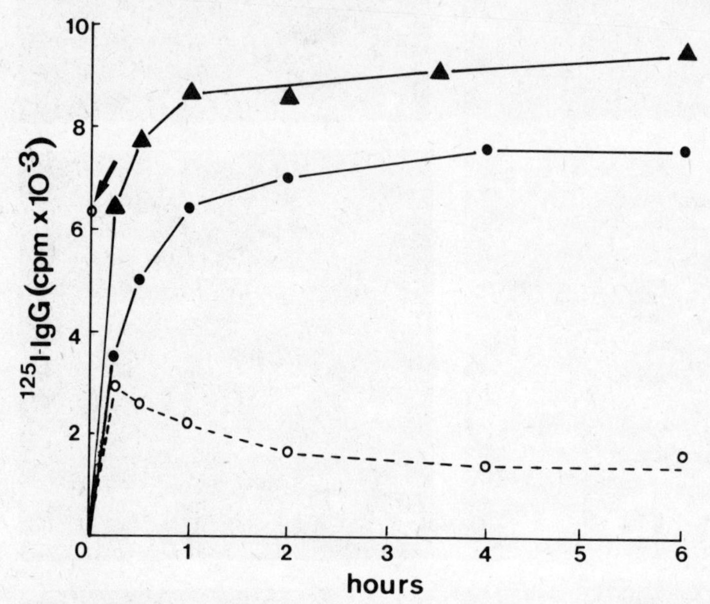

FIG. 4. Binding and uptake of immune complexes by J774 cells. Cells were incubated at 37 °C with 20 μg ml^{-1} ^{125}I-labelled rabbit IgG immune complexes. At the indicated time intervals, cells were either harvested (total cell-associated ^{125}I, solid triangles) or treated with 1 mg ml^{-1} subtilisin at 0 °C for 2 h. Subtilisin-releasable radioactivity (cell-surface, open circles) and subtilisin-resistant radioactivity (intracellular, solid circles) were determined, the sum of which closely approximated the amount of total cell-associated ^{125}I. Arrow indicates amount of [^{125}I]IgG bound at 0 °C.

bound to surface Fc receptors at the time this new equilibrium was reached was almost four-fold lower than the amount specifically bound to control cells at 0 °C (Fig. 4). That this decrease in surface-bound ligand was a reflection of a decreased number of Fc receptors was confirmed by using ^{125}I-labelled 2.4G2 Fab. After a 2-h incubation in unlabelled complexes at 37 °C a three-fold reduction in the number of 2.4G2 binding sites was found.

The fact that the amount of cell-associated ^{125}I-labelled IgG remained constant for long periods (up to 24 h) indicates that a steady state is reached in which the rate of degradation of IgG (which is continuous for 24 h) is balanced by the continuous internalization of complexes. Thus, the apparent redistribution of Fc receptors may not be an instance of 'down-regulation' or irreversible removal of receptors from the cell surface. However, it is not yet clear whether this continued uptake of complexes results from receptor recycling or from new synthesis of receptors.

Interestingly, some interiorized receptors seem to be able to return to the PM. When cells were incubated in complex-free medium after a 2-h exposure at 37 °C, almost 50% of the internalized 2.4G2 binding sites returned to the cell surface within 2 h. However, it is possible that these Fc receptors were recruited from a pre-existing intracellular pool, or were the result of *de novo* synthesis.

We have recently begun examining the mechanism of IgG-immune complex pinocytosis in J774 cells. Our preliminary results with colloidal gold conjugates suggest that uptake occurs via the formation of clathrin-coated pits and coated vesicles. We have bound rabbit immune complexes to cells at 0 °C (for 1 h) and then visualized them by using protein A–gold (17 nm diameter) complexed with affinity purified goat anti-rabbit IgG (for 1 h at 0 °C). This complex was prepared by incubating protein A–gold (1:10 dilution) with the antibody (50 μg/ml) in phosphate-buffered saline that contained 0.2% polyethylene glycol (1 h, 0 °C). The complexes were then pelleted at 100 000 **g**, and the pellets were resuspended in phosphate-buffered saline devoid of added antibody. At 0 °C, gold particles were observed singly or in pairs apparently at random over the entire cell surface. Most of this binding could be inhibited by 2.4G2. Following warming to 37 °C for 4 min, most of the gold was present intracellularly in irregularly shaped vacuoles (endosomes) or on the surface in clusters of 3–6 particles at coated invaginations of the plasma membrane. A few particles were also noted, at the same time, in coated vesicles and dense bodies (which may be lysosomes).

In any event, these observations are consistent with a ligand-induced alteration of the rates of Fc receptor interiorization, or recycling, or both. We are currently investigating this problem by study Fc receptor turnover during the internalization of soluble complexes. In addition, we are examining in detail the morphological aspects of the internalization of both soluble complexes and the receptor itself, using our anti-Fc receptor reagents.

Acknowledgements

I should like to thank Professor Zanvil A. Cohn for his advice and support; many of the studies described here were initiated in his laboratory. I am also grateful for the collaborative assistance of Jay C. Unkeless, Ralph M. Steinman and Helen Plutner. This work is supported by grants from the

National Institutes of Health (GM29765 and BRSG RR05358), by a Junior Faculty Research Award from the American Cancer Society (JFRA–26) and by an award from the Swebilius Fund.

REFERENCES

Aggeler J, Heuser J, Werb Z 1982 The distribution of clathrin in phagocytosing macrophages. J Cell Biol, in press

Besterman JM, Airhart JA, Woodworth RC, Low RB 1981 Exocytosis of pinocytosed fluid in cultured cells: kinetic evidence for rapid turnover and compartmentation. J Cell Biol 91:716-727

Bretscher MS, Thomson JN, Pearse BMF 1980 Coated pits act as molecular filters. Proc Natl Acad Sci USA 77:4156-4159

Cohn ZA, Steinman RM 1982 Phagocytosis and fluid-phase pinocytosis. In: Membrane recycling. Pitman Books Ltd, London (Ciba Found Symp 92) p 15-34

Gefter MC, Margulies DH, Scharff MD 1977 A simple method for polyethylene glycol-promoted hybridization of mouse myeloma cells. Somat Cell Genet 3:231-236

Helenius A, Marsh M 1982 Endocytosis of enveloped animal viruses. In: Membrane recycling. Pitman Books Ltd, London (Ciba Found Symp 92) p 59-76

Hubbard AL, Cohn ZA 1975a Externally disposed plasma membrane proteins. I: Enzymatic iodination of mouse L cells. J Cell Biol 64:438-460

Hubbard AL, Cohn ZA 1975b Externally disposed plasma membrane proteins. II: Metabolic fate of iodinated peptides in mouse L cells. J Cell Biol 64:461-479

Kaplan G, Unkeless JC, Cohn ZA 1979 Insertion and turnover of macrophage plasma membrane proteins. Proc Natl Acad Sci USA 76:3824-3828

Leslie RGQ 1980 The binding of soluble immune complexes of guinea pig IgG_2 to homologous peritoneal macrophages. Determination of avidity constants at 4 °C. Eur J Immunol 10:317-322

Marsh M, Helenius A 1980 Adsorptive endocytosis of Semliki Forest virus. J Mol Biol 142:439-454

Mellman IS, Unkeless JC 1980 Purification of a functional mouse Fc receptor through the use of a monoclonal antibody. J Exp Med 152:1048-1069

Mellman IS, Steinman RM, Unkeless JC, Cohn ZA 1980 Selective iodination and composition of pinocytic vesicles. J Cell Biol 86:712-722

Mellman IS, Unkeless JC, Steinman RM, Cohn ZA 1981 Internalization and fate of Fc receptors during endocytosis. J Cell Biol 91:124a

Muller WA, Steinman RM, Cohn ZA 1980 The membrane proteins of the vacuolar system. II: Bidirectional flow between secondary lysosomes and the plasma membrane. J Cell Biol 86: 304-314

Pearse BMF, Bretscher MS 1981 Membrane recycling by coated vesicles. Annu Rev Biochem 50:85-107

Steinman RM, Cohn ZA 1972 The interaction of soluble horseradish peroxidase with mouse peritoneal macrophages in vitro. J Cell Biol 55:186-204

Steinman RM, Silver JM, Cohn ZA 1974 Pinocytosis in fibroblasts. Quantitative studies in vitro. J Cell Biol 63:949-969

Steinman RM, Brodie SE, Cohn ZA 1976 Membrane flow during pinocytosis. A stereologic analysis. J Cell Biol 68:665-687

Steinman RM, Nogueira N, Witmer MD, Tydings JD, Mellman IS 1980 Lymphokine enhances the expression and synthesis of Ia antigens on cultured mouse peritoneal macrophage. J Exp Med 152:1248-1261
Storrie B, Dreesen TD, Maurey KM 1981 Rapid cell surface appearance of endocytic membrane protein in Chinese hamster ovary cells. Mol Cell Biol 1:261-268
Unkeless JC 1979 Characterization of a monoclonal antibody directed against mouse macrophage and lymphocyte Fc receptors. J Exp Med 150:580-596
Unkeless JC, Fleit H, Mellman IS 1981 Structural aspects and heterogeneity of immunoglobulin Fc receptors. Adv Immunol 31:247-270

DISCUSSION

Rothman: Have you tried to conjugate a particle like low density lipoprotein (LDL), which you know enters the cell via coated pits, to lactoperoxidase? One could then study the same issues with coated vesicles specifically; by iodinating inside the cell, one can test whether the vesicle membrane is representative of the plasma membrane as a whole.

Mellman: This is an excellent idea. Initially, however, I intend to do these experiments by using lactoperoxidase coupled to Semliki Forest virus (see Helenius & Marsh, this volume, p 59-76). The virus provides an excellent morphological marker so that, at least in principle, one has the ability to see where the lactoperoxidase is.

Geisow: How do you interpret the sudden disappearance of the fluorescence in your slides of labelled immunoglobulin G (IgG)? Is it because the endocytic vesicles become acid very quickly and quench the fluorescence? (You said that the fluorescence returned with Triton treatment.)

Mellman: These were indirect immunofluorescence experiments in which the rabbit IgG–dinitrophenol–bovine serum albumin immune complexes were visualized using a rhodamine-conjugated second antibody. Thus, these findings cannot be explained by invoking acidification; after 2 h at 37 °C in immune complexes, most of the ligand was simply inaccessible to the fluorescent probe unless the macrophages were permeabilized. The point here is that very little of the ligand was on the cell surface, although the ligand was present continuously during the 2 h incubation. Fc receptors had apparently been redistributed from the plasma membrane to some intracellular site(s).

We have looked at the rapidity of this redistribution by examining the internalization at 37 °C of immune complexes pre-bound at 0 °C. This internalization is quite rapid, with a $t_{\frac{1}{2}}$ of 3–4 min (I. S. Mellman, unpublished). We know that J774 macrophages have approximately 500 000 receptors per cell. So an enormous number can be internalized very quickly.

Geisow: I have looked spectroscopically at the uptake by macrophages of fluoresceinated ovalbumin in soluble complexes with immunoglobulin. I found that almost immediately after warm-up of cells with bound fluorescent complexes the environment of the resulting endosomes was acidic (M. J. Geisow, unpublished results). Either the endosome itself becomes acid or it fuses with an acidic vesicle almost instantaneously. In my experiments there was also a rapid recovery of fluorescence because, presumably, degradation occurs in the lysosomes to which the fluorescent complexes are transferred. The label then comes out of the lysosomes, so causing the fluorescence to build up again. The feature here is that an acid environment quenches the fluorescence. The acid environment observed after warm-up therefore quickly converts to an alkaline environment, which represents the cytoplasm.

Mellman: It does not surprise me that acidification occurs soon after endocytosis, especially in view of the findings of Helenius & Marsh (this volume, p 59-76) on the kinetics of virus infection. While I favour the possibility that the endosome itself may develop a low pH, one must keep in mind that in macrophages (and fibroblasts) endosomes and lysosomes may fuse within a few minutes of internalization (e.g. Steinman et al 1976). Although I have done only limited electron microscopy with conjugates of colloidal gold and immune complexes, I was surprised to find evidence for the localization of some gold particles in morphologically identifiable lysosomes (dense, multivesicular bodies) 4 min after endocytosis (unpublished).

Goldstein: In your experiments with the monoclonal ^{125}I-labelled IgG directed against the Fc receptor, only about 25% of internalized monoclonal IgG was degraded. Why was all of it not degraded?

Mellman: Because it is degraded relatively slowly, with a $t_{\frac{1}{2}}$ of about 2.5 h. The rate of degradation of the soluble IgG complex therefore appears to be similar to that for IgG bound to a phagocytic ligand. Given enough time, complete degradation will occur.

Goldstein: Negatively charged LDL, such as acetylated LDL, binds with high affinity to the cell surface of the J774 cells as well as to mouse peritoneal macrophages. In pulse–chase experiments, virtually 100% of the receptor-bound ^{125}I-labelled acetyl-LDL that is internalized is rapidly degraded in lysosomes; [^{125}I]monoiodotyrosine begins to appear in the culture medium within five minutes, and all the cell-bound acetyl-LDL is degraded within 15–30 min (Goldstein et al 1979). It seems that these kinetics are quite different from those you are describing for IgG coupled to red cells or to soluble monoclonal IgG complexes involving the Fc receptor.

Mellman: This difference may just reflect the varying susceptibilities of different proteins to intracellular proteolysis. For example, the albumin portion of these complexes seems to be degraded much more rapidly than is the IgG (I. S. Mellman, unpublished; Leslie 1980). IgG degradation by fibroblasts also appears to be quite slow (Schneider et al 1979). In macrophages, the proteolysis of IgG seems to be initiated quite rapidly by a clip that occurs around the hinge region. The Fab and/or Fc fragments so generated, however, persist for some time (I. S. Mellman, unpublished).

Goldstein: What is the half-time for internalization? That would tell you whether there is a difference in the internalization process or in the resistance of the IgG to degradation. If you bind the antibody to the Fc receptor at 4 °C, wash away unbound antibody, and then warm the cells to 37 °C, is all the receptor-bound antibody inside the cells after 5–10 min?

Mellman: Yes.

Goldstein: So that is similar to the acetylated LDL.

Widnell: When you use IgG-coated erythrocytes to provoke the internalization of receptor and deplete the number of surface receptors by about 50%, what happens if you then expose the cells to additional IgG-coated erythrocytes? Can you eventually internalize all the receptors, leaving none on the surface?

Mellman: I have not tried that, but I would suspect not. 'Rosetting' of erythrocytes around the macrophage is a highly cooperative process. If one measures the numbers of erythrocytes bound per cell as a function of the number of IgG molecules present on the erythrocytes one observes a steep sigmoidal relationship. Optimal conditions for rosetting of erythrocytes around the macrophage are in the middle of that curve so a slight decrease in the number of IgG molecules or receptors for IgG may decrease disproportionately the number of erythrocytes that can be bound (D. Yoshie, I. S. Mellman and P. Lengyel, unpublished results).

Cohn: There are ways of removing most of the Fc receptors from given surfaces of the macrophage. Michl et al (1979) have shown that by plating macrophages on immune complexes one can modulate all the Fc receptors, and by plating them on immune complexes with complement one can modulate the complement receptor, leaving the Fc receptor on the surface. These approaches might be useful for studying receptor movement.

Mellman: When Michl et al (1979) measured the presence of apical Fc receptors by rosetting they found that the disappearance of the receptors was more rapid than the disappearance of 2.4G2 binding sites. This suggests that slight decreases in the concentration of receptor have relatively great effects on rosetting.

Widnell: So if the cell no longer rosettes, and it still has half the Fc receptors on the surface, does that influence the capacity of the cell to take up immunocomplexes by receptor-mediated pinocytosis?

Mellman: I don't think it has any influence. We measured the internalization of labelled complexes by cells previously exposed to unlabelled complexes. The remaining receptors (about 33% of control) were still capable of mediating uptake at an apparently normal rate. Thus, the capacity for complex pinocytosis seems to be a function of the number of receptors present on the cell surface.

Hubbard: Your pulse–chase experiments indicated the reappearance of the Fc receptor after washing out excess soluble immune complexes. What is your evidence that this is the reappearance of pre-existing receptors rather than the expression of new receptors?

Mellman: Kinetically, the return of receptors to the cell surface happens too rapidly for new synthesis; however, this interpretation is complicated by two points. First, I tried a few experiments using cycloheximide only to find that receptor reappearance was blocked, but the inhibitor also killed the cells (I. S. Mellman, unpublished). In addition, I cannot say whether we are seeing the return of previously internalized receptors, or the recruitment to the cell surface of receptors from a pre-existing intracellular pool.

Baker: When you talk about turnover of all membrane components, seemingly in parallel, what is the evidence that trans-membrane proteins like sodium–potassium pumps, and proteins at the inner face of the plasma membrane, like cyclase, are going through the same system at the same rate? Have you labels for these proteins?

Mellman: Most of the polypeptides labelled on macrophages by lactoperoxidase-catalysed iodination turn out to be intrinsic membrane proteins (I. S. Mellman, unpublished); this is also true for the Fc receptor (Mellman & Unkeless 1980). I expect a number of these will be shown to span the bilayer as well. While I have not yet studied the internalization and fate of sodium–potassium pumps in particular, I suspect that they are internalized, recycled, and turned over in a manner similar to most other plasma membrane proteins. My suspicion is based on the overall similarity between the polypeptide constituents of endosome and plasma membranes. Acidification of endosomes may result from the interiorization of a plasma membrane proton pump (or exchanger), which would then pump protons into the lumen of the endocytic vacuole. I cannot speak about the behaviour of any integral or peripheral membrane proteins that are not externally disposed (i.e. that face the cytoplasm) because none of the probes I have used can label such proteins.

Baker: So although you say that the membrane as a whole is turning over, it is really for just a small subset of labelled proteins that you have evidence of turnover, and it is not necessarily the whole membrane that turns over.

Mellman: Obviously I do not have information about all plasma membrane proteins. However, most components are probably subject to this general pathway of membrane turnover and, if anything, only a small subset may be excluded. Iodination seems to label most, if not all, of the 'major' membrane polypeptides, since the electrophoretic pattern obtained with this label is quite similar to patterns obtained if cells are labelled metabolically with [^{35}S]methionine or [^{3}H]mannose and if cell-surface proteins are isolated after derivatization with trinitrobenzene sulphonic acid (Kaplan et al 1981). Similarly, these [^{35}S]methionine- and [^{3}H]mannose-labelled proteins turn over at slow and fairly synchronous rates (Kaplan et al 1979, Baumann & Doyle 1980).

In addition to examining only major membrane proteins by one-dimensional sodium dodecyl sulphate gel electrophoresis, we have increased our level of resolution by using the monoclonal anti-macrophage antibodies that I described. Many of these antigens cannot be recognized in the gel pattern of the whole cell lysate and they represent both highly redundant and relatively minor polypeptides. In every case, these antigens are present on internalized membrane, sometimes in increased amounts.

Jourdian: Has anyone looked at the effect of pH on the different types of endocytosis (fluid-phase, solid, receptor-mediated)? Receptor-mediated endocytosis, at least in lysosomal enzymes carrying the phosphomannosyl recognition marker, is under fine pH control (Rome et al 1979, Gonzalez-Noriega et al 1980, Sahagian et al 1981).

Helenius: We have looked at the pH dependence of both fluid-phase uptake and virus uptake, and they are different. The rate of virus uptake decreases by about 30% when the pH is dropped from 7.0 to 6.4. Fluid-phase uptake, by contrast, remains the same at both pH values (M. Marsh, E. Bolzau, J. White and A. Helenius, unpublished work).

Bretscher: Is the binding of virus similar at both pHs?

Helenius: The virus appears to bind to the same sites at both pHs, and the binding actually increases somewhat at lower pH. Since the viruses are not binding into the coated pits, they must be transported along the membrane to reach the coated pits. This 'gliding' of the virus along the surface of the cell may be slowed down at lower pH.

Goldstein: It would be interesting to determine directly whether the slow lysosomal degradation of the monoclonal antibody against the Fc receptor is due to some inherent structural property of this particular antibody molecule. One could conjugate mannose to the monoclonal IgG and have it taken up by the mannose receptor of macrophages (Stahl et al 1980). When mannose is conjugated to albumin, the uptake and degradation of the mannose–albumin complex is rapid, as for acetylated LDL. One could do a similar experiment with a mannose conjugate to your monoclonal antibody. This approach would tell us whether there are different routes for the degradative pathway of various receptor-bound ligands or whether there is an inherent resistance to lysosomal degradation of the particular IgG molecule that you are studying.

Dean: We have done an experiment of the kind you just described. We compared the degradation of mannose-bovine serum albumin with that of human serum albumin taken up in macrophages. The degradative half-lives of the two polypeptides were different, in spite of their molecular similarity, and this is consistent with the possibility of different uptake routes (Faghihi Shirazi et al 1982).

Cohn: We know very little about the cytoplasmic face of these vesicles. It is difficult to label these structures preferentially, partly because isolated vesicles do not survive long enough. If one tries to iodinate the external face of isolated phagolysosomes, one often labels an internal marker in the phagolysosome, which prevents us from following this up.

Sly: Is the accelerated half-life of Fc caused by some Fc being internalized from the medium, compared to the other markers?

Mellman: That is possible. The medium is a low percentage (10%) fetal calf serum which contains little complexed IgG. For instance, one cannot use the serum to inhibit the binding of labelled complexes. It is a poor competitive inhibitor.

Sly: If you had Fc receptors that were free of ligand you might increase their half-life to a level comparable to the half-life for the other ligands that are not being occupied and internalized.

Mellman: I have never tried to deplete the serum of all immunoglobulin and to maintain cells in it for long periods of time to see if there is an increase in half-time for the receptor.

Rodewald: There should be virtually no IgG in your fetal calf serum, though, because in this species maternal IgG is transferred to the young wholly after birth.

Mellman: That's what my measurements have suggested, but I have not used a sensitive method such as radioimmunoassay to test and confirm it.

Palade: How good is the evidence that coated vesicles are involved in the uptake of quasi-representative samples of the plasmalemma? For example, in your experiments, what was the rate of uptake of marked antigen–antibody complexes taken up by coated vesicles, as compared to the rate of uptake by other means?

Mellman: Because I have examined only a small number of sections, I cannot say for certain. However, I suspect that most of the IgG–gold complexes enter macrophages in coated vesicles. Soon after warming cells to 37 °C, almost all the surface-bound gold particles were clustered at coated pits. At 0 °C, such accumulations were not observed. However, I have no way of determining the contribution of smooth vesicles to uptake.

With respect to your first question, there is no direct evidence that coated vesicles internalize quasi-representative samples of plasma membrane. My results are not directly relevant to this issue unless, of course, fluid-phase and adsorptive uptake turn out largely to reflect the same coated-vesicle-mediated process (as suggested by Marsh & Helenius 1980). My data support the concept of an intracellular compartment that contains a largely representative sample of plasma membrane proteins. This compartment, which we are now calling endosomes, would thus be in communication with the plasma membrane via coated vesicles, at least in one direction. Thus, coated vesicles may deliver plasma membrane proteins to endosomes. If IgG-complexes are internalized primarily in coated vesicles, this view is supported by one of the intracellular iodination experiments that I discussed earlier (Fig. 4, p 48). Lactoperoxidase internalized in the fluid phase in the presence of receptor-bound complexes labelled *both* the IgG and the representative plasma membrane components intracellularly.

However, the coated vesicle membrane is specialized in some ways. Certain receptor-bound ligands seem to localize preferentially at coated pits. In addition, Dr Bretscher (see this volume, p 266-282) has some results showing that two membrane antigens may be excluded from coated regions. It remains to be seen whether these specializations are causes or consequences of coat formation.

Bretscher: Coated pits contain a 50-fold higher concentration of LDL receptors than the rest of the plasma membrane (Anderson et al 1976). Presumably this is also true for many other specific receptors. It is therefore unlikely that the composition of the coated pit can be similar to that of the plasma membrane, irrespective of any experiments that have been done.

Rothman: One should try to distinguish whether the apparently random sampling of membrane that is internalized is diluted or concentrated with regard to lipid.

Mellman: That is possible, but our work is now moving towards subcellular fractionation. We have developed a very sensitive radioimmunoassay for membrane proteins, which we hope to use to analyse coated-vesicle fractions (Mellman & Galloway 1982). We want to correlate the amount of protein present with the amount of lipid that can be measured either by isotopic means or by direct analysis.

REFERENCES

Anderson RGW, Goldstein JL, Brown MS 1976 Localisation of low density lipoprotein receptors on plasma membrane of normal human fibroblasts and their absence in cells from a familial hypercholesterolemia homozygote. Proc Natl Acad Sci USA 73:2434-2438

Baumann H, Doyle D 1980 Metabolic fate of cell-surface glycoproteins during immunoglobulin-induced internalization. Cell 21:897-907

Faghihi Shirazi M, Aronson NN, Dean RT 1982 Temperature dependence of integrated membrane functions in macrophages. J Cell Sci, in press

Goldstein JL, Ho YK, Basu SK, Brown MS 1979 Binding site on macrophages that mediates uptake and degradation of acetylated low density lipoprotein, producing massive cholesterol deposition. Proc Natl Acad Sci USA 76:333-337

Gonzalez-Noriega A, Grubb JH, Talkad V, Sly WS 1980 Chloroquine inhibits lysosomal enzyme pinocytosis and enhances lysozyme enzyme secretion by impairing receptor recycling. J Cell Biol 85:839-844

Kaplan G, Unkeless JC, Cohn ZA 1979 Insertion and turnover of macrophage plasma membrane proteins. Proc Natl Acad Sci USA 76:3824-3828

Kaplan G, Plutner H, Mellman I, Unkeless JC 1981 Studies on externally disposed plasma membrane proteins. Exp Cell Res 133:103-114

Leslie RGQ 1980 Macrophage handling of soluble immune complexes: ingestion and digestion of surface-bound complexes at 4 °C, 20 °C and 37 °C. Eur J Immunol 10:323-333

Marsh M, Helenius A 1980 Adsorptive endocytosis of Semliki Forest virus. J Mol Biol 142:439-454

Mellman I, Galloway C 1982 Selective labeling and quantitative analysis of internalized plasma membrane. Methods Enzymol, in press

Mellman IS, Unkeless JC 1980 Purification of a functional mouse Fc receptor through the use of a monoclonal antibody. J Exp Med 152:1048-1069

Michl J, Pieczonka MM, Unkeless JC, Silverstein SC 1979 Effects of immobilized immune complexes on Fc- and complement-receptor function in resident and thioglycollate-elicited mouse peritoneal macrophages. J Exp Med 150:607-621

Rome LH, Weissmann B, Neufeld EF 1979 Direct demonstration of bonding of a lysosomal enzyme α-L-iduronidase to receptors on cultured fibroblasts. Proc Natl Acad Sci USA 76:2331-2334

Sahagian GG, Distler J, Jourdian GW 1981 Characterization of a membrane-associated receptor from bovine liver that binds phosphomannosyl residues of bovine testicular β-galactosidase. Proc Natl Acad Sci USA 78:4289-4293

Schneider Y-J, Tulkens P, de Duve C, Trouet A 1979 Fate of plasma membrane during endocytosis. II: Evidence for recycling (shuttle) of plasma membrane constituents. J Cell Biol 82:466-474

Stahl P, Schlesinger PH, Sigardson E, Rodman JS, Lee YC 1980 Receptor-mediated pinocytosis of mannose glycoconjugates by macrophages: characterization and evidence for receptor recycling. Cell 19:207-215

Steinman RM, Brodie SE, Cohn ZA 1976 Membrane flow during pinocytosis. A stereologic analysis. J Cell Biol 68:665-687

Endocytosis of enveloped animal viruses

ARI HELENIUS and MARK MARSH

Section of Cell Biology, Yale University School of Medicine, P.O. Box 3333, 333 Cedar Street, New Haven, Connecticut 06510, USA

Abstract After attaching to the cell surface, virus particles are rapidly internalized by endocytosis and channelled into the lysosomal compartment. The endocytosis occurs by a pinocytic process involving coated pits and coated vesicles. Intermediate pre-lysosomal vacuoles, termed endosomes, are recognized as a part of the intracellular pathway. Our studies have shown that for several of the enveloped viruses (toga viruses, orthomyxoviruses and rhabdoviruses) the endocytic pathway is essential for productive infection. In these cases the viral genome penetrates from the lysosomes where the virus membrane fuses with the lysosomal membrane. The low pH in the lysosomes triggers membrane fusion by causing a conformational change in the virus spike glycoproteins, which results in the expression of potent fusion activity. As a result of the fusion reaction the nucleocapsids are transferred into the cytoplasm. In this paper we review some work in which Semliki Forest virus (SFV) has been used to probe the adsorptive endocytosis pathway in baby hamster kidney (BHK-21) cells. In addition, we present new data on the kinetics by which the contents of the endocytic vacuoles become acidified. In these studies the pH-dependent penetration by SFV has been used as an indicator of pH.

Viruses depend totally on their host cells for replication. Virus genomes code only for the structural proteins of the virus particles themselves and for some key factors not present in the cell, such as RNA-dependent RNA-polymerases and reverse transcriptases. This extensive reliance on cellular functions renders viruses quite resistant to the defence mechanisms of their host organism and to therapeutic strategies. The entry into the host cell is one of the few stages in the replication cycle where the virus must express active functions of its own; the virus must attach to the plama membrane of relevant host cells and deliver its nucleocapsid into the cytoplasm. In enveloped animal viruses both of these functions are associated with the viral envelope and its spike glycoproteins.

Our studies with Semliki Forest virus (SFV, a togavirus), influenza A virus

1982 Membrane recycling. Pitman Books Ltd, London (Ciba Foundation symposium 92) p 59–76

(fowl plague virus, an orthomyxovirus) and vesicular stomatitis virus (a rhabdovirus) have shown that during the entry stages these viruses also depend extensively on host cell functions. In this paper we shall discuss some of the data on SFV endocytosis and its relation to normal cellular pinocytosis. Moreover, experiments will be described where the virus has been used as a probe to characterize the endocytic pathway.

Adsorptive endocytosis of SFV and other viruses

From early studies with electron microscopy it is known that virus particles can be internalized into cells by endocytosis (see Dales 1973). This phenomenon, called *viropexis* in published work, has long been thought to play a role in virus infection. That virus endocytosis is a prerequisite for productive infection by many viruses has been verified experimentally only recently. The best example so far is SFV. Our studies (Helenius et al 1980, Marsh & Helenius 1980) have shown that after attaching to cell-surface glycoproteins the SFV particles are recruited into coated pits and internalized in coated vesicles. They are then routed as passive ligands along a common endocytic pathway until they reach the lysosomal compartment. There, in the lysosome, the low pH triggers a conformational change in the spike glycoproteins which results in membrane fusion between the virus membrane and the lysosomal membrane; the nucleocapsids are then released into the cytoplasm for replication (Helenius et al 1980, White & Helenius 1980, White et al 1980, Helenius et al 1982). The importance of endocytosis lies thus in passing the virus into a compartment of sufficiently low pH to permit fusion. Similar pathways of entry have now also been demonstrated for orthomyxoviruses (Matlin et al 1981, Maeda & Ohnishi 1980), rhabdoviruses (Miller & Lenard 1980, Matlin et al 1982) and other toga viruses (Talbot & Vance 1980). The strategy used by these viruses to gain entry into the cell is very similar to the 'Horse of Troy': not before the viruses have been carried into the cell and centrally located in the cytoplasm are the deadly contents of the virus particle released.

Our studies indicate that the endocytic process by which viruses are internalized is analogous to that observed in the receptor-mediated endocytosis of physiological ligands such as low density lipoprotein, asialoglycoproteins, antibodies, lysosomal enzymes and polypeptide hormones, which involves the use of specific receptors (see Goldstein et al 1979). One important difference, however, is in the nature and specificity of the receptors. It seems unlikely that cells should carry on their surfaces specific and unique surface structures designed to bind various lethal viruses. In some cases viruses may use receptor systems that take up physiological molecules.

In other cases they seem to bind relatively non-specifically: for example, influenza virus, which requires sialic acid for binding (see Schulze 1975), can attach to a variety of cell-surface molecules; and vesicular stomatitis virus is apparently capable of attaching quite non-specifically (Miller & Lenard 1980). We know that binding of SFV occurs to surface glycoproteins and, in some cells, the involvement of the major histocompatibility antigens has been implicated (Helenius et al 1978), though these are probably not the only functional receptors.

Although non-physiological and 'opportunistic', the SFV particles offer certain advantages as ligands in studies of the pinocytic process. The particles are easily recognized by electron microscopy in thin sections, both on the cell surface and intracellularly. Specific antibodies and other reagents are available for analysis of the fate of the individual components. The viruses are structurally very well characterized. Metabolically labelled viruses, with the label located in the viral lipid, carbohydrate, RNA or protein, are easily prepared. Replication of progeny virus, moreover, greatly amplifies detection, which simplifies studies at low ligand concentrations and provides simple assays of successful penetration.

Uptake capacity

The rate-limiting step in SFV uptake is the binding of the viruses to the cell surface. At low temperature (0–4 °C) half-maximal binding to monolayers of baby hamster kidney (BHK-21) cells takes 45 min, and at 37 °C the binding is only marginally faster (Fries & Helenius 1979). To determine the rate and the capacity of internalization we allowed radioactively labelled viruses to bind at low temperature, and followed the uptake and further fate of the viruses after warming to 37 °C (Marsh & Helenius 1980). The results were as follows:

(a) Uptake is rapid (the average residence time on the cell surface is approximately 10 min).

(b) The rate of uptake is not dependent on multiplicity over a range of multiplicities from 0.1 to 10^5 viruses per cell.

(c) The viruses are routed into the lysosomes. Cell fractionation studies, by free-flow electrophoresis (Marsh et al 1982), have shown that a large proportion (30%) of the radioactively labelled intracellular viruses are present in the lysosomes.

(d) Uptake of prebound viruses is not inhibited by lysosomotropic weak bases (such as methylamine, chloroquine, and ammonium chloride) nor by carboxylic ionophores (such as monensin or nigericin). These agents are potent inhibitors of SFV-penetration (Helenius et al 1980, Helenius et

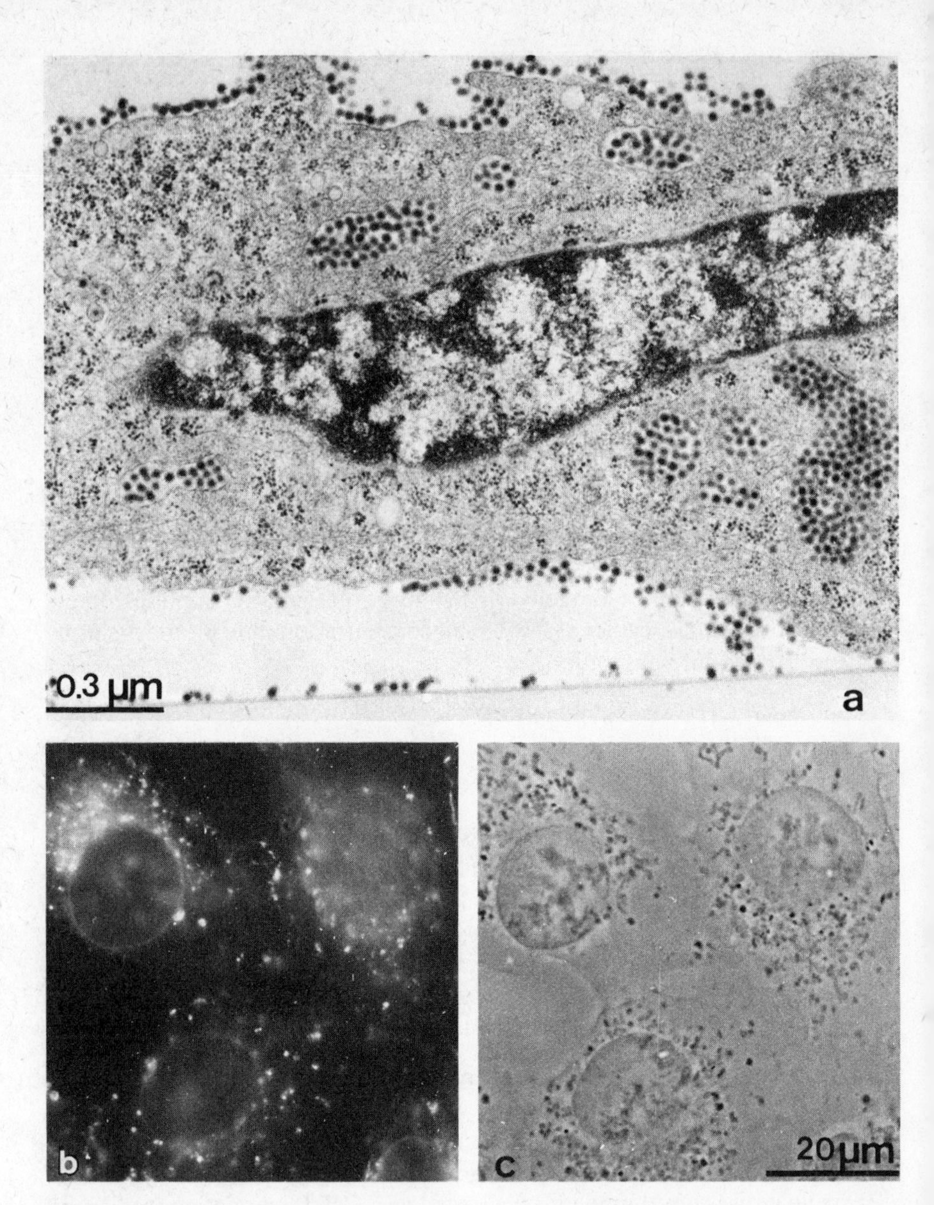

FIG. 1. Morphology of internalized Semliki Forest virus (SFV) in BHK-21 cells. (a) Electron microscopy of thin section. Cells grown on glass cover-slips were allowed to bind SFV for 1 h at a very high multiplicity (4 μg SFV per 10^6 cells) in the cold. After washing, the cells were warmed to 37 °C and fixed after 20 min. A vertical section of a cell is shown, and several large virus-filled vacuoles can be seen. The majority of these are probably endosomes. Viruses are also seen on the cell surface and a few are attached to the glass. (b) Immunofluorescence. SFV (1.0 μg per 10^6 cells) were allowed to bind as above. One hour after warming to 37 °C the cells were fixed with formaldehyde and the membranes permeabilized with 0.1% Triton X-100. Immunofluorescence staining, with affinity-purified rabbit anti-spike IgG and rhodamine-conjugated goat anti-rabbit

al 1982, Marsh et al 1982). They inhibit penetration by increasing the lysosomal pH above the threshold value needed to activate the fusion reaction.

We found that after warming to 37 °C each cell was capable of ingesting up to 3000 virus particles per minute. Electron microscopy of cells fixed shortly after warming showed an average of 1.3 virus particles per coated vesicle. This suggested that, in the conditions used, at least 2300 coated vesicles were formed at the cell surface every minute. A calculation based on 90 nm internal diameter, $2.5 \times 10^{-2} \mu m^2$ surface area and $3.8 \times 10^{-13} \mu l$ volume for a coated vesicle indicated that an area of $50 \mu m^2$ and a volume of $0.9 \times 10^{-9} \mu l$ per cell could be internalized by coated vesicles every minute. This corresponds to about 1% of the cell surface area and 0.04% of the cell volume—values which in turn correspond quite closely to the estimates by Steinman et al (1976) for pinocytosis in fibroblasts. The measured value for fluid-phase uptake in our BHK-21 cells, obtained with $[^3H]$sucrose and horseradish peroxidase as fluid-phase markers, was $0.5 \times 10^{-9} \mu l$ per min per cell. Taken together these values, although rough estimates, indicate that endocytosis of coated vesicles plays a major role in fluid-phase pinocytosis and in membrane recycling in BHK-21 cells.

We also measured the effect of virus uptake on fluid-phase pinocytosis (Marsh & Helenius 1980). Because a virus occupies about one quarter of the internal volume of an endocytic coated vesicle, should the viruses induce the formation of the coated vesicles which carry them into the cytoplasm we would expect to record a transient 70% increase in fluid-phase uptake during massive virus uptake. If the viruses used a vesicle system not involved in fluid-phase uptake we would expect no change. However, if the viruses used the same vesicles as the ongoing fluid-phase uptake we would expect a transient decrease of 20–25% in the uptake of our markers due to the displacement of fluid, equivalent to the virus volume, from the vesicles. This third alternative is what we observed, and it suggested that the virus exploits the continuous pinocytic uptake for entry into the cell. That SFV uptake was by pinocytosis and not by phagocytosis was supported by the relative insensitivity to various inhibitors, the involvement of small coated vesicles, the kinetic properties of uptake and the lack of induction by the particle itself.

Fig. 1a shows a BHK-21 cell which has been exposed to massive amounts of SFV, and incubated for 20 min at 37 °C. Viruses are seen on the top and bottom surface of the substrate-attached cell and densely packed within

IgG, was used to visualize the distribution of viral antigens. The majority of the stain was observed in cytoplasmic vacuoles distributed in the perinuclear region. (c) Phase-contrast microscopy. Same cells as in b. Comparison of (b) and (c) indicates a partial overlap of phase-dense vacuoles and fluorescently labelled vacuoles.

smooth-surfaced vacuoles in the cytoplasm. The overall distribution of virus-containing vacuoles observed after immunofluorescence (Fig. 1b) is similar, but not identical, to that of phase-dense vacuoles observed by light microscopy (Fig. 1c). On the basis of kinetic and morphological evidence we have suggested that, before delivery into the lysosomes, the virus particles dwell for some time in pre-lysosomal, or 'endosomal', vacuoles devoid of lysosomal hydrolases. The most convincing evidence for the existence of such pre-lysosomal organelles comes from recent cell fractionation studies, by Ann Hubbard and co-workers (see p 109-115, this volume).

Kinetics of acidification

Lysosomes have two features that distinguish them from other cellular organelles—a low pH and a high concentration of degradative hydrolytic enzymes. The low pH (4.5 to 5.0) is probably generated by an ATP-dependent proton pump (Schneider 1981) and the enzymes are glycoproteins synthesized in the endoplasmic reticulum and transported via the Golgi apparatus into the lysosomal compartment.

Most substances internalized into cells by endocytosis are routed into the lysosomes. The degradation usually starts 20–60 min after uptake, with some variation observed for different substances and different cell types. Fig. 2 shows the kinetics of ^{35}S-labelled SFV uptake in BHK-21 cells and the degradation as measured by the appearance of acid soluble [^{35}S]methionine radioactivity (solid squares). In this case degradation begins about 45 min after internalization and proceeds thereafter linearly for at least two hours. The lag time includes the time taken for the viruses to be internalized via the coated pits (half-life is about 10 min), the residence time in coated vesicles (probably a minute or less), the residence time in the 'endosomes' and the time required in the lysosomes before the proteins are digested to an acid-soluble form. The time for the degradation to be completed in the lysosomes is difficult to estimate. The lack of intermediate degradation products indicates that, when attacked by hydrolases, a virus protein is quite rapidly degraded to the final stage; however, we do not know how long virus particles can stay intact before digestion begins.

When SFV enters the cell the nucleocapsids are transferred into the cytoplasm. The RNA thereby becomes accessible to the ribosomes and the initial translation takes place; however, it also becomes accessible to RNase if the cell is opened by mild lysis and the enzyme is added (Helenius et al 1982). We have taken advantage of this to determine the kinetics of penetration (Fig. 2). [^{3}H]uridine-labelled virus was used at a multiplicity of 20 plague-forming units per cell and the RNA degradation into an acid soluble form was

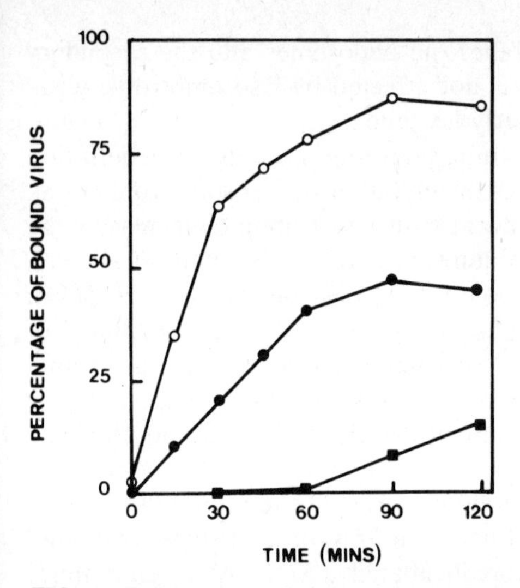

FIG. 2. Endocytosis, uncoating and degradation of SFV. [^{3}H]uridine- or [^{35}S]methionine-labelled SFV at low multiplicity was bound to BHK-21 cells for 1 h at 0 °C. The unbound activity was washed away, the cells overlaid with fresh medium and warmed to 37 °C. After the indicated time period the plates were returned to 0 °C. The media was assayed for degraded activity by trichloroacetic acid (TCA) precipitation (Marsh & Helenius 1980). The cells were assayed for internalized activity using Proteinase K (Marsh & Helenius 1980). After incubation with [^{3}H]uridine SFV the cells were isotonically lysed and the lysate assayed for uncoated RNA: uncoated RNA was defined as that which became TCA soluble after degradation by added RNase (Helenius et al 1982). The points are expressed as percentages of initial bound virus. Open circles, endocytosed virus; solid circles, TCA-soluble [^{3}H]uridine activity (uncoated RNA); solid squares, TCA-soluble [^{35}S]methionine activity in medium (degraded activity).

measured in RNase-containing lysates. It can be seen, by this criterion, that penetration occurred throughout the first hour of virus–cell contact and that about 50% of the cell-associated virus penetrated during the first hour (Fig. 2, solid circles). These results suggest that at least half the viruses must have reached organelles with low pH within one hour. No lag between internalization and penetration was observed.

Using the low-pH dependent membrane fusion activity of SFV as a monitor of pH it is possible to gain some insight into the kinetics of the acidification process intrinsic to the endocytic pathway. The fusion activity, which is a prerequisite for infection, has the following essential features (Helenius et al 1980, White & Helenius 1980, White et al 1980): (1) it has a sharp pH threshold at 6.0 to 6.1; (2) it occurs very rapidly after the pH is lowered; (3) it is highly efficient; (4) it is non-discriminating as to the nature and the origin of the target membrane (no protein besides the virus spikes are needed and the only specific lipid requirement is that for cholesterol; cholesterol is known as a component of the plasma membrane, the endosomes and the secondary lysosomes; Bode et al 1975); (5) it is not affected by lysosomotropic weak bases, provided that the pH is properly lowered.

A more exact time course for the minimum time needed for penetration was obtained by studying the kinetics of inhibition by lysosomotropic weak bases. These lysosomal inhibitors prevent virus penetration by increasing the lysosomal pH above the threshold required for fusion (Helenius et al 1980, Helenius et al 1982). Ohkuma & Poole (1978) have shown that the pH in lysosomes increases to beyond pH 6 in less than 30 s after addition of 10 mM-NH_4Cl or 0.1 mM-chloroquine. As mentioned above, we have found that these and other lysosomotropic weak bases do not affect virus binding or virus endocytosis, nor do they affect any of the biosynthetic or morphogenic steps in the replication cycle.

Our experimental strategy was as follows: viruses were allowed to bind to the cell surface either for a very short period at 37 °C or in the cold. Unbound viruses were removed and the cells were incubated at 37 °C. At different times 15 mM-NH_4Cl was added and, after a single infection cycle, the extent of infection was determined either by plaque titration of the progeny virus or by incorporation of [^3H]uridine into virus RNA. We found that if the inhibitor was added together with the virus inoculation the infection was efficiently blocked. When the inhibitor was added later, the time of addition determined the efficiency of infection (Fig. 3). Thus, addition within the first 4 min of virus uptake gave virtually full inhibition whereas after 6 min the inhibitor was already largely ineffective in intercepting the virus penetration. If we assume, in accordance with the results of Ohkuma & Poole (1978), that the increase in lysosomal pH is almost instantaneous, we can conclude that the first virus particles reach a low pH compartment (pH less than 6.1) within 3–4 min. The experiment was done with high and low multiplicity of virus (see Fig. 3). In separate experiments with chloroquine, where progeny virus production rather than uridine incorporation was monitored, a very similar time course of inhibition was observed (Helenius et al 1980).

In view of the rapid penetration by the virus particles, the question arises of whether they can penetrate through the plasma membrane before being endocytosed. It can be argued that a local pH drop on the plasma membrane may trigger penetration. Our results (White et al 1980) indicate that fusion of SFV with the plasma membrane is, indeed, possible when the pH of the medium is lowered, and that this can result in infection of the cells. However, the results also clearly exclude cell-surface fusion as a pathway of entry in normal tissue culture conditions. We have found that the low pH-induced infection, through fusion with the cell surface, differs fundamentally from the natural entry process because it is not inhibited by the lysosomotropic weak bases, nor by carboxylic ionophores (White et al 1980, Marsh et al 1982). On

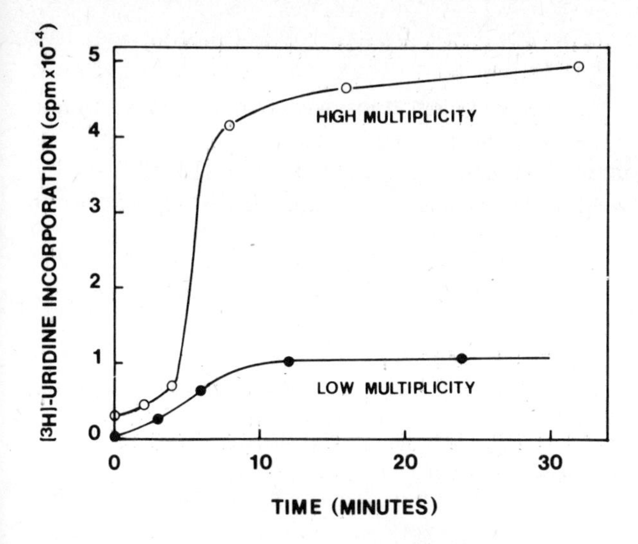

FIG. 3. Kinetics of ammonium chloride inhibition. Viruses were added in 50 μl medium to cells grown in 24-well plates either at low or high multiplicity (1 or 260 plague-forming units per cell). After 2 min the unbound viruses were removed and fresh medium added. At the indicated times the medium was aspirated and replaced with 15 mM-NH_4Cl medium. [^{3}H]uridine incorporation into virus RNA was measured during a 2-h period beginning either at 4 h (for the low multiplicity case) or 2 h (high multiplicity) (Helenius et al 1982). The whole experiment was done at 37 °C and the zero time denotes when SFV was added.

the basis of this difference we can conclude that the penetration of the early as well as the later penetrating viruses must take place in an intracellular compartment. Further evidence supporting this point has been elaborated in our recent paper (Helenius et al 1982).

Our studies with SFV give the following overall picture of acidification in the endocytic pathway of BHK-21 cells. Acidification of the contents of incoming endocytic vesicles occurs within a few minutes after leaving the cell surface. Acidification does not occur at the plasma membrane, nor to any significant extent in the endocytic coated vesicles as these exist only for a minute or less. The most likely explanation is that the pH drops in the endosomes.

The available data on the time of acidification in other pinocytic systems suggest a relatively early event. Steinman et al (1976) observed that horseradish peroxidase began to enter the lysosomes (and thus presumably an acid compartment) within 5 min after addition to the cells. Maxfield & Tycko (1981) have recently followed the acidification more directly, using fluorescent α_2-macroglobulin. Their conclusion that acidification takes place in endosomes within 20 min is compatible with our results.

Acknowledgements

We thank Eva Bolzau for electron and fluorescence microscopy, Linda Herzman for typing and Ira Mellman for helpful discussions. The work was supported by a grant from the National Institutes of Health (AI 18582) and by a BRSG grant RR 05358. Mark Marsh is a recipient of a EMBO postdoctoral fellowship.

REFERENCES

Bode F, Baumann K, Kinne R 1975 Analysis of the pinocytic process in rat kidney. II: Biochemical composition of pinocytic vesicles compared to brush border microvilli, lysosomes and basolateral plasma membranes. Biochim Biophys Acta 433:294-310

Dales S 1973 Early events in cell–animal virus interactions. Bacteriol Rev 37:103-135

Fries E, Helenius A 1979 Binding of Semliki Forest virus and its isolated glycoprotein to cells. Eur J Biochem 97:213-220

Goldstein JL, Anderson RG, Brown MS 1979 Coated pits, coated vesicles, and receptor-mediated endocytosis. Nature (Lond) 279:679-685

Helenius A, Morein B, Fries E et al 1978 Human (HLA-A and HLA-B) and murine (H-2K and H-2D) histocompatibility antigens are cell surface receptors for Semliki Forest virus. Proc Natl Acad Sci USA 75:3846-3850

Helenius A, Kartenbeck J. Simons K, Fries E 1980 On the entry of Semliki Forest virus into BHK-21 cells. J Cell Biol 84:404-420

Helenius A, Marsh M, White J 1982 Inhibition of Semliki Forest virus penetration by lysosomotropic weak bases. J Gen Virol 58:47-62

Maeda T, Ohnishi S 1980 Activation of Influenza virus by acidic media causes hemolysis and fusion of erythrocytes. FEBS (Fed Eur Biochem Soc) Lett 122:283-287

Marsh M, Helenius A 1980 Adsorptive endocytosis of Semliki Forest virus. J Mol Biol 142:439-454

Marsh M, Wellstead J, Kern H, Harms E, Helenius A 1982 Monensin inhibits Semliki Forest virus penetration into cultured cells. Proc Natl Acad Sci USA, in press

Matlin KS, Reggio H, Helenius A, Simons K 1981 Infectious entry pathway of Influenza virus in a canine kidney cell line. J Cell Biol 91:601-613

Matlin KS, Reggio H, Helenius A, Simons K 1982 Pathway of Vesicular Stomatitis virus entry leading to infection. J Mol Biol 156:609-631

Maxfield FR, Tycko B 1981 Rapid acidification of endocytic vesicles containing α_2-macroglobulin. J Cell Biol 91:212a (abstr)

Miller DK, Lenard J 1980 Inhibition of Vesicular Stomatitis virus infection by spike glycoprotein. J Cell Biol 84:430-437

Ohkuma S, Poole B 1978 Fluorescence probe measurement of the intralysosomal pH in living cells and the perturbation of pH by varying agents. Proc Natl Acad Sci USA 75:3327-3331

Schneider DL 1981 ATP-dependent acidification of intact and disrupted lysosomes. Evidence for an ATP-driven proton pump. J Biol Chem 258:3858-3864

Schulze IT 1975 The biologically active protein of Influenza virus: the hemagglutinin. In: Kilbourne ED (ed) The influenza viruses and influenza. Academic Press, New York, p 53-82

Steinman RM, Brodie SE, Cohn ZA 1976 Membrane flow during pinocytosis. J Cell Biol 68:665-687

Talbot PJ, Vance DE 1980 Evidence that Sindbis virus infects BHK-21 cells via a lysosomal route. Can J Biochem 58:1131-1137

White J, Helenius A 1980 pH-dependent fusion between the Semliki Forest virus membrane and liposomes. Proc Natl Acad Sci USA 77:3273-3277

White J, Kartenbeck J, Helenius A 1980 Fusion of Semliki Forest virus with the plasma membrane can be induced by low pH. J Cell Biol 87:264-272

DISCUSSION

Sly: Do you conclude that acidification occurs after about four minutes?

Helenius: Yes. We can exclude the possibility of acidification occurring on the cell surface. As I mentioned we can acidify the environment at the cell surface simply by applying a medium of low pH. We then see virus fusing with the cell surface, and the cell becomes infected (White et al 1980). This infection, unlike that occurring at physiological pH, cannot be inhibited by monensin or by lysosomotropic weak bases. These results indicate that entry cannot normally occur by fusion at the cell surface; the fusion must be at some intracellular site.

Hubbard: Can you use temperature or something else to inhibit fusion of the endocytic vesicle with the lysosome in order to find out if the pump is in the endosome?

Helenius: It could be done. We could allow viruses to interact with cells at various temperatures in the 10–20 °C range and then warm the cells to 37 °C, in the presence of ammonium chloride, and determine infection and lysosomal degradation. If there is a temperature at which internalization into acid endosomes does occur, but no further transport into lysosomes takes place, we should find infection but no lysosomal degradation of viral proteins.

Sabatini: Does the Sendai virus fuse directly with the plasma membrane wherever it binds, or does it have to glide to a particular place of fusion?

Helenius: Like influenza virus it binds to sialic acid-containing receptors, but little is known about the fusion mechanism or possible movement along the membrane. We have found that Sendai virus fusion is less efficient than that for Semliki Forest virus and influenza virus (J. White, unpublished results), and it is independent of pH (Väänänen & Kääriäinen 1980). However, Miller & Lenard (1981) have shown that Sendai virus infection is inhibited by certain lysosomotropic weak bases. It is possible that although fusion can take place on the cell surface, the infective entry actually requires the endocytic pathway.

Sabatini: Can the influenza virus and the Sendai virus, as well as the Semliki Forest virus, fuse with the liposomes?

Helenius: Yes. A morphological study by Anne Haywood (1974) showed that liposomes and Sendai viruses seem to be fused. Using quantitative biochemical methods, we have shown fusion of influenza virus and Semliki Forest virus with liposomes (White & Helenius 1980, White et al 1982). In all cases we have found that fusion is very efficient.

Palade: Could you elaborate in some detail on the functions of the spike glycoproteins in influenza? Why does the virus need a receptor-destroying activity, and what causes the cleavage between the HA1 and HA2 fragments of the influenza virus spike glycoprotein?

Helenius: The cleavage usually occurs in the host cell, and the location and identity of the enzyme is not exactly known. In certain tissue-culture systems the virus emerges without cleavage, and it can be induced by addition of trypsin (Lazarowitz & Choppin 1975). The viruses bind quite firmly to red blood cells in the cold, when the activity of the viral neuraminidase is very low. When the cells are warmed up, a large fraction of the viruses is released back into the medium. In cell lines from the canine kidney we have similar results except that some of the viruses are endocytosed into the cell before they are released (Matlin et al 1981). The neuraminidase is probably necessary because the virus has such a ubiquitous receptor. Sialic acid residues are present on basal membranes, red blood cells, mucous surfaces, etc, where irreversible virus binding would lead to a dead-end unless a releasing enzyme were present. So if the virus is not endocytosed within a given period it will be released and free to bind to another site. It can thus 'test' several surfaces until it finds a metabolically active cell willing to endocytose it.

Cohn: Margaret Kielian studied fusion between phagocytic vacuoles and lysosomes and noted that one could block the system, as you have, by low temperature (see Kielian & Cohn 1980). She could maintain a phagosome unfused for very long periods of time at 4–10 °C. By rapidly raising the temperature and adding an agent such as acridine orange, which accumulates only in acidified structures, she found that the phagocytic vacuoles were not acidified because they did not take up the dye. This is an indirect approach but suggests that endocytic vacuoles that have been residing in the cytoplasm for some time, albeit at low temperature, do not have a low vacuolar pH.

Cuatrecasas: I would nevertheless suggest that the pH in the endosomes might indeed be lower. There really seems to be a discrepancy between the rate of delivery to the lysosomes and the viral entry into the cytoplasm of the cell. Do you suspect that it is reaching the lysosome earlier, or that at some stage the endosome is becoming acidified?

Helenius: I think we may have been mistaken in some of our earlier work in emphasizing the lysosome as the vacuole where the virus penetrates. In fact the kinetic data on penetration indicate that acidification occurs before the virus enters the lysosomal compartment, and probably in the endosomes, which suggests that they have a pH below 6.

Baker: What happens to the virus membrane that becomes incorporated? Does it change the permeability of the compartments that it enters? For instance, if the compartment has a low pH does the fusion of virus membrane permit hydrogen ions to leak out into the cytosol?

Helenius: We have permeability data only for fusion with liposomes. In these cases the membrane does not become leaky to proteins (White & Helenius 1980, White et al 1982), but we don't know whether it becomes leaky to small molecules.

Baker: But when Sendai virus fuses to cells, I believe it makes them permeable to small molecules (see Poste & Pasternak 1978).

Helenius: Yes; that is quite possible. We have found with influenza that it makes a difference how the virus is prepared. With standard preparations, up to 80% of the fusion can be lytic. But if the virus is prepared more carefully and if fresh liposomes are used the fraction of lytic fusion is only 5% of the total. The extensive lysis observed with Sendai virus is probably an artifact due to damage during virus isolation.

Hubbard: We should bear in mind that all pH measurements on lysosomes have been done on secondary lysosomes. We do not actually know the pH of the primary lysosome. Maybe the primary lysosome does *not* contribute to the lowering of pH that we have discussed.

Have you compared the surface area of the endosome compartment with that of the incoming coated vesicle to see whether there is a pre-existing compartment with which the coated vesicle fuses?

Helenius: No, but the fact that the viruses accumulate in the endosomes seems to indicate that the endosomes are formed by fusion of the primary endocytic vesicles with each other. It cannot be ruled out, however, that a primary lysosome or some other vacuole, which is not of endocytic origin, serves as the initial vesicle where endosome formation begins.

Rothman: Are purified spike glycoproteins from these viruses, after reconstitution into lipid bilayers, capable of pH-dependent fusion with other lipid bilayers that are free from protein? In other words, in your *in vitro* fusion reaction can you replace the intact virus with a reconstituted viral envelope?

Helenius: Yes, to some extent. We have prepared reconstituted vesicles by octylglucoside dialysis (Helenius et al 1977). Mark Marsh (unpublished results) has recently shown that such vesicles do have some fusion activity at low pH, which forms about 25% of the fusion activity of a comparable amount of intact virus.

Rothman: Why is it less efficient?

Helenius: In the Semliki Forest virus the spikes are ordered in a defined array owing to an interaction of each spike glycoprotein with the nucleocapsid. In the reconstituted vesicles they are not fixed, and can probably move laterally in the membrane. The defined spatial arrangement and the high density of spikes in the virus may be necessary for optimal fusion activity.

Sabatini: Are the spikes in both sides of those vesicles?

Helenius: In the vesicles that we have used 95% are facing outwards.

Mellman: Do you think that the many viruses that are internalized via coated vesicles each bind to specific cell-surface receptors?

Helenius: Viruses and toxins belong to what Dr Goldstein has called opportunistic ligands for receptor-mediated endocytosis (Goldstein et al 1979). It would not make sense for cells to provide specific unique receptors for such ligands especially if they are toxic. Unless it is completely non-specific, their uptake by the coated vesicle pathway must depend on receptors for other ligands. But to account for the binding of Semliki Forest virus, half a million such receptors would be needed, because that is approximately the number of binding sites that we can measure by binding studies (Fries & Helenius 1979). Influenza viruses bind to highly non-specific structures—cell-surface sialic acid residues—and although all the molecules to which they bind cannot be receptors, the virus particles are efficiently cleared from the cell surface by receptor-mediated endocytosis in coated vesicles (Matlin et al 1981). When considered from the point of view of such opportunistic ligands there seems to be a large degree of non-specificity in adsorptive uptake by coated vesicles!

Rothman: One need not conclude that the coated vesicle does not select its content. One need only suggest that the virus doesn't have a specific receptor, but that it binds to a multiplicity of components, some of which could be selectively concentrated in the coated vesicles.

Goldstein: Do the receptors for these viruses recycle?

Helenius: The receptors for Semliki Forest virus do seem to recycle. Even with massive uptake of viruses we never see any decrease in the binding capacity (Marsh et al 1982a). We can cleave the receptor off the surface by proteases, but when we warm the cells up again, binding activity is rapidly restored.

Raff: Isn't it possible that the virus binds to many different plasma membrane proteins but that only a small proportion of these are responsible for directing the virus to the coated pits? The other proteins that bind the virus might be dragged along into the coated pits and vesicles.

Helenius: I agree. However, vesicular stomatitis virus (VSV) doesn't seem to have specific receptors (Miller & Lenard 1980), and it is still cleared from the surfaces in coated vesicles. This indicates again that cells may be able to internalize, by pinocytosis, practically any small particle that binds to the cell surface, and not only those that have specific unique receptors.

Bretscher: I believe Semliki Forest virus attaches to the surface of the cell by many interactions, each of which is rather weak. So the virus isn't just attached statically to glycoproteins but is probably rolling all over the surface of the cell by Brownian motion. At any moment it may find itself in a coated pit, in which case it may be endocytosed. The same is probably true for vesicular stomatitis virus.

Helenius: This is what I mean about the non-specificity of the uptake.

Raff: Rather than rolling along from protein to protein, the virus is surely more likely to cross-link a variety of different glycoproteins, some of which will change their conformation and localize in coated pits. Even if only a small fraction of the glycoproteins behaves in that way, all the virus particles will eventually end up in coated pits.

Sly: What is the evidence for transfer of the virus to the lysosome, and what are the kinetics in the presence of agents that block viral replication?

Helenius: Together with E. Harms and H. Kern in Heidelberg we have used the technique of free-flow electrophoresis to isolate, from a cell homogenate, a fraction enriched in lysosomes (Marsh et al 1982b). It is about 25-fold enriched in lysosomal enzymes over the starting homogenate. We have done the homogenization only at one time point so far. When we isolate the lysosomes one hour after addition of ^{32}P-labelled virus, about 30–40% of the intracellular radioactivity is in the lysosomal fractions. If this is repeated, after pretreatment with 10 μM-monensin or 15 mM-ammonium chloride, there is no difference in the amount of viral radioactivity in the lysosomal fractions. This implies that the transport of virus into the lysosomes is not blocked at these concentrations of the inhibitors.

Cohn: Could the specificity of these hydrophobic runs be related to the mechanism by which macromolecular nutrients or factors escape from the vacuolar system and into the cytosol?

Helenius: I believe that the viruses have not invented the mechanism of membrane fusion but, rather, during evolution they have picked it up from existing cellular mechanisms. There may therefore be cellular processes in which similar mechanisms operate.

Cuatrecasas: In some of Dr Hubbard's experiments on glycosylated proteins or orosomucoid receptor systems, the protein was initially associated with the inner aspect of the smooth vesicle endosomes but later the material appeared to be within the endosomal lumen. Would that suggest association between the endocytosed ligand and its receptor?

Hubbard: That's right. We have been preferentially labelling this intermediate compartment (see p 109-115) and find that receptors for asialoglycoproteins are present in the compartment after pulse–chase of labelled ligand into that compartment (D. A. Wall & A. L. Hubbard, unpublished work). However we don't know whether any dissociation occurs in the compartment. We don't have a probe like Dr Helenius's virus to indicate the pH within the isolated vesicles that we are examining. But the lactosaminated ferritin to which you were referring is a low-affinity ligand and it could therefore give a different picture from a high-affinity ligand such as asialo-orosomucoid. Low-affinity ligands for this receptor system appear to be exocytosed after they are endocytosed (Connolly et al 1982, Tolleshaug et al 1981).

Geisow: Some of these receptor proteins (particularly glycoproteins containing sialic acid) that are being pulled in to endosomes have an acid isoelectric point. So the pH in the endosome may simply be determined by a Gibbs-Donnan equilibrium; endocytosis of very negatively charged molecules is taking place. The effect of an asymmetrical distribution of acid–base equivalents across a membrane is described by Goldman & Rottenberg (1973). One does not need to envisage a proton pumping mechanism. This might be particularly so for coated pits: selective uptake of receptors into the coated pit might produce the acid environment physicochemically by virtue of the charge on the molecules that are being taken in.

Palade: The endosome membrane is supposed to be essentially plasma membrane, according to Dr Mellman's paper (p 35-58). But the endosomes apparently have a proton pump, according to the results reported by Dr Helenius here and by Dr Geisow (Geisow et al 1981). What could be the source of this proton pump? We need to know if there is any evidence for a proton pump in the plasmalemma.

Raff: Given that virus escape occurs only a minute or so before you can see virus particles in lysosomes, isn't it likely that the fusion of a small number of primary lysosomes produces sufficient acidification to allow the virus to escape but insufficient to produce a structure recognizable as a secondary lysosome?

Helenius: We do not have enough information to exclude that possibility. The question of acidification kinetics becomes even more complicated if we consider the lack of difference in penetration kinetics observed at high and low virus multiplicities. We know that uptake has a half-life of about 10 min, regardless of multiplicity (Marsh & Helenius 1980). As we measure infection in our assays, we only score the penetration of the first virus. At low multiplicity, where less than one virus per cell is added, we should observe the average time that it takes for viruses to reach the acid compartment. The observed time includes the average residence time on the surface and the average time taken for the virus to move through the endocytic pathways to the low-pH compartment. At high multiplicity we should record the rate for the fastest viruses rather than for the average virus. We were therefore expecting that we would see a shorter time of inhibition at high than at low multiplicity. But the values were identical (see Fig. 3, p 67). We don't understand this result.

Rothman: Wouldn't it just mean that everything is synchronous?

Helenius: If we look biochemically and morphologically at the uptake we see that it's not synchronous. There may, however, be a very rapid and synchronous uptake underlying the non-synchronous uptake that we normally measure.

Palade: The fusogen HA1 fragment of the influenza virus spike protein provides an impressive example of protein directly involved in membrane fusion. A similar fusogenic protein may exist on the cytoplasmic aspect of vesicles involved in vesicular transport.

REFERENCES

Connolly DT, Townsend RR, Kawaguchi K, Bell WR, Lee YC 1982 Binding and endocytosis of cluster glycosides by rabbit hepatocytes. J Biol Chem 257:939-945

Fries E, Helenius A 1979 Binding of Semliki Forest virus and its isolated glycoprotein to cells. Eur J Biochem 97:213-220

Geisow MJ, D'Arcy Hart P, Young MR 1981 Temporal changes of lysosomes and phagosomes during phagolysosome formation in macrophages: studies by fluorescence spectroscopy. J Cell Biol 89:645-652

Goldman R, Rottenberg H 1973 Ion distribution in lysosomal suspension. FEBS (Fed Eur Biochem Soc) Lett 33:233-238

Goldstein JL, Anderson RG, Brown MS 1979 Coated pits, coated vesicles, and receptor-mediated endocytosis. Nature (Lond) 279:679-685

Haywood AM 1974 Characteristics of Sendai Virus receptors in a model membrane. J Mol Biol 83:427-436

Helenius A, Fries E, Kartenbeck J 1977 Reconstitution of Semliki Forest virus membrane. J Cell Biol 75:866-880

Kielian MC, Cohn ZA 1980 Phagosome–lysosome fusion. Characterization of intracellular membrane fusion in mouse macrophages. J Cell Biol 85:754-765

Lazarowitz SG, Choppin PW 1975 Enhancement of the infectivity of influenza A and B viruses by proteolytic cleavage of the hemagglutinin polypeptide. Virology 68:440-454

Marsh M, Helenius A 1980 Adsorptive endocytosis of Semliki Forest virus. J Mol Biol 142:439-454

Marsh M, Matlin K, Simons K, Reggio H, White J, Kartenbeck J, Helenius A 1982a Are lysosomes a site of enveloped virus penetration? Cold Spring Harbor Symp Quant Biol 46:835-843

Marsh M, Wellstead J, Kern H, Harms E, Helenius A 1982b Monensin inhibits Semliki Forest virus penetration into baby hamster kidney cells. Proc Natl Acad Sci USA, in press

Matlin KS, Reggio H, Helenius A, Simons K 1981 Infectious entry pathway of Influenza virus in a canine kidney cell line. J Cell Biol 91:601-613

Miller DK, Lenard J 1980 Inhibition of Vesicular Stomatitis virus infection by spike glycoprotein. J Cell Biol 84:430-437

Miller DK, Lenard J 1981 Antihistaminics, local anesthetics, and other amines as antiviral agents. Proc Natl Acad Sci USA 78:3605-3609

Poste G, Pasternak CA 1978 Virus-induced cell-fusion. Cell Surf Rev 5:305-367

Tolleshaug H, Chindemi PA, Regoeczi E 1981 Diacytosis of human asialotransferrin Type 3 by isolated rat hepatocytes. J Biol Chem 256:6526-6528

Väänänen P, Kääriäinen L 1980 Fusion and haemolysis of erythrocytes caused by three togaviruses: Semliki Forest, Sindbis and rubella. J Gen Virol 46:467-475

White J, Helenius A 1980 pH-dependent fusion between the Semliki Forest virus membrane and liposomes. Proc Natl Acad Sci USA 77:3273-3277

White J, Kartenbeck J, Helenius A 1980 Fusion of Semliki Forest virus with the plasma membrane can be induced by low pH. J Cell Biol 87:264-272

White J, Kartenbeck J, Helenius A 1982 Membrane fusion activity of Influenza virus. EMBO J 1:217-222

Receptor-mediated endocytosis and the cellular uptake of low density lipoprotein

JOSEPH L. GOLDSTEIN*†, RICHARD G. W. ANDERSON‡ and MICHAEL S. BROWN*†

Departments of Molecular Genetics, Cell Biology‡, and Internal Medicine†, University of Texas Health Science Center at Dallas, 5323 Harry Hines Boulevard, Dallas, Texas 75235, USA*

Abstract During receptor-mediated endocytosis various extracellular nutritional and regulatory molecules bind to plasma membrane receptors and rapidly enter target cells. In many systems (including those for certain plasma transport proteins, protein hormones, glycoproteins, toxins and viruses, and other plasma proteins) the receptors cluster in discrete regions of the surface membrane called coated pits, which invaginate into the cell to form endocytic vesicles. The extracellular ligand enclosed in the endocytic vesicle is delivered to intracellular sites, frequently to lysosomes, where it is degraded.

In one system of receptor-mediated endocytosis, namely the one for plasma low density lipoprotein (LDL), the receptor functions to internalize LDL. The LDL is delivered to lysosomes where it is degraded and its cholesterol is released for use in the synthesis of membranes, steroid hormones and bile acids. Three recent advances in the LDL receptor system are reviewed: (1) the development of a method for purifying the receptor to apparent homogeneity and the demonstration that the LDL-binding site is contained within a glycoprotein of relative molecular mass 164 000 and an acidic isoelectric point of 4.6; (2) the production of monoclonal antibodies directed against the receptor and the use of these antibodies as probes for receptor-mediated endocytosis; and (3) the use of monovalent carboxylic ionophores (such as monensin) to demonstrate by immunofluorescence that the LDL receptor enters the cell together with LDL, after which it recycles to the surface.

Receptor-mediated endocytosis is a general process by which nucleated animal cells ingest highly selected proteins from the extracellular fluid (for reviews, see Goldstein et al 1979, Pastan & Willingham 1981, Kaplan 1981). The first step is the binding of the protein to its receptor on the surface of the plasma membrane. The protein enters cells when the membrane to which it is bound folds inwards and pinches off to form an endocytic vesicle. Receptor-mediated endocytosis is one mechanism by which protein-bound nutrients

1982 Membrane recycling. Pitman Books Ltd, London (Ciba Foundation symposium 92) p 77-95

(such as cholesterol, iron and vitamin B_{12}) are delivered to cells. In addition, this mechanism accounts for the cellular uptake and degradation of certain plasma proteins (such as glycoproteins and α_2-macroglobulin). Receptor-mediated endocytosis also operates on protein hormones (such as insulin, chorionic gonadotropin and epidermal growth factor) that bind to receptors on cell surfaces. Finally, receptor-mediated endocytosis is a route by which some protein toxins (such as *Pseudomonas* toxin and diphtheria toxin) and viruses (such as Semliki Forest virus and certain strains of influenza virus) gain entry to animal cells. The relevance of receptor-mediated endocytosis to human disease is emphasized by the observation that a genetic defect in one of the receptors leads to a failure of cells to take up cholesterol-carrying plasma lipoproteins, thus producing a common disease, familial hypercholesterolaemia.

General characteristics of receptor-mediated endocytosis

Receptor-mediated endocytosis has four characteristics that distinguish it from other cellular uptake mechanisms:

(1) The receptors are proteins that are embedded in the bilayer of the plasma membrane; they bind extracellular proteins with high affinity and specificity.

(2) Internalization of the ligand is efficiently coupled to binding. The half-time for internalization of receptor-bound ligands is less than 10 min.

(3) In all cases for which adequate ultrastructural data are so far available, the receptor-bound proteins enter cells through coated pits, which are specialized regions of the surface membrane that invaginate rapidly into the cell during endocytosis to form coated endocytic vesicles (see below).

(4) The internalized proteins are usually, but not always, delivered to lysosomes, where they are degraded completely to amino acids.

So far, at least 22 types of protein and macromolecular complex are recognized to be internalized by receptor-mediated endocytosis. These ligands are listed in Table 1.

The coated pit as a mechanism for receptor-mediated endocytosis

In all the cases analysed in detail so far, receptor-mediated endocytosis takes place in discrete regions of the plasma membrane called coated pits, which occupy about 2% of the surface of most cells (Anderson et al 1976). The pits are regions in which the plasma membrane is indented and coated on its cytoplasmic surface (Roth & Porter 1964) by a lattice-like structure composed of a protein called *clathrin* (Pearse & Bretscher 1981). By electron microscopy the coated pits appear to be constantly folding inwards and pinching off to form coated vesicles, which are membrane-bounded sacs surrounded by clathrin. Ligands are internalized when their receptors migrate laterally in the plane of the membrane and become trapped in the membrane of a coated pit. Certain receptors appear to move to coated pits spontaneously, even when no ligand is bound. This is particularly true of the receptors for plasma low density lipoprotein (Anderson et al 1976, 1977a) and plasma asialoglycoproteins (Wall et al 1980). Other receptors seem to occupy random locations on the plasma membrane. After the ligand attaches to these receptors, the receptor–ligand complex moves laterally towards coated pits. Such ligand-induced movement has been noted in the case of receptors for epidermal growth factor, insulin and certain viruses (Pastan & Willingham 1981, Helenius et al 1980).

After the coated pit pinches off from the plasma membrane, the clathrin coat rapidly falls off the vesicle, and the uncoated endocytic vesicle migrates through the cytoplasm (Anderson et al 1977a, Wall et al 1980, Pastan & Willingham 1981). Each vesicle appears to contain sufficient information to direct it to a specific cellular site. By electron microscopy many endocytic vesicles appear to fuse together creating larger, irregularly shaped vesicles. These have been variously called intermediate endocytic vesicles, sorting vesicles, endosomes and receptosomes. Eventually the contents of the endocytic vesicles reach lysosomes. This may occur by the fusion of small primary lysosomes with the endosomes, which exposes the ligands to an acid pH and to the hydrolytic action of multiple lysosomal hydrolase enzymes. Table 1 lists those internalized ligands that are delivered to lysosomes.

In some cases endocytic vesicles are directed to subcellular structures other than lysosomes (Table 1). For example, endocytic vesicles containing nerve growth factor are transported retrogradely in neuronal axons, to the nucleus. Plasma lipoproteins that are taken up by coated pits and coated vesicles in chicken oocytes are delivered to the yolk granules where the lipids are liberated from the proteins for storage. Maternal immunoglobulins are transported in endocytic vesicles that completely traverse the trophoblasts of the placenta and the epithelial cells of the neonatal gut, being discharged intact on the opposite side. In each case the uptake mechanism appears to be similar to receptor-mediated endocytosis, but the vesicles are targeted to specific sites other than lysosomes.

In many of the systems listed in Table 1 the receptors are not destroyed in lysosomes but, together with clathrin, they are returned to the plasma membrane where they again bind another molecule of ligand. This process is termed *receptor recycling* and is discussed in more detail below.

Fig. 1 illustrates schematically the sequential steps in receptor-mediated endocytosis.

TABLE 1 Proteins that undergo receptor-mediated endocytosis

Protein	Major target cell	Internalization via coated pits and coated vesicles	Fate of internalized protein	
			Degraded in lysosomes	*Other*
Transport proteins				
Low density lipoprotein (LDL)	Fibroblasts, hepatocytes, adrenocortical cells, lymphocytes, other non-macrophage cells	Yes	Yes; cholesterol retained by cells	—
Chylomicron remnants	Hepatocytes	Data not available	Yes; cholesterol retained by cells	
Yolk proteins (lipovitellin)	Oocytes (chicken, mosquito)	Yes	No	Delivered to yolk granules
Transcobalamin II	Kidney cells, hepatocytes, fibroblasts	Yes	Yes; vitamin B_{12} retained by cells	—
Transferrin	Erythroblasts, reticulocytes, fibroblasts	Yes	Iron dissociates from transferrin in lysosomes; apotransferrin is discharged intact from cells, while iron is retained	—
Protein hormones				
Epidermal growth factor	Fibroblasts, hepatocytes	Yes	Yes	—
Nerve growth factor	Sympathetic ganglion cells	Data not available	Data not available	Carried in vesicles retrogradely up the axon
Insulin	Hepatocytes, lymphocytes, adipocytes, fibroblasts	Yes (for some cell types)	Yes	Also may be delivered to Golgi region
Chorionic gonadotropin	Leydig cells, ovarian luteal cells	Data not available	Yes	—
β-Melanotropin	Melanoma cells	Data not available	Data not available	Delivered to Golgi region and melanosomes

Prolactin	Breast cells, hepatocytes	Data not available	Yes	Also may be delivered to Golgi region
Glycoproteins				
Asialoglycoproteins (galactose)[a]	Hepatocytes	Yes	Yes	—
Lysosomal enzymes (mannose 6-phosphate)[a]	Fibroblasts	Yes	No	Delivered to lysosomes and Golgi-region; enzymes remain active for many days
Glycoproteins (mannose or glucose)[a]	Macrophages	Data not available	Yes	—
Other plasma proteins				
α_2-Macroglobulin	Macrophages, fibroblasts	Yes	Yes	
Maternal immuno-globulins (IgG)	Placenta, fetal yolk sac, neonatal intestinal epithelial cells	Yes	No	Transferred intact in coated vesicles to basal surface of cells where IgG is discharged into fetal or neonatal circulation
Immune complexes (Fc domain)[b]	Macrophages	Data not available	Yes	—
Chemotactic peptide	Neutrophils, macrophages	Data not available	Yes	—
Viruses and toxins				
Semliki Forest Virus	Fibroblasts	Yes	Acidic pH of lysosome allows fusion between viral membrane and lysosomal membrane, releasing viral nucleocapsid into cytosol	—
Diphtheria toxin	Fibroblasts, kidney cells	Data not available	Yes	Some of the A chains escape lysosomal degradation and inhibit protein synthesis in cytosol
Pseudomonas toxin	Fibroblasts	Yes	Yes	
Ricin toxin	Fibroblasts	Data not available	Yes	

[a]Carbohydrate moiety that is recognized by cell surface receptor. [b]Portion of antibody component of immune complex that is recognized by cell surface receptor.

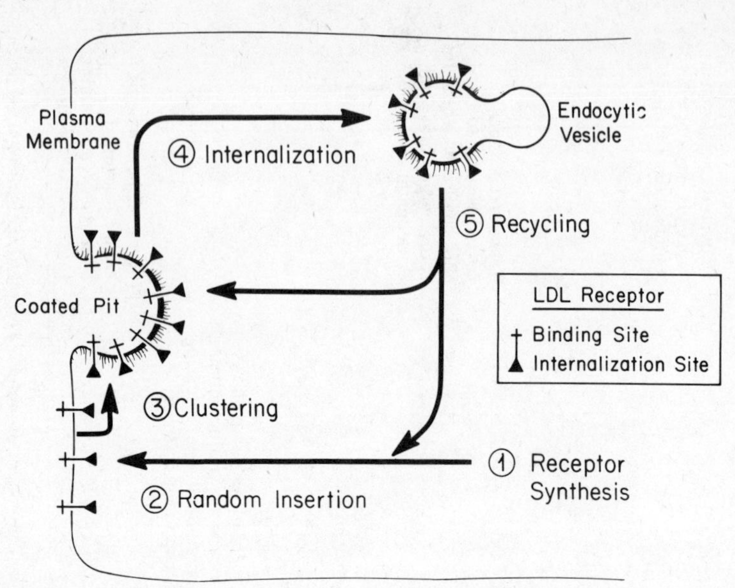

FIG. 1. Schematic illustration of the proposed pathway by which certain cell-surface receptors (such as LDL receptors) become localized to clathrin-containing coated pits on the plasma membrane of cells. The sequential steps are as follows: (1) synthesis of receptors on polyribosomes; (2) insertion of receptors as integral membrane proteins at random sites along non-coated segments of plasma membrane; (3) clustering together of receptors in coated pits; (4) internalization of receptors and their bound ligands when coated pits invaginate to form coated endocytic vesicles, followed by delivery of the ligands to lysosomes; and (5) recycling of internalized receptors and clathrin back to the plasma membrane.

The LDL receptor system as an example of receptor-mediated endocytosis

Low density lipoprotein (LDL) is the major cholesterol transport protein in human plasma. The bulk of cholesterol in LDL is contained in a central core in which each cholesterol molecule is esterified with a long-chain fatty acid. The LDL particle, containing a core of approximately 1500 molecules of cholesteryl ester, is internalized by cells after binding to the LDL receptor. Within lysosomes the protein component of LDL is degraded completely to amino acids. The cholesteryl esters of LDL are also hydrolysed by a lysosomal acid lipase enzyme. The liberated cholesterol crosses the lysosomal membrane and enters the cytosol where it is used for several metabolic purposes, e.g. the synthesis of plasma membranes in most tissues, the synthesis of steroid hormones in adrenal cortex and ovarian corpus luteum, and the synthesis of bile acids in hepatocytes (Brown et al 1981).

Several allelic mutations affecting the LDL receptor have been identified in humans who have a genetic disease called familial hypercholesterolaemia. The most common mutant gene specifies a receptor that is either unable to bind to LDL or missing altogether (*receptor-negative* allele). Another mutant gene specifies an altered receptor that cannot be incorporated into coated pits and therefore cannot mediate internalization of LDL (*internalization-defective* allele) (Goldstein & Brown 1979). This latter mutation has proved especially valuable in delineating the role of coated pits in receptor-mediated endocytosis and in verifying the occurrence of receptor recycling (Basu et al 1981).

A summary of the characteristics of the LDL receptor is presented in Table 2.

TABLE 2 Characteristics of the LDL receptor system

Distribution Present on nearly all human and animal cells
Function Facilitates uptake and degradation of cholesterol-rich lipoproteins, supplying cholesterol to cells
Structure Binding site consists of an acidic glycoprotein of M_r 164 000 (isoelectric point, 4.6)
Specificity Binds lipoproteins containing apoprotein B (LDL) or apoprotein E (cholesterol-rich remnants)
Mechanism of uptake Receptors located in clathrin-coated pits that invaginate to form coated vesicles
Regulation Number of receptors increases when cellular requirement for cholesterol increases
Genetics Patients with familial hypercholesterolaemia have defective gene for LDL receptor. LDL accumulates in plasma owing to impaired cellular degradation, and severe atherosclerosis follows

Purification of the LDL receptor

The LDL receptor has been most extensively studied in intact cultured cells and in membranes from animal tissues. In the body the highest concentration of receptors occurs in the cortex of the adrenal gland, which uses these receptors to supply cholesterol for steroid hormone synthesis (Brown et al 1981).

Wolfgang Schneider in our laboratory has recently developed a method for the purification of the LDL receptor from bovine adrenal cortex membrane. This purification scheme is relatively rapid and provides an apparently homogeneous receptor preparation in high yield (Schneider et al 1982). After solubilization with non-ionic detergents, the receptor adheres tightly to a DEAE–cellulose column at pH 6. After elution from DEAE–cellulose, the detergent is removed, leaving the receptor in the form of a soluble aggregate. The receptor is then subjected to affinity chromatography on a column containing LDL coupled to Sepharose 4B. The receptor is eluted with suramin, a newly found inhibitor of LDL–receptor interactions (Schneider et al 1982). This procedure yields a single protein with a relative molecular mass (M_r) of 164 000. The same protein is also isolated when the crude DEAE–cellulose fraction is applied to an immunoaffinity column containing a monoclonal antibody that is known to bind to the receptor (see below). The 164 000 M_r receptor protein has an acidic isoelectric point of 4.6, which rises to 4.8 after extensive treatment with neuraminidase (EC 3.2.1.18). The purified receptor retains all the binding properties of the receptor on intact cells and on crude membranes.

Monoclonal antibodies to the LDL receptor

Ulrike Beisiegel in our laboratory has recently prepared monoclonal antibodies against the LDL receptor by immunizing mice with partially purified receptor from bovine adrenal cortex (Beisiegel et al 1981b). The most extensively studied monoclonal antibody, designated immunoglobulin G–C7 (IgG–C7), reacts with human and bovine LDL receptors but not with LDL receptors from the mouse, rat, Chinese hamster, rabbit or dog. ^{125}I-labelled monoclonal antibody binds to monolayers of intact human fibroblasts and to the purified bovine adrenal receptor in amounts that are equimolar to ^{125}I-labelled LDL (Beisiegel et al 1981b, Schneider et al 1982). This observation suggests that each LDL receptor has one antibody-binding site per LDL-binding site.

After binding to the receptor of human fibroblasts, monoclonal IgG-C7 is rapidly internalized by the cells and degraded in lysosomes in a manner similar to the receptor-mediated uptake of LDL. The monoclonal antibody has thus proved useful as a probe for the study of the receptor-mediated endocytosis of LDL (see below).

Recycling of the LDL receptor

Ever since the earliest studies of receptor-mediated endocytosis, the importance of receptor recycling has been evident (Goldstein & Brown 1974). Even though the internalized ligand is destroyed in lysosomes, in many cases the receptors survive and are recycled back to the plasma membrane and re-utilized. Recycling appears to be the rule in those systems in which receptor-mediated endocytosis functions to transport plasma proteins other than hormones into cells. These include receptor systems for LDL in fibroblasts (Goldstein et al 1976, Anderson et al 1977a); asialoglycoproteins in hepatocytes (Steer & Ashwell 1980, Tolleshaug & Berg 1979); mannose 6-phosphate-containing lysosomal enzymes in fibroblasts (Gonzalez-Noriega et al 1980); mannose-conjugated proteins in macrophages (Stahl et al 1980); and α_2-macroglobulin in macrophages (Kaplan 1981) and fibroblasts (Van Leuven et al 1980). Many protein hormones also undergo receptor-mediated endocytosis after binding to their surface receptors, but the receptors for most protein hormones are not recycled. Rather, the binding and uptake lead to the loss of receptors from the surface (for reviews, see Goldstein et al 1979, Kaplan 1981).

From the standpoint of cellular economy, the value of receptor recycling is obvious: one receptor molecule can facilitate the internalization of many ligands during its lifetime. The number of cycles that each receptor can undergo may be large. For example, it has been estimated that each LDL receptor is used more than 150 times during its 30-h lifespan (Goldstein et al 1979).

The earliest evidence for receptor recycling, albeit indirect, came from experiments in which the synthesis of new receptors was blocked with the protein synthesis inhibitor, cycloheximide. When human fibroblasts are incubated with LDL and cycloheximide, they continue to internalize and degrade the ligand at a steady rate for at least 8 h, even though all the receptors transfer their LDL into the cell within 10–12 min (Goldstein et al 1976, 1979). If recycling does not occur, all the receptors would disappear by 12 min and internalization would cease. Similar observations have been made with other receptor systems.

Additional indirect evidence for receptor recycling has emerged from experiments with lysosomal inhibitors. These agents, including chloroquine and methylamine, are weak bases that diffuse into cells and become protonated. As a result they are trapped within the lysosome and raise the lysosomal pH, which in turn leads to an inhibition of lysosomal enzyme activity (de Duve et al 1974). In several receptor systems, including those for asialoglycoproteins (Tolleshaug & Berg 1979), lysosomal enzymes (Gonzalez-Noriega et al 1980) and α_2-macroglobulin (Van Leuven et al 1980), incubation of cells with chloroquine or methylamine plus the exogenous ligand leads to a decline in the number of receptors on the surface. This decline has been attributed to the trapping of receptors within the lysosomes, owing to the raised pH of the organelle.

To obtain more direct evidence for receptor recycling, we have recently developed a new approach to the problem in the LDL receptor system. In this approach we use two types of agent: (1) monovalent carboxylic ionophores such as monensin and nigericin; and (2) monoclonal and polyclonal antibodies against the LDL receptor (Basu et al 1981, Beisiegel et al 1981a, b). The studies have shown that monensin and nigericin trap internalized LDL receptors within the cell and prevent their return to the surface. The shift of receptors into cells has been confirmed directly by making the cells permeable to an antibody to the LDL receptor, followed by indirect immunofluorescence. We discuss this subject further in a recent review article (Brown et al 1982).

An unexpected finding in our studies of recycling of the LDL receptor was made with the monoclonal antibody IgG-C7. After the IgG-C7 is bound to the receptor and internalized during endocytosis, the monoclonal antibody apparently can be detached from the receptor within the cell, allowing the receptor to undergo its usual recycling process in which it returns to the surface and binds another molecule of antibody. This conclusion follows from kinetic data showing that fibroblasts continue to bind, internalize and degrade ^{125}I-labelled IgG-C7, just like ^{125}I-labelled LDL, at a steady rate without any depletion of receptors for at least 6 h at 37 °C (Beisiegel et al 1981b).

The ability of the LDL receptor to recycle in the presence of the monoclonal antibody may depend on the rapid dissociation of the antibody from the receptor at 37 °C. Previous studies have shown that the affinity of the fibroblast receptor for LDL is approximately 10-fold lower at 37 °C than at 4 °C (Goldstein et al 1976). An even more striking difference was observed in the apparent affinity of the ^{125}I-labelled IgG-C7 for the receptor at the two temperatures. Half-maximal binding at 4 °C was achieved at an ^{125}I-labelled IgG-C7 concentration of approximately 1 nM, whereas a concentration of 75 nM was required for half-maximal uptake at 37 °C, suggesting that the affinity of the receptor for monoclonal antibody is approximately 75-fold lower at 37 °C than at 4 °C (Beisiegel et al 1981b). The low affinity of the antibody at 37 °C was shown to be due largely to a more rapid dissociation at that temperature.

If such rapid dissociation were to occur at 37 °C within the cell, it would liberate the receptor so that it could return to the surface and bind another molecule of antibody. We have previously studied a polyclonal rabbit antibody to the LDL receptor (Beisiegel et al 1981a). This antibody does not rapidly dissociate from the receptor at 37 °C (R. G. W. Anderson, M. S. Brown & J. L. Goldstein, unpublished observations). The lack of dissociation apparently prevents the receptor from returning to the surface. When fibroblasts are incubated with this polyclonal antibody, the number of receptors on the surface immediately declines and remains low for several hours. In these conditions, the receptors can be shown by indirect immunofluorescence to be localized within the cell in an antigenically intact form (Brown et al 1982).

Unanswered questions

The selectivity problem: How are receptors recognized as components of coated pits, so that they can be internalized with high efficiency while other membrane proteins remain on the surface? It has been suggested that the LDL receptor is a transmembrane protein whose cytoplasmic segment contains a sequence that binds to clathrin, either directly or through interaction with some other protein that itself is bound to clathrin (Anderson et al 1977b, Goldstein et al 1979). If this mechanism is correct, it implies that other receptors are also capable of interacting with clathrin or with clathrin-associated protein(s). This in turn raises the possibility that all receptors that migrate to coated pits have common structural features that allow this recognition. The information for such recognition must be contained in the primary sequence of the receptors, but it might operate by making the receptors susceptible to some specific modification such as glycosylation or phosphorylation, which then serves as the recognition signal. It seems likely that coated pit-associated membrane proteins may contain a common structural feature analogous to the amino-terminal signal sequence of many membrane proteins.

The recycling problem: How are receptors separated from their ligands so that the receptors can return to the surface while the ligands are delivered to other organelles such as lysosomes or the Golgi? Many ligands dissociate from their receptors at acid pH. Such dissociation might occur in an endocytic vesicle even before it fuses with a lysosome, if the pH of the fluid of the endocytic vesicle becomes acidic. The evaluation of this hypothesis requires that we measure the pH of endocytic vesicles at different stages after their invagination. Alternatively, many ligands (such as LDL, asialoglycoproteins and α_2-macroglobulin) show an absolute requirement for Ca^{2+} ions in order to bind to their respective cell-surface receptors. These ligands might separate from their receptors if the Ca^{2+} concentration were lowered as the coated pit is converted into an endocytic vesicle. To evaluate this idea, we need to learn the Ca^{2+} concentration of endocytic vesicles at various stages after their invagination.

Acknowledgement

The research described in this article was supported by a grant from the National Institutes of Health (HL 20948).

REFERENCES

Anderson RGW, Goldstein JL, Brown MS 1976 Localization of low density lipoprotein receptors on plasma membrane of normal human fibroblasts and their absence in cells from a familial hypercholesterolemia homozygote. Proc Natl Acad Sci USA 73:2434-2438

Anderson RGW, Brown MS, Goldstein JL 1977a Role of the coated endocytic vesicle in the uptake of receptor-bound low density lipoprotein in human fibroblasts. Cell 10:351-364

Anderson RGW, Goldstein JL, Brown MS 1977b A mutation that impairs the ability of lipoprotein receptors to localize in coated pits on the cell surface of human fibroblasts. Nature (Lond) 270:695-699

Basu SK, Goldstein JL, Anderson RGW, Brown MS 1981 Monensin interrupts the recycling of low density lipoprotein receptors in human fibroblasts. Cell 24:493-502

Beisiegel U, Kita T, Anderson RGW, Schneider WJ, Brown MS, Goldstein JL 1981a Immunologic cross-reactivity of the LDL receptor from bovine adrenal cortex, human fibroblasts, canine liver and adrenal gland, and rat liver. J Biol Chem 256:4071-4078

Beisiegel U, Schneider WJ, Goldstein JL, Anderson RGW, Brown MS 1981b Monoclonal antibodies to the low density lipoprotein receptor as probes for study of receptor-mediated endocytosis and the genetics of familial hypercholesterolemia. J Biol Chem 256:11923-11931

Brown MS, Kovanen PT, Goldstein JL 1981 Regulation of plasma cholesterol by lipoprotein receptors. Science (Wash DC) 212:628-635

Brown MS, Anderson RGW, Basu SK, Goldstein JL 1982 Recycling of cell surface receptors: observations from the LDL receptor system. Cold Spring Harbor Symp Quant Biol 46:713-721

de Duve C, DeBarsy T, Poole B, Trouet A, Tulkens P, Van Hoof F 1974 Lysosomotropic agents. Biochem Pharmacol 23:2495-2534

Goldstein JL, Brown MS 1974 Binding and degradation of low density lipoproteins by cultured human fibroblasts: comparison of cells from a normal subject and from a patient with homozygous familial hypercholesterolemia. J Biol Chem 249:5153-5162

Goldstein JL, Brown MS 1979 The LDL receptor locus and the genetics of familial hypercholesterolemia. Annu Rev Genet 13:259-289

Goldstein JL, Basu SK, Brunschede GY, Brown MS 1976 Release of low density lipoprotein from its cell surface receptor by sulfated glycosaminoglycans. Cell 7:85-95

Goldstein JL, Anderson RGW, Brown MS 1979 Coated pits, coated vesicles, and receptor-mediated endocytosis. Nature (Lond) 279:679-685

Gonzalez-Noriega A, Grubb JH, Talkad V, Sly WS 1980 Chloroquine inhibits lysosomal enzyme pinocytosis and enhances lysosomal enzyme secretion by impairing receptor recycling. J Cell Biol 85:839-852

Helenius A, Kartenbeck J, Simons K, Fries E 1980 On the entry of Semliki Forest virus into BHK-21 cells. J Cell Biol 84:404-420

Kaplan J 1981 Polypeptide-binding membrane receptors: analysis and classification. Science (Wash DC) 212:14-20

Pastan IH, Willingham MC 1981 Receptor-mediated endocytosis of hormones in cultured cells. Annu Rev Physiol 43:239-50

Pearse BMF, Bretscher MS 1981 Membrane recycling by coated vesicles. Annu Rev Biochem 50:85-101

Roth TF, Porter KR 1964 Yolk protein uptake in the oocyte of the mosquito *Aedes aegypti* L. J Cell Biol 20:313-332

Schneider WJ, Beisiegel U, Goldstein JL, Brown MS 1982 Purification of the low density lipoprotein receptor, an acidic glycoprotein of 164,000 molecular weight. J Biol Chem 257:2664-2673

Stahl P, Schlesinger PH, Sigardson E, Rodman JS, Lee YC 1980 Receptor-mediated pinocytosis of mannose glycoconjugates by macrophages: characterization and evidence for receptor recycling. Cell 19:207-215

Steer CJ, Ashwell G 1980 Studies on a mammalian hepatic binding protein specific for asialoglycoproteins: evidence for receptor recycling in isolated rat hepatocytes. J Biol Chem 255:3008-3013

Tolleshaug H, Berg T 1979 Chloroquine reduces the number of asialoglycoprotein receptors in the hepatocyte plasma membrane. Biochem Pharmacol 28:2912-2922
Van Leuven F, Cassiman J-J, Van Den Berghe H 1980 Primary amines inhibit recycling of α_2M receptors in fibroblasts. Cell 20:37-43
Wall DA, Wilson G, Hubbard AL 1980 The galactose-specific recognition system of mammalian liver: the route of ligand internalization in rat hepatocytes. Cell 21:79-93

DISCUSSION

*J. L. Gowans**: I don't know the extent to which decreased density of low density lipoprotein (LDL) receptors pedisposes to arterial disease in conditions other than familial hypercholesterolaemia, but there might be therapeutic advantages if one could re-equip cells that genetically lack LDL receptors. If, for example, it were possible to prepare large amounts of the receptor by recombinant DNA techniques, would administration of the receptor in a suitable form lead to its uptake by cells and incorporation with their membranes? Is this possible *in vitro* and is it likely to be a feasible strategy *in vivo*?

Goldstein: In this disease, one in 500 people in most populations is heterozygous for an LDL receptor mutation and about one in a million people has the homozygous form of the disease (Goldstein & Brown 1979). The number of LDL receptors is normally regulated through a feedback mechanism that responds to the cholesterol content of the cell. One can take advantage of this receptor regulation in designing a therapy for the disease in heterozygotes. By administering an inhibitor of cholesterol synthesis (which deprives the liver of cholesterol) together with a bile acid-binding resin (which removes cholesterol from the liver by promoting the excretion of bile acids from the body), one can markedly reduce plasma LDL levels (Brown & Goldstein 1981). We have been able to show in animal studies that this lowering of plasma LDL occurs as a result of an increase in LDL receptors in the liver, which in turn causes a removal of LDL from the plasma (Kovanen et al 1981). If these drugs turn out to be non-toxic in humans, this would be the therapy of choice for familial hypercholesterolaemia. Since this approach is based on the concept of receptor regulation, it should work in heterozygotes, who have half the normal number of receptors, but not in homozygotes, who lack receptors.

Branton: Why is it that when you apply the fluorescent monoclonal antibody to the untreated cell which had been permeabilized you see no

---

**Observer, Medical Research Council, Park Crescent, London W1.*

fluorescence but if you permeabilize the cells after monensin treatment, you do see the fluorescence. Does that imply a difference in the anchorage of the receptor?

Goldstein: We wipe out the surface binding sites for the fluorescent monoclonal antibody by permeabilization of untreated cells. The internalized binding sites in the monensin-treated cells are not affected by the permeabilization.

Jourdian: Is the internalized receptor still intact? Does it have the same molecular weight?

Goldstein: We have not yet tested that point.

Sabatini: Does trifluoperazine dihydrochloride have any effect?

Goldstein: Yes. This and related drugs inhibit either the binding or internalization of LDL, depending on the concentration of drug used. Of numerous drugs that we have tested in the last few years, monensin has been the only one that has not affected the binding and internalization of LDL. It therefore allows us to assess specifically the recycling of the receptor.

Meldolesi: Is it possible that monensin is effective because it exchanges Na^+ or K^+ with H^+, and therefore raises pH in intracellular acidic compartments?

Goldstein: Yes. The simplest interpretation of the data is that monensin works by raising the pH of intracellular vesicles, although we have no direct evidence for this speculation.

Cohn: Is there any possibility that monensin interacts with the receptor itself, or can you wash it out of the system?

Goldstein: Monensin has no effect on the binding of [^{125}I]LDL to the receptor at 4 °C. If we add the monensin for only 15 min in the absence of LDL at 37 °C, its effect on recycling is reversible, but after longer periods of time at 37 °C it is not completely reversible.

Sabatini: Do the receptors that are interiorized without the ligand in the presence of monensin come back to the cell surface? It is possible that those receptors pass through a different or less complete route from the one they use when they bind the ligand?

Goldstein: That's possible. We have some preliminary results that might be relevant. We initially thought that the 50% of receptors that recycled in the presence of monensin and in the absence of LDL were the ones that were spontaneously clustered in coated pits. We also believed that the remaining 50% that did not recycle in the presence of monensin alone were the ones that were not in coated pits but which migrated there after addition of LDL. However, our preliminary electron microscopic studies with R. G. W. Anderson and M. S. Brown (unpublished) do not support this interpretation, but show that receptors remaining on the surface after monensin treatment *are* localized in coated pits.

Rothman: Is the monoclonal antibody going in with the receptor and recycling with it or is it separated from the receptor in the cell?

Goldstein: It is separated from the receptor within the cell and degraded in lysosomes, just like LDL.

Raff: One could hypothesize simply that the acidification is needed for the ligand to dissociate from the receptor. If so, then one would predict that the monoclonal anti-receptor antibody would dissociate from receptor at low pH while the polyclonal antibodies would not. Have you studied this?

Goldstein: ^{125}I-labelled LDL dissociates from the receptor at acid pH. Preliminary studies suggest that the ^{125}I-labelled monoclonal antibody also dissociates from the receptor at acid pH. We haven't tested acid-mediated dissociation with the polyclonal antibody because we can't obtain 'clean' results when we radiolabel the polyclonal antibody.

Mellman: I agree with Dr Raff's point about your observation on the polyclonal antibody. We find similar results with one of our monoclonal antibodies (2.4G2) against the Fc receptor: we don't see a regeneration of new 2.4G2 binding sites, apparently because the antibody never comes off the receptor even after internalization; it seems to bind without being delivered to lysosomes (I. S. Mellman, unpublished). The receptor–antibody complex is apparently recycling intact, as most receptors would in the absence of ligand. Your result indicates that the polyclonal antibody interferes with recycling.

Goldstein: Yes, I agree.

Geisow: Is there a possible distinction here between non-recycling and recycling receptors? The affinity seems to remain too high for dissociation under the vacuole conditions, whatever they turn out to be, and thus we get trapping of the receptor. Therefore epidermal growth factor and Fc receptors don't recycle because they can't dissociate in the vesicular conditions of the endosome or lysosome.

Mellman: In principle that is possible. We were confused for some time because the isolated receptor had a pH optimum for binding which made it look like a good delivery system; it was reversibly inactivated at pHs that approximated to intralysosomal pH (Mellman & Unkeless 1980). But we have never been able to duplicate those results using receptor in intact membranes.

Geisow: Of course, in the endosome, the ligand concentration may be relatively high, and this may encourage continued association of the ligand with the receptor, and make it particularly difficult in some circumstances for dissociation to occur, even at low pH.

Hopkins: Have you coupled ferritin to either the monoclonal or the polyclonal anti-receptor antibody and followed where it goes after it leaves the coated pit?

Goldstein: Yes. When the monoclonal antibody is coupled to ferritin, we can see the ferritin particles in lysosomes.

Hopkins: Is there a pre-lysosomal stage?

Goldstein: Yes. It appears to follow the same pathway as LDL.

Hopkins: I wondered whether it was different from LDL, which is so big that it is difficult to relate the ferritin to where the receptor might be. But if there was only a monoclonal antibody between the ferritin and the receptor one would have a better idea of localizing it. With antibody–ferritin, is the ferritin more closely associated with the membrane?

Goldstein: We have not looked closely enough at that point. At one minute at 37 °C, we have seen the LDL–ferritin lining the membrane of the endocytic vesicle. After 1–2 min the LDL–ferritin begins to move into the lumen of the vesicle, suggesting that dissociation between ligand and receptor has occurred within 1–2 min. That would be consistent with our not finding a large internal pool of LDL receptors. It would also be consistent with an estimated life-time for the receptor inside the cell of about one minute.

Kornfeld: Have you done a similar experiment with the ferritin-labelled monoclonal antibody to see if the antibody is still bound to membrane when it reaches the lysosomes?

Goldstein: No.

Bretscher: If the lifetime of the LDL receptor on the surface of cells is 6 min and the time inside is 1 min, the one-seventh of the molecules, or 15%, must be inside the cells at any moment. If only 5% are inside the cell, then the transit time through the cell is only one-third of a minute, which is incredibly short!

Helenius: Our recent results with monensin show that it blocks Semliki Forest virus infection in the same way as does ammonium chloride. It inhibits the penetration of the virus from intracellular vacuoles by increasing the vacuolar pH above the pH needed to trigger fusion (Marsh et al 1982). We have no data on the recycling of receptors but the drug does not affect the internalization.

Raff: If the dissociation of the ligand from the receptor is pH-related one would expect, in the monensin-treated cells, not to see the LDL–ferritin dissociating from the membrane in the vesicles. Have you looked at that specifically?

Goldstein: No.

Helenius: Neither have we.

Widnell: I think there may be differences in the effects of monensin, depending on the concentration added to the medium. We have been looking (Wilcox et al 1982) at effects on fluid-phase endocytosis in rat fibroblasts as judged both by the uptake of horseradish peroxidase (HRP) and also by the internalization and return to the surface of 5′-nucleotidase (which seems to follow the kinetics that one would expect if it were a constituent of the membranes involved in fluid-phase endocytosis). If cells are treated with the concentrations of monensin (i.e. 0.5–1.0 μM) that have been used to inhibit secretion (Tartakoff & Vassalli 1978), there is a significant lag phase of about three hours before any inhibition of HRP uptake or other membrane flow can be detected. However, if cells are treated with 25 μM-monensin, there is a rapid effect: 50% inhibition within about one hour. We have also found that galactosyltransferase is inactivated following monensin treatment, presumably as the result of an effect on the membranes of the Golgi complex. In contrast to the inhibition of pinocytosis, the kinetics of inactivation of galactosyltransferase are exactly the same as either 1 μM or 25 μM-monensin (R. P. Kitson, D. K. Wilcox & C. C. Widnell, unpublished results). It thus seems possible that different concentrations of monensin may result in effects at different cellular sites. At high concentrations, for example, one may see a preferential effect of monensin on ion exchange in pinocytic vesicles whereas at low concentrations it may be acting not even as an ionophore, and at another site in the cell. One thus has to be cautious about the concentration of this drug that one uses in relation to any interpretation of effects that it may have.

Palade: Can the results of the monensin experiments be interpreted in terms of imperfect synchronization or lack of synchronization within the short time (less than 1 min) spent by the receptor in the cell? Suppose, for example, that during the experiment half the receptors move from the cell surface to a compartment from which they can no longer be recruited back to the surface; this could give you the figure of 50%. Suppose that after binding of LDL, the time the receptor spends on the cell surface is shortened; this may explain the results with monensin plus LDL.

Goldstein: That idea would be consistent with the experimental data.

Mellman: Do none of the familial hypercholesterolaemia homozygotes have any cross-reacting material detectable by your anti-receptor antibodies?

Goldstein: Of 30 familial hypercholesterolaemia homozygotes who have no binding activity for [^{125}I]LDL, five have significant amounts of cross-reacting material, as determined by binding studies with the ^{125}I-labelled monoclonal antibody.

Mellman: And presumably the internalization-defective mutant does have cross-reactive material?

Goldstein: Yes.

Rothman: Is the cross-reacting material of the same relative molecular mass as wild-type receptor?

Goldstein: Yes. The receptor in the internalization-defective mutant ('JD') has a relative molecular mass of 164 000, similar to that observed for the receptor of normal fibroblasts. This has been done by one-dimensional polyacrylamide gel electrophoresis of fibroblast extracts, followed by electrophoretic transfer and immunoblotting with the monoclonal antibody.

Sly: Does the polyclonal antibody bind the internalization-defective receptors?

Goldstein: The monoclonal antibody does bind, but we haven't looked at whether or not the polyclonal antibody binds.

Palade: What are the latest figures on the distribution of receptors on the plasmalemma proper and in coated pits?

Goldstein: The average values from numerous studies done over the last six years show that about 70% of the LDL receptors of fibroblasts are located within the coated pits and 30% are scattered at random on the cell surface. The coated pits occupy 2% of the cell surface in fibroblasts.

Palade: Does the receptor bind in any system to clathrin or to any other protein of the geodetic cage?

Goldstein: We are just beginning studies designed to identify different domains of the purified receptor.

Cohn: What is the lipoprotein specificity of the isolated receptor?

Goldstein: It's exactly the same as the native receptor in monolayer culture. The purified receptor binds lipoproteins that contain either apoprotein B (such as LDL) or apoprotein E (such as apo E-HDL_c) (Schneider et al 1982). Apo E-HDL_c is a unique lipoprotein isolated from the plasma of dogs fed high amounts of cholesterol (Mahley 1979). Neither the purified LDL receptor nor the LDL receptor of intact fibroblast monolayers binds typical high density lipoprotein (HDL), which contains apoproteins A-I and A-II.

Cohn: Would you expect that because of its highly anionic nature it would preferentially bind cationic polypeptides on the surface?

Goldstein: Probably not specifically, but they would bind non-specifically. Are you referring to any particular proteins?

Cohn: I was thinking of cationized ferritin or something like that. Would that be a driving force, a non-specific ligand, that might modify the receptor?

Goldstein: I am certain that those cationic proteins *would* bind, but I don't know how one could assess the specificity.

Cohn: Would it have anything to do with the aggregation of the receptor?

Goldstein: The internalization of receptor-bound LDL proceeds at a normal rate for hours in phosphate-buffered saline that contains calcium but no exogenous proteins. This suggests that exogenous proteins are not required for either endocytosis or recycling of LDL receptors.

REFERENCES

Brown MS, Goldstein JL 1981 Lowering plasma cholesterol by raising LDL receptors (editorial). N Eng J Med 305:515-517

Goldstein JL, Brown MS 1979 The LDL receptor locus and the genetics of familial hypercholesterolemia. Annu Rev Genet 13:259-289
Kovanen PT, Bilheimer DW, Goldstein JL, Jaramillo JJ, Brown MS 1981 Regulatory role for hepatic low density lipoprotein receptors *in vivo* in the dog. Proc Natl Acad Sci USA 78:1194-1198
Mahley RW 1979 Dietary fat, cholesterol, and accelerated atherosclerosis. Atheroscler Rev 5:1-34
Marsh M, Wellstead J, Kern H, Harms E, Helenius A 1982 Monensin inhibits Semliki Forest virus penetration into baby hamster kidney cells. Proc Natl Acad Sci USA, in press
Mellman IS, Unkeless JC 1980 Purification of a functional mouse Fc receptor through the use of a monoclonal antibody. J Exp Med 152:1048-1069
Schneider WJ, Beisiegel U, Goldstein JL, Brown MS 1982 Purification of the low density lipoprotein receptor, an acidic glycoprotein of 164,000 molecular weight. J Biol Chem 257:2664-2673
Tartakoff, A, Vassalli P 1978 Comparative studies of intracellular transport of secretory proteins. J Cell Biol 79:694-707
Wilcox DK, Kitson RP, Widnell CC 1982 Inhibition of pinocytosis in rat embryo fibroblasts treated with monensin. J Cell Biol 92:859-864

Epidermal growth factor: uptake and fate

PEDRO CUATRECASAS

Department of Molecular Biology, Wellcome Research Laboratories, Burroughs Wellcome Co., 3030 Cornwallis Road, Research Triangle Park, North Carolina 27709, USA

Abstract Lateral diffusion of epidermal growth factor (EGF) receptors along the plane of the cell membrane can be measured using fluorescently labelled analogues of EGF and the fluorescence photobleaching recovery method in cultured cells. With the aid of high image-intensified fluorescent microscopy, the receptors, which are initially distributed diffusely, form patches and undergo endocytosis at 37 °C. These gross processes may not be critical in mediating the initial, rapid actions of the hormone. The processes of uptake and endocytosis correspond biochemically to the loss of surface receptors ('down-regulation') and degradation of the receptor and hormone via lysosomes. The EGF receptors are not apparently recycled or re-utilized, and they are continuously internalized, even in the absence of ligand. Since all manoeuvres that interfere with intracellular degradation or processing block mitogenesis, it is proposed that one or both of these may be essential processes, although in such a case they must be continuous and protracted functions. Slow nuclear accumulation of the complex of hormone and receptor may be an important process. In addition, evidence suggests that limited (submicroscopic) receptor aggregation (dimerization) at the cell surface may be necessary and sufficient to trigger the long-term effects (but not the immediate effects), and thus this aggregation may be required for endocytosis. The ligand itself may not be an essential structural component of the action of the receptor since anti-receptor antibodies can elicit mitogenic responses. Recent results suggest that EGF receptors normally exist in a low affinity state which is rapidly converted by EGF (at 37 °C but not at 4 °C) to a high affinity state by a process that requires prior intact protein synthesis. Furthermore, the accumulation of a special, stable intracellular pool of the complex may be related to the control of cellular growth (and tumour promotion).

Epidermal growth factor (EGF) promotes the growth and differentiation of epithelial and fibroblastic cells both *in vitro* and *in vivo*. Although it is generally accepted that EGF and other mitogenic peptides initiate a cascade of intracellular anabolic events through specific interactions with cell-surface receptors, it is not known what secondary processes are required for establishing their ultimate mitogenic effect (King & Cuatrecasas 1981).

1982 Membrane recycling. Pitman Books Ltd, London (Ciba Foundation symposium 92), p 96-108

The lateral diffusion of receptors along the plane of the membrane has been measured using fluorescently labelled analogues of EGF and the fluorescence photobleaching recovery method in cultured cells (Schlessinger et al 1978, Shechter et al 1978a,b). The diffusion constant measured is about 4×10^{10} cm^2 s^{-1}, a value near that of other macromolecular markers on the cell membrane, such as concanavalin A, acetylcholine receptors and insulin receptors. The lipid probe, rhodamine-stearic acid, has shown a diffusion constant of about 10^{-8} cm^2 s^{-1} in both artificial bilayers and intact cells. Although detailed interpretations of such studies must be guarded, it is safe to say that EGF receptors *can* diffuse in the plane of the membrane, and that at least 40 to 50% of the receptors are mobile. The lateral mobility initially increases markedly but then decreases rapidly when the temperature is increased from 4 °C to 37 °C. The latter effect has been attributed to the aggregation and subsequent internalization of receptors. Very recently, Zidovetzki et al (1981) have used a phosphorescence emission and anisotropy method for measuring the rotational diffusion of EGF receptors in epidermoid carcinoma cells (A-431). In contrast to lateral diffusion, rotational diffusion measurements became longer as the time and temperature of the incubation were increased, presumably reflecting the immobilization attendant on the progressive formation of microclusters (which probably contain 10–50 receptors per cluster). The rotational correlation times became shorter when the clusters were internalized, suggesting a decrease in the size of the dynamic units.

By high image-intensified fluorescence microscopy, EGF–rhodamine is initially distributed diffusely on the cell surface (Schlessinger et al 1978). Upon warming from 4 °C to 24 or 37 °C the fluorescence rapidly forms a patchy pattern which can be removed by trypsin. With further incubation larger patches are seen, and subsequently the material is seen in endocytic vesicles which demonstrate the saltatory motion characteristic of intracellular vesicles. Visualization by electron microscopy using EGF–ferritin conjugates in A-431 cells, which are enriched in EGF receptors (about 2×10^6 per cell), demonstrates microclusters of 5 to 10 molecules within 1 to 2 min at 37 °C (Haigler et al 1979, McKanna et al 1979). The internalization requires metabolic energy.

The rates at which these processes of gross aggregation and internalization occur are much too slow to explain the immediate, almost instantaneous effects of hormones, such as increased glucose or amino acid transport, phospholipid breakdown, and so on. However, *limited* (i.e. submicroscopic) mobility, rearrangements and aggregation may still be prerequisites for such rapid (as well as more protracted) biological effects. For example, if the receptors are assumed to be randomly distributed and to have a diffusion constant of 4×10^{-10} cm^2 s^{-1}, two receptors could collide within 50 ms.

Thus, limited aggregation is fast enough for such processes. There is considerable evidence, to which I shall return later, to suggest that submicroscopic receptor microaggregation is very important (if not mandatory) for receptor activation.

What could be the significance of these dramatic processes of gross patching and internalization? Clearly, this is another example of receptor-mediated endocytosis, by which ligands undergo adsorptive concentration into cells through coated pits. The functions to be served by peptide hormones such as EGF and insulin, however, are very different from those of low density lipoprotein (LDL), lysosomal enzymes and asialoglycoproteins and, therefore, it would not be surprising if some major mechanistic differences were found between these various systems. For mitogenic hormones (e.g. EGF, insulin, concanavalin A, and even oestrogen hormones) a legitimate question is whether their long-term mitogenic effects, such as cell division, are related to endocytosis or to the regulation of hormonal responsiveness (desensitization, adaptation).

The biochemical correlate of the fluorescence studies described above is the process of 'down-regulation' (Carpenter & Cohen 1976a). Whereas at 4 °C the association of ^{125}I-labelled EGF proceeds slowly to a stable steady-state level, at 37 °C the cell-bound radioactivity increases rapidly (during the first hour) to reach a level greater than an amount equivalent to the total number of cell receptors, after which there is a rapid fall to a level that is 10 to 20% of the peak level (by about two hours). During the first hour the hormone binds to its receptor and is internalized and delivered to lysosomes (Aharonov et al 1978, Carpenter & Cohen 1976a, Vlodavsky et al 1978). Subsequently the hormone and the receptor (Fox & Das 1979, Fox et al 1980) are degraded, and the products are secreted into the medium. There is a fall in the number of surface receptors (to 10–20%). The EGF receptors are internalized after migration to coated pits (Haigler et al 1979, McKanna et al 1979), and their presence in intracellular 'receptosomes' precedes delivery to lysosomes, presumably after processing in the Golgi–Golgi endoplasmic reticulum lysosome (GERL) complex (Pastan & Willingham 1981). Minimal degradation occurs during the first 30 to 60 min, reflecting the lag period for processing of the receptosomes.

Several studies have clearly demonstrated that for EGF to elicit a mitogenic response it has to be exposed to cells (in the medium) continuously for at least 8–10 h (Aharonov et al 1978, Carpenter & Cohen 1976b, Shechter et al 1978a). Removal of the hormone by washing and exposure to anti-EGF antibodies abrogates the response during the first 10 h (Shechter et al 1978a). Thus, there is a need for continuous and prolonged exposure to EGF. Therefore, the initial, dramatic delivery of hormone into the cells during the first two hours (the 'down-regulation' period) is certainly not a sufficient event for mitogenesis. The simplest explanation would be a requirement for continuous occupation of surface receptors to generate signals which when sustained would produce some significant and irreversible intracellular event (Shechter et al 1978a). Although the need for continuous occupation of surface receptors is clear, this does not exclude a possible requirement for a continual process of receptor-mediated endocytosis. For example, this could in theory produce a threshold, critical concentration of receptor in some intracellular organelle. Recent studies have cast doubt on the possible role of endocytosis in the mitogenic action because primary and tertiary amines, which were thought to block internalization, were found to be without effects on the biological effects of EGF (Maxfield et al 1979a,b). However, the results of subsequent detailed studies alter these interpretations (King et al 1980a, 1981).

Primary and tertiary amines such as CH_3NH_2, chloroquine and dansylcadaverine, clearly do not block internalization (King et al 1980a, McKanna et al 1979, Michael et al 1980). In their presence, the radioactive hormone accumulates in the cell and the degradation of the hormone and receptor is prevented. EGF–ferritin complexes also can be shown to internalize and accumulate intracellularly (in large, multivesicular structures) in the presence of chloroquine and CH_3NH_2 (Haigler et al 1979). Very importantly, mitogenesis was blocked under all conditions that interfered with endocytic processing, suggesting the possible interrelationship of these processes (King et al 1981). It is impossible to say whether the degradation process itself or some other aspect of intracellular processing (e.g. delivery of a component to another cell organelle) may be related to the biological effects. For nerve growth factor (NGF), there is evidence that the intact hormone and the receptor accumulate in the cell nucleus with prolonged incubation (Andres et al 1977, Marchisio et al 1980, Yankner & Shooter 1979). For EGF, there is apparent association of the labelled hormone with the nucleus in the presence of compounds that decrease degradation and increase intracellular accumulation of the intact hormone (Savion et al 1981). It is, however, under these conditions that the mitogenic effects of the hormone are blocked; it is conceivable that this hormone ‘association’ with the nucleus could be an artifact of isolation of nuclei under conditions where large multivesicular structures accumulate intracellularly.

Unlike in a number of other receptor-mediated endocytic systems, such as those for LDL and for asialoglycoproteins, receptors for EGF are not re-utilized and do not recycle in cultured cells (King & Cuatrecasas 1981, King et al 1980b). The receptors are internalized and degraded continually at a relatively rapid rate ($t_{\frac{1}{2}}$ about four to five hours) even in the absence of EGF. The receptors, once internalized (with or without EGF) are degraded and must be synthesized *de novo* to replenish the normal surface complement and to allow turnover. It is by this mechanism that the cell can accumulate or process EGF in quantities larger than those equal to the total number of receptors that exist at any one time. In the presence of EGF the rate of internalization increases about 10-fold, but this effect disappears after about one to two hours (the period for down-regulation), and there is no concurrent increase in the rate of synthesis of the receptor. After this initial phase (about two hours) the EGF–receptor complexes continue to be internalized, degraded and replenished through *de novo* synthesis at a steady rate in the continued presence of the hormone.

On the basis of studies using affinity labelling of the EGF receptor, it has been suggested that degradation of the ligand–receptor complex may produce an intracellular second messenger (Fox & Das 1979, Fox et al 1980). Although this hypothesis is consistent with the observed specific cleavage of EGF receptors (Fox et al 1980), and with the above data implicating a requirement for endocytosis and processing (possible 'degradation'), it is difficult to reconcile the need for a product of EGF proteolysis with the recent observations (Schreiber et al 1981b) that monoclonal receptor antibodies are capable of simulating the action of EGF in A431 cells. Therefore, if a proteolytic fragment is involved in transmitting information intracellularly, and if this is to arise from the EGF–receptor complex, it is likely to be a fragment of the receptor molecule.

The tumour-promoting phorbol ester, phorbol myristate acetate (PMA), which is known to enhance the mitogenic effects of EGF, has been used as a tool to study the process of EGF processing by cultured cells (King & Cuatrecasas 1982). PMA has an immediate but transient effect on the affinity of cell receptors for EGF, apparently by inhibiting the conversion of EGF receptors from low to high affinity. This process normally occurs very rapidly on exposure to EGF, and it requires increased temperatures (37 °C) and depends completely on the immediately preceding *de novo* protein synthesis. There is some evidence that this effect, which is not observed at 4 °C, may depend on a very rapidly-turning-over effector (rather than on the receptor itself being synthesized). It is pertinent that for NGF there is evidence for the rapid peptide-induced transformation of low affinity receptors to ones of high affinity (Sutter et al 1979). With continued exposure of cells to EGF and PMA, the effect on affinity appears to be lost, and instead there is a slow accumulation of hormone–receptor complexes which after two hours reaches a level higher than that originally observed at one hour in the absence of PMA. The usual falls in surface receptors and in cell-associated hormone (down-regulation) are not observed. The achievement of levels higher than those seen without PMA is blocked by cycloheximide. The accumulated intracellular pool of EGF–receptor complexes is not readily susceptible to degradation through the lysosomal system. Thus, PMA appears to prevent the high affinity transformation of receptors induced by EGF, but it favours the uptake of the EGF–receptor complex into a stable intracellular pool. The latter may be related to EGF-induced mitogenesis, and to the mechanism by which PMA enhances EGF-initiated mitogenesis. Notably, in the presence of PMA, EGF can induce mitogenesis with lag phases much shorter than the usual periods of about 20 h.

Earlier, I mentioned evidence suggesting that receptor microaggregation may be an important or essential feature in the hormone activation of immediate as well as long-term biological effects. An analogue of EGF (CBrN–EGF), prepared by treating the hormone with cyanogen bromide, is devoid of biological activity yet it retains specific receptor affinity (Shechter et al 1979b). This inactive analogue, which is basically a competitive antagonist, can be converted to a fully active hormone by adding anti-EGF antibodies. This effect depends on the bivalence of the antibodies, and is not seen with Fab fragments. This suggests that CBrN–EGF may be deficient in an ability to induce receptor cross-linking, and that this cross-linking can be achieved effectively by passive induction with antibody. Fluorescent microscopy using rhodamine-labelled CBrN–EGF confirms that the distribution on the cell surface is diffuse even after incubation at 37 °C for two hours; in the presence of anti-EGF antibodies gross patches and aggregates are seen, and internalization with endocytic vesicles is observed.

It has also been possible to enhance greatly the mitogenic activity of native EGF itself in cells with antibodies to the hormone (Shechter et al 1979a,b). This is especially dramatic if the cells used are relatively 'resistant' to EGF. In this case extremely low concentrations of EGF (100 to 1000 times lower than the ED_{50}) can be shown to elicit activity with the proper concentration of anti-EGF antibody. In this case the bivalence of the antibody is also essential. Importantly, it has recently been shown that bivalent (but not univalent) EGF receptor antibodies can induce mitogenesis in the absence of EGF (Zidovetzki et al 1981).

All these studies suggest that receptor aggregation (or dimerization) can induce a hormone-like response, and that perhaps the hormone may ordinarily function simply to induce such aggregation. However, the receptors may also exist normally in an equilibrium state between a reservoir of inactive, low affinity dispersed receptors and a state of high affinity pre-aggregated receptors in which a subtle change induced by EGF binding triggers a minor conformational change and, thus, activity. In this case the use of external agents to cross-link the receptors passively would artificially recruit and activate receptors that would ordinarily not be susceptible to EGF stimulation. On the other hand, the transformation may indeed be mediated by EGF, and this effect could be the mechanism for explaining the temperature-dependent conversion from low to high affinity receptor, described already for the PMA studies. It is interesting that for insulin and for immunoglobulin E (in mast cells) there is also strong evidence for a role of receptor dimerization and aggregation in activation.

It has recently been described that the immediate ability of EGF to induce tyrosine phosphorylation in membranes (perhaps the receptor itself being the substrate) can be produced by CBrN–EGF (Schreiber et al 1981a). At least some of the very rapid, immediate effects of the hormone may therefore not require microaggregation, and the latter may be peculiar to the slower processes that require endocytosis and processing. Furthermore, those receptors mediating immediate effects may actually escape endocytosis (by not aggregating). Thus, one might separate and distinguish between those very different biological effects while utilizing a single, unique receptor macromolecule for all the effects. In any case, the immediate and late effects are likely to be mediated by different chemical signals arising from the same receptor.

There is no evidence that EGF-induced endocytosis and degradation act in any important way simply to 'regulate' the cell-surface concentration of receptors. The term 'down-regulation' may thus be a misnomer. This process occurs physiologically and may be essential (rather than inhibitory) to the biological response. Furthermore, it is observed virtually only during the first few hours. Endocytosis appears to be, rather, a mechanism for directing the hormone–receptor complex into the cell interior where specific degradation and processing lead, over protracted periods, to the generation of specific biochemical signals.

REFERENCES

Aharonov A, Pruss RM, Herschmann HR 1978 Insulin and epidermal growth factor: human fibroblast receptors related to deoxyribonucleic acid synthesis and amino acid uptake. J Biol Chem 253:3970-3977

Andres RY, Jeng I, Bradshaw RA 1977 Nerve growth factor receptors: identification of distinct classes in plasma membranes and nuclei of embryonic dorsal root neurons. Proc Natl Acad Sci USA 74:2785-2789

Carpenter G, Cohen S 1976a ^{125}I-labeled human epidermal growth factor: binding, internalization, and degradation in human fibroblasts. J Cell Biol 71:159-171

Carpenter G, Cohen S 1976b Human epidermal growth factor and the proliferation of human fibroblasts. J Cell Physiol 88:227-237

Fox CF, Das M 1979 Internalization and processing of the EGF receptor in the induction of DNA synthesis in cultured fibroblasts: the endocytic activation hypothesis. J Supramol Struct 10:199-214

Fox CF, Wrann M, Linsley P, Vale R 1980 Hormone-induced modification of EGF receptor proteolysis in the induction of EGF action. J Supramol Struct 12:517-531

Haigler HT, McKanna JA, Cohen S 1979 Direct visualization of the binding and internalization

of a ferritin conjugate of epidermal growth factor in human carcinoma cells A-431. J Cell Biol 81:382-395
King AC, Cuatrecasas P 1981 Peptide hormone-induced receptor mobility, aggregation, and internalization. N Eng J Med 305:77-88
King AC, Cuatrecasas P 1982 Resolution of high and low affinity epidermal growth factor receptors: inhibition of high affinity component by low temperature, cycloheximide and phorbol esters. J Biol Chem 257:3053-3060
King AC, Hernaez LJ, Cuatrecasas P 1980a Lysosomotrophic alkylamines cause intracellular accumulation of receptors for epidermal growth factor. Proc Natl Acad Sci USA 77:3283-3287
King AC, Willis RA, Cuatrecasas P 1980b Accumulation of epidermal growth factor within cells does not depend on receptor recycling. Biochem Biophys Res Commun 97:840-845
King AC, Davis LH, Cuatrecasas P 1981 Lysosomotrophic alkylamines inhibit mitogenesis induced by growth factors. Proc Natl Acad Sci USA 78:717-721
Marchisio PC, Naldini L, Calissano P 1980 Intracellular distribution of nerve growth factor in rat pheochromocytoma PC12 cells: evidence for a perinuclear and intranuclear location. Proc Natl Acad Sci USA 77:1656-1660
Maxfield FR, Davies PJA, Klempner L, Willingham MC, Pastan I 1979a Epidermal growth factor stimulation of DNA synthesis is potentiated by compounds that inhibit its clustering in coated pits. Proc Natl Acad Sci USA 76:5731-5735
Maxfield FR, Willingham MC, Davies PJA, Pastan I 1979b Amines inhibit the clustering of α_2-macroglobulin and EGF on the fibroblast cell surface. Nature (Lond) 277:661-663
McKanna JA, Haigler HT, Cohen S 1979 Hormone receptor topology and dynamics: morphological analysis using ferritin-labeled epidermal growth factor. Proc Natl Acad Sci USA 76:5689-5693
Michael HJ, Bishayee S, Das M 1980 Effect of methylamine on internalization, processing and biological activation of epidermal growth-factor receptor. FEBS (Fed Eur Biochem Soc) Lett 117:125-130
Pastan IH, Willingham MC 1981 Receptor-mediated endocytosis of hormones in cultured cells. Annu Rev Physiol 43:239-250
Savion N, Vlodavsky I, Gospodarowicz D 1981 Nuclear accumulation of epidermal growth factor in cultured bovine corneal endothelial and granulosa cells. J Biol Chem 256:1149-1154
Schlessinger J, Shechter Y, Willingham MC, Pastan I 1978 Direct visualization of binding, aggregation, and internalization of insulin and epidermal growth factor on living fibroblastic cells. Proc Natl Acad Sci USA 75:2659-2663
Schreiber AB, Yarden Y, Schlessinger J 1981a A non-mitogenic analogue of epidermal growth factor enhances the phosphorylation of endogenous membrane proteins. Biochem Biophys Res Commun 101:517-523
Schreiber AB, Lax I, Yarden Y, Eshhar Z, Schlessinger J 1981b Monoclonal antibodies against receptor for epidermal growth factor induce early and delayed effects of epidermal growth factor. Proc Natl Acad Sci 78:7535-7539
Shechter Y, Hernaez L, Cuatrecasas P 1978a Epidermal growth factor: biological activity requires persistent occupation of high affinity cell surface receptors. Proc Natl Acad Sci USA 75:5788-5791
Shechter Y, Schlessinger J, Jacobs S, Chang K-J, Cuatrecasas P 1978b Fluorescent labeling of hormone receptors in viable cells: preparation and properties of highly fluorescent derivatives of epidermal growth factor and insulin. Proc Natl Acad Sci USA 75:2135:2139
Shechter Y, Chang K-J, Jacobs S, Cuatrecasas P 1979a Modulation of binding and bioactivity of insulin by anti-insulin antibody: relation to possible role of receptor self-aggregation in hormone action. Proc Natl Acad Sci USA 76:2720-2724

Shechter Y, Hernaez L, Schlessinger J, Cuatrecasas P 1979b Local aggregation of hormone-receptor complexes is required for activation by epidermal growth factor. Nature (Lond) 278:835-838

Sutter A, Riopelle RJ, Harris-Warrick RM, Shooter EM 1979 Nerve growth factor receptors: characterization of two distinct classes of binding sites on chick embryo sensory ganglia cells. J Biol Chem 254:5972-5982

Vlodavsky I, Brown KD, Gospodarowicz D 1978 A comparison of the binding of epidermal growth factor to cultured granulosa and luteal cells. J Biol Chem 253:3744-3750

Yankner BA, Shooter EM 1979 Nerve growth factor in the nucleus: interaction with receptors on the nuclear membrane. Proc Natl Acad Sci USA 76:1269-1273

Zidovetzki R, Yarden Y, Schlessinger J, Jovin TM 1981 Rotational diffusion of epidermal growth factor complexed to cell surface receptors reflects rapid microaggregation and endocytosis of occupied receptors. Proc Natl Acad Sci USA 78:6981-6985

DISCUSSION

Goldstein: I believe that the inhibitors you use to prevent processing of epidermal growth factor (EGF) in endocytic vesicles would also inhibit the lysosomal degradation of EGF. So, how can you separate an inhibitory effect on processing in endocytic vesicles from an inhibitory effect on the lysosomal generation of some active metabolite that is required for the action of EGF?

Cuatrecasas: We cannot separate them, and we must consider both alternatives.

Goldstein: Is it correct that no small peptide fragment of EGF has ever been isolated after incubation of intact EGF with cells?

Cuatrecasas: That's right. You might know that Fox and his colleagues (1980) have proposed that there may be an important product of proteolysis of EGF and/or the receptor. This hypothesis is based on several studies, including the use of affinity labelling of the EGF receptors. Their work has shown that after exposure to EGF a macromolecule can be isolated in the cytosol with a relative molecular mass (M_r) of 60–80 000. This molecule can stimulate DNA synthesis *in vitro*, in isolated nuclei from frog liver. Some fragment of the receptor is proposed to be generated.

Hubbard: What's the evidence for continued internalization of the receptor in the absence of ligand?

Cuatrecasas: By the addition of cycloheximide at various times, we can show a decrease in the number of surface receptors, and this decrease is enhanced very significantly by addition of EGF. Unlike Dr Goldstein's results (p 77-95) with low density lipoprotein (LDL), we can effectively 'turn off' the appearance of new EGF receptors. We presume that in addition to being internalized, the receptors could be removed onto the exterior. However, if we add cycloheximide, methylamine or chloroquine to prevent degradation, we can recover the receptors intracellularly, in the absence of ligand.

Hubbard: So the receptors don't seem to reappear on the cell surface, and the process is one-way?

Cuatrecasas: Yes. With cycloheximide we begin to lose the receptors from the cell surface, but if one adds chloroquine or dansylcadaverine one still loses receptors from the cell surface, but they can be recovered inside the cell.

Raff: Isn't it time to bury the idea that protein ligands have to enter cells in order to signal them? For example, antibodies against the EGF receptor can stimulate all the early and late events that EGF itself induces (Schreiber et al 1981). Even in the case of nerve growth factor (NGF) the evidence seems overwhelming: injecting anti-NGF antibodies into the cytosol doesn't block the effect of extracellular NGF and injecting NGF itself into the cytosol is ineffective (Heumann et al 1981).

Cuatrecasas: I would tend to agree with you, but one hypothesis which is difficult to exclude is that the NGF has been translocated in vesicles and is not accessible to antibodies because it is in the inner nuclear membrane.

Hubbard: What is responsible for the 'down-regulation' of the EGF receptor in the presence of ligand if, in the absence of ligand, the receptor still enters?

Cuatrecasas: The apparent entry of the receptor into a coated pit in the absence of ligand is considered by many people not to need any special mechanisms. The hormone does not seem to be required to induce a conformational change that takes the receptor into a coated pit. The process may simply be determined by diffusion. If one considers that in fibroblasts the coated pits constitute about 2% of the cell surface, one can calculate that random movement of receptors into coated pits would be sufficiently fast for the receptors to be captured at the rates known to occur during internalization. When a receptor containing a hormone reaches a coated pit it will stay there longer than when it is unoccupied. Perhaps a given number of them stay there long enough to be internalized even in the absence of hormone, since in such a case the receptor disappears with a fast half-life of about 4–5 hours. In the presence of EGF the *rate* of disappearance increases 10-fold, and new synthesis of receptor does not compensate for that.

Sly: That is analogous to what Dr Mellman said earlier (see this volume, p 35-58) about the Fc receptors, which enter continuously with a half-life of about 9.5 h, but whose entry is greatly accelerated when the receptors interact with ligand.

Mellman: I cannot say for sure how fast entry occurs in the absence of ligand, except that the receptor is internalized at the same rate as most other membrane polypeptides. If one assumes that internalized membrane has the same protein concentration (i.e. with respect to lipid) as the cell surface, arguing from Dr Cohn's stereological data (see Cohn et al, this volume p 15-34), the macrophage Fc receptor should be continuously internalized with a $t_{\frac{1}{2}}$ of 15 min. This rate is accelerated some four-fold after interaction with soluble immune complexes. I have not yet examined receptor turnover with this ligand. However, receptors internalized selectively during the phagocytosis of immunoglobulin G (IgG)-coated erythrocyte ghosts turn over with a $t_{\frac{1}{2}}$ of 2.5 h, compared with 9–10 h in control cells.

Dr Cuatrecasas, do you measure receptor turnover in the absence of ligand by the decrease in EGF binding sites, either in the presence or absence of cycloheximide, or by immunoprecipitation of labelled receptor?

Cuatrecasas: By the disappearance of receptor in the presence of cycloheximide. The monoclonal antibodies have only recently been obtained, so the details of these hypotheses have yet to be confirmed, with the aid of the antibodies.

Mellman: How about the turnover of the receptor in the presence of ligand? Were those data derived mostly from the affinity labelling experiments of Das & Fox (1978)?

Cuatrecasas: No, these are our own data, but they correspond quite closely to theirs.

Kornfeld: You stated that if you bound EGF to cells at 4 °C and then lowered the pH to 4.5, a mitogenic response could be detected. How long were the cells exposed to that pH, and how does this observation correlate with the experiments in which you added anti-EGF antibodies up to 8 h after EGF addition and still saw an effect on mitogenesis? Secondly, can you detect any changes in EGF in solution at that pH?

Cuatrecasas: The exposure to pH values of about 6 can be as short as 5 min, at 24–37 °C (King & Cuatrecasas 1982). At 4 °C the cells detach from the dish, and such experiments are more difficult. The low pH at 4 °C has also been done with diphtheria toxin, another important molecule to consider by analogy to Dr Helenius's work on viruses and EGF (see p 59-76). We can wash the cells with antibody to EGF and then we don't need any additional exogenous EGF for DNA response. We can see effects on DNA synthesis after periods as short as 10 h after such treatments. We don't know how much earlier the effects could be seen; ordinarily, of course, one sees nothing before about 20 h.

Helenius: Can you by-pass the chloroquine block?

Cuatrecasas: Yes, as in your work. It is incredible that this block can be overcome by such treatment.

Incidentally, the comparison of viral entry with diphtheria toxin, and perhaps with EGF, is quite important. As Dr Helenius said (p 72), viruses and toxins are probably not using normal receptors for those ligands but they may be using processes that are ordinarily used by some other physiological system. Diphtheria toxin is a single polypeptide containing two sub-units with a single disulphide. A critical proteolytic cleavage occurs between the sub-units. The proteolysis must occur before the toxin reaches the cell surface. The toxin is taken into lysosomes and, presumably, it is then reduced, and the B subunit, which recognizes the receptor only, apparently changes its conformation. The B subunit is very hydrophobic, and it apparently enters the membrane and produces an ionophore which allows the hydrophilic A subunit (a catalytic protein that catalyses the NAD-dependent inactivation of an elongation factor) to pass into the cytoplasm. The lysosome itself is not absolutely essential since if the toxin is bound to the cell surface and the pH is lowered, it is possible to by-pass everything, and to go directly to the cytosol and inhibit protein synthesis.

Raff: But something doesn't fit. If antibody against EGF receptors can do everything that EGF can do, the information for signalling the cell must be in the receptor. What effect do you think lowering the pH has? It seems inconceivable that a fall in pH will get EGF into the cytosol and that this somehow mimics the effect of receptor activation. It is really very different from the case of diphtheria toxin, where the toxin itself mediates the intracellular response.

Cuatrecasas: Perhaps in this case a part of the receptor rather than EGF is cleaved and internalized; the hormone may facilitate one of these early events.

Palade: We need to know what happens to the EGF at low pH. Is it accessible to proteases at the cell surface; is it accessible to the antibody at the cell surface; or is it no longer at the cell surface because it has been interiorized?

Cuatrecasas: Those are crucial points. The problem so far has been that the amount of EGF required is extraordinarily small.

Palade: Considering the immediate effects of phosphorylation of tyrosine residues in the receptor and (as I understand it) in other membrane proteins, can you obtain the same effects with the antibody and with lectins?

Cuatrecasas: With lectin, that hasn't been done, but a monoclonal EGF receptor antibody *will* induce mitogenesis and will produce the immediate phosphorylation (tyrosine) effects.

Palade: Is the rate of phosphorylation of these tyrosine residues increased as you increase the size of the aggregates by using successive layers of appropriate antibodies?

Cuatrecasas: That has not been done.

Mellman: Fc receptors for IgG, have traditionally been considered to function only as scavengers for immune complexes. However, they should be considered as effector receptors as well. Rouzer et al (1980) looked at some of the immediate effects of Fc receptor binding to IgG. They coated Sephadex beads with IgG which were too large to be internalized. They measured the production of prostaglandins by the cells and found that it was greatly potentiated soon after these particles were bound. A similar experiment on Sephadex beads coated with Fab fragments of our anti-receptor antibody (2.4G2) showed that the potentiation of prostaglandin release was not nearly as great. This suggests that the signal is not simply the clustering or cross-linking of receptors, but that something inherent in the nature of the ligand is required.

Cuatrecasas: Those are important experiments. Although the bead containing Fab is multivalent, it is not the same as actually physically cross-linking the receptors in very close proximity, as could occur with a soluble multivalent ligand.

Palade: Are the diphtheria toxin effects now believed to be independent of the activity of lysosomotropic agents?

Helenius: No; the effects are inhibited by all the lysosomotropic weak bases and monensin. Marnell et al (1982) have shown recently that monensin and lysosomotropic weak bases block the same step in the entry pathway.

Cuatrecasas: But with exposure to low pH, one can by-pass this blockage.

REFERENCES

Das M, Fox CF 1978 Molecular mechanism of mitogen action: processing of receptor induced by epidermal growth factor. Proc Natl Acad Sci USA 75:2644-2648

Fox CF, Wrann M, Linsley P, Vale R 1980 Hormone-induced modification of EGF receptor proteolysis in the induction of EGF action. J Supramol Struct 12:517-531

Heumann R, Schwab M, Thoenen H 1981 A second messenger required for nerve growth factor biological activity? Nature (Lond) 292:838-840

King AC, Cuatrecasas P 1982 Exposure of cells to an acidic environment reverses the inhibition by methylamine of the mitogenic response to epidermal growth factor. Biochem Biophys Res Commun 106:479-485

Marnell MH, Stookey M, Draper RK 1982 Monensin blocks the transport of diphtheria toxin to the cell cytoplasm. J Cell Biol 93:57-62

Rouzer CA, Scott WA, Kempe J, Cohn ZA 1980 Prostaglandin synthesis by macrophages requires a specific receptor–ligand interaction. Proc Natl Acad Sci USA 77:4279-4282

Schreiber AB, Lax I, Yarden Y, Eshhar Z, Schlessinger J 1981 Monoclonal antibodies against receptor for epidermal growth factor induce early and delayed effects of epidermal growth factor. Proc Natl Acad Sci USA 78:7535-7539

General discussion I

Receptor-mediated endocytosis of asialoglycoproteins in the hepatocyte

Hubbard: I would like to describe our recent work on the asialoglycoprotein receptor system in liver. Ashwell & Morell (see 1974 review) first reported that desialylated glycoproteins were rapidly cleared from the circulation by the liver and degraded. The receptor which initially binds these galactose (or galactosamine)-terminating ligands is localized exclusively to the major cell of the liver, the parenchymal cell.

We have studied the ligand pathway both *in vivo* and in an isolated perfused liver system (Hubbard & Stukenbrok 1979, Wall et al 1980, Wall &

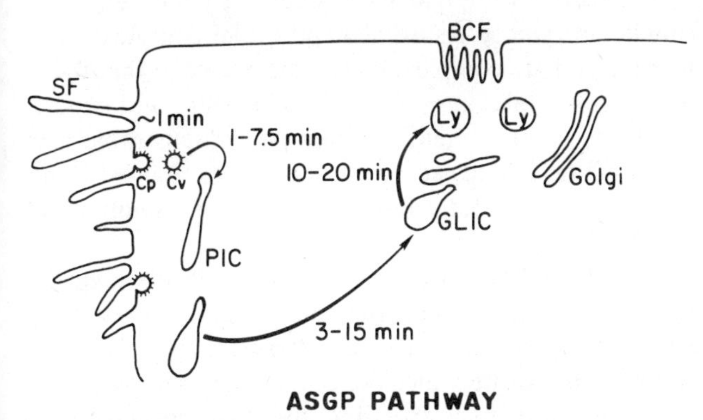

FIG. 1. (*Hubbard*) The proposed pathway taken by asialoglycoproteins (ASGP) through the hepatocyte. SF, sinusoidal front; BCF, bile canalicular front; Cp, coated pit; Cv, coated vesicle; PIC, peripheral intermediate compartment; GLIC, Golgi–lysosome intermediate compartment; Ly, lysosome.

Hubbard 1981). We have used iodinated ligands as well as ligands conjugated to electron microscopic tracers, such as lactosaminated ferritin (Lac–Fer) or asialo-orosomucoid (ASOR) coupled to horseradish peroxidase (HRP). Fig. 1 shows schematically our current picture of the pathway taken by asialoglycoproteins (ASGP), on the basis of these studies.

The first step in the pathway, that of ligand binding to receptor at the hepatocyte cell surface, has been studied in the isolated perfused liver where

1982 Membrane recycling. Pitman Books Ltd, London (Ciba Foundation symposium 92) p 109-119

temperature, medium composition, or both, can be manipulated easily but where the *in vivo* cell–cell associations are maintained. In the absence of internalization (at 4 °C or formaldehyde prefixation), we find that 60% of the ASGP ligand (Lac–Fer) is concentrated in coated pits along the blood sinusoidal surface while the remaining 40% is randomly distributed along this same surface.

Approximately one minute after injection *in vivo*, or after warming perfused livers that were previously exposed to ligand at 4 °C, we find ligand (ASOR–HRP or Lac–Fer) in coated vesicles. Ruthenium red staining indicates that at least half the profiles are coated vesicles, and not pits, in continuity with the cell surface. Shortly thereafter, the ligand is found in uncoated, large tubular structures (which we call the *peripheral intermediate compartment*) still near the hepatocyte periphery. This compartment (which is pre-lysosomal) begins to fill as early as one minute after injection *in vivo*, and continues to be filled up to 7.5 min. From 3–15 min we find ligand appearing in another pre-lysosomal compartment deeper in the cell, in a region near the bile canaliculus where lysosomes and Golgi complexes are found (Golgi–lysosome intermediate compartment). The evidence that these two compartments are not lysosomal comes from histochemistry and experiments at 16 °C. That is, these compartments contain no demonstrable aryl sulphatase and can accumulate ligand at 16 °C for up to 8 h with no degradation. (Lysosomes *in vitro* are fully functional at 16 °C; Dunn et al 1980).

From 10–20 min we find ligand in structures that are morphologically recognizable as lysosomes. Degradation commences at about 15 min, both *in vivo* and also in the isolated perfused liver, as measured by release of acid-soluble products into the perfusate which we periodically sample.

Doris Wall and I are currently characterizing the two pre-lysosomal compartments biochemically in order to determine whether they are similar or different, whether receptor is present in either of them and, if so, where and when ligand–receptor dissociation occurs (D. A. Wall & A. L. Hubbard, unpublished results 1982). We have approached these questions using the isolated perfused liver system and temperature-shift experiments. Iodinated ligand or ASOR–HRP is bound at 4 °C, the excess is washed out and then the liver is warmed to 32 °C for varying periods of time. After rapid cooling of the liver, all the cell-surface ligand can be dissociated by introduction of ethyleneglycolbis(aminoethylether)tetra-acetate (EGTA) into the perfusion medium (this ligand–receptor complex requires calcium for binding). We have subsequently processed the liver for electron microscopy or homogenized the tissue and done subcellular fractionation.

Using such a protocol we can show morphologically that 2.5 min after warming, virtually all the ASOR–HRP is present in the peripheral pre-lysosomal compartment. After 7–10 min at 32 °C ligand accumulates in *both* pre-lysosomal compartments. Thus, the contents of different intracellular vesicles in the ligand pathway can be 'marked' and the vesicles can be followed by subcellular fractionation.

Using [^{125}I]ASOR to mark the peripheral intermediate compartment, we have homogenized livers after 2.5 min at 32 °C, prepared several fractions by differential centrifugation, and found that 90% of the radioactivity sediments between 10 000 and 100 000 g (a microsomal fraction). Ninety-five per cent of the radioactivity is sequestered, as measured by accessibility to EGTA. That is, when microsomes are incubated with 10mM-EGTA, centrifuged (at 100 000 $g \times 60$ min), and the pellet and supernatant are analysed, 95% of the [^{125}I]ASOR is sedimentable, indicating its presence in an inaccessible compartment. We have evidence from preliminary experiments that ligand and receptor are in the same vesicle at 2.5 min (the peripheral intermediate compartment). We don't know whether they are bound to each other, owing to the conditions used during fractionation (no Ca^{2+}).

Unlike some of the other receptor systems that we have heard about at the symposium, in the asialoglycoprotein system approximately 10–20% of the receptor is on the cell surface and 80–90% is elsewhere, inside the cell, according to most reports (e.g. Steer & Ashwell 1980, Stockert et al 1980a,b, Pricer & Ashwell 1976). Therefore, in addition to studying the pre-lysosomal vesicles using the approach outlined above, we have also been measuring the distribution of the ASOR receptor in these same subcellular fractionation experiments. First, the majority of the cellular receptor is found in a microsomal fraction (75–80%). The bulk of the receptor in this fraction is 'latent'; that is, it can be measured only by adding detergent. We find that the membrane vesicles containing the 'latent' receptors are very similar biochemically to those comprising the peripheral intermediate compartment. For example, when we incubate microsomes, from livers labelled with [^{125}I]ASOR at 2.5 min, with low concentrations of digitonin (0.01%) (a glycoside that forms a complex with cholesterol; Amar-Costesec et al 1974) the density of the peripheral intermediate compartment on a sucrose gradient increases, demonstrating that it contains cholesterol. By the same criterion, we find that the internal pool of receptor also resides in a cholesterol-rich compartment. In fact, the distributions of sequestered ligand and latent receptor in both control *and* digitonin-treated microsomes coincide on sucrose gradients. A 'shoulder' of binding activity in the control curve coincides with the distribution of 5′-nucleotidase, a plasma-membrane marker enzyme. These biochemical experiments suggest that the peripheral intermediate compartment and the internal receptor pool may be one and the same entity. Clearly, ultrastructural localization studies are now required.

Other work that we have just begun with Y. C. Lee and R. Townsend is based on a recent report from Lee's lab describing a phenomenon called 'short-circuiting' (Connolly et al 1982) or *diacytosis* (Tolleshaug et al 1981). Certain ASGP ligands (low affinity and/or 'mutant' ligands) appear to be internalized by cells into an EGTA-inaccessible compartment, but are subsequently exocytosed intact into the medium. Two ligands have been used in the perfused liver: [^{125}I]ASOR, a molecule with five oligosaccharide chains, and a glycopeptide derived from α-antitrypsin with one 'triantennary' oligosaccharide chain. ASOR has only a 10-fold greater affinity for the receptor than does the triantennary glycopeptide (1 nM cf. 10nM, R. R. Townsend & Y. C. Lee, unpublished observations), despite the multivalent nature of ASOR relative to the glycopeptide. However, their fates can be different. For instance, when livers are exposed to [^{125}I]ASOR at 37 °C for 20 min and the excess and surface-bound is washed out, and then the system is perfused at 37 °C in the absence of ligand, but in the presence of a molecule that will dissociate accessible ligand from receptor (e.g. EGTA or *N*-acetylgalactosamine), more than 95% of the internalized [^{125}I]ASOR is degraded whether or not a dissociating agent is in the perfusate.

On the other hand, when livers are perfused with iodinated triantennary peptide under the same conditions as those used with [^{125}I]ASOR, and the excess is removed and the system perfused at 37 °C with *N*-acetylgalactosamine to dissociate any returning ligand, we find a different result. Of the radioactivity that appears in the medium over a one-hour period, 60% is undegraded and 40% is degraded. Thus, the triantennary ligand is 'short-circuited', but ASOR is not. Why? How? What happens to each ligand at shorter loading times? We are currently synthesizing HRP conjugates of the triantennary glycopeptide in order to localize the 'short-circuit' pathway. Perhaps it is the pathway of receptor recycling.

Meldolesi: Did you calculate the size and concentration of ASOR receptors in the first pre-lysosomal compartment? I understand that most receptors are located intracellularly. Thus, either they are very concentrated in the membranes of the pre-lysosomal compartment, or those membranes constitute a very large pool.

Hubbard: So far we have found that, biochemically, the internal pool of receptor and the peripheral intermediate compartment behave similarly. Dr Widnell, I believe, has some data from fibroblasts on the 5′-nucleotidase-containing compartments, which are distinct geographically from one another despite their biochemical similarities. We don't know where the internal pool is, inside the cell. If the density of receptors in the internal pool is similar to that at the cell surface (number per unit length of membrane), then you are right—there may be a huge membrane pool. However, we find the ASGP receptors at the cell surface both concentrated in coated pits and diffusely distributed, so we cannot predict the packing density in the internal pool.

Nonetheless, thee is a large amount of smooth surface membrane underneath the sinusoidal front membrane.

Meldolesi: So you are suggesting that the concentration of ASOR receptors might be very high in the membranes of pre-lysosomal compartment(s)?

Hubbard: The *amount* is high but I don't know what the *concentration* is.

Goldstein: Ann Hubbard's results on rat liver showed that 50% of the cytoplasmic vesicles that were identified as being coated and as containing a ferritin-labelled asialoglycoprotein ligand were stained with ruthenium red (Wall et al 1980). This is probably the strongest functional evidence to indicate that coated vesicles exist. The similarities between the asialoglycoprotein receptor system and the low density lipoprotein (LDL) receptor system are really remarkable: (1) both receptors require calcium for binding; (2) the time-course for internalization of the two receptor-bound ligands is virtually the same; and (3) the electron microscopic appearance of the two endocytic pathways is virtually identical. The one difference is that there is a large internal pool of receptors for asialoglycoproteins, but not for LDL. Several lines of evidence indicate that there is recycling of the asialoglycoprotein receptor, just like the LDL receptor. Dr Hubbard, which pool of asialoglycoprotein receptors do you think is recycling? Is it the external receptor that is moving into the cell, or the receptor in the first internal pool, or in the second internal pool, or all three? You have the ideal system for trying to dissect out recycling.

Hubbard: One group of workers suggests that the internal pool of receptor is not in exchange with that at the cell surface (see Stockert et al 1980a,b). This is evidence from work on neuraminidase treatment to the whole surfaces of isolated cells. The technique is difficult to interpret because one does not know how high is the local concentration of asialoglycoproteins, or how that might affect binding of either the active or the altered receptor. We have no direct evidence that the internal pool is ever expressed at the cell surface, except that in no conditions can one ever reduce the number of receptors at the cell surface without using trypsin or collagenase in order to destroy the receptors. In those conditions we see the appearance of receptors in the absence of protein synthesis, so the internal pool in some circumstances may be expressed at the cell surface (Zeitlin & Hubbard 1982). We don't know whether it is always in equilibrium.

Goldstein: Is there any evidence that asialoglycoprotein receptors in the internal and external pools are the products of the same gene?

Hubbard: No.

Cohn: Have you induced this peripheral compartment with the large bolus of asialoglycoprotein or do you find it normally in the absence of ligand?

Hubbard: Unfortunately we don't have a way of labelling this compartment in the absence of ligand, but the profiles are similar and one sees a lot of smooth surface membrane with this configuration in the periphery of the hepatocyte, in the absence of any internalization of specific ligands.

Palade: Is there any difference, in addition to their positions inside the cell, between those two pre-lysosomal compartments?

Hubbard: Yes. We have run out the microsomal fraction on metrizamide gradients and found that in a homogenate made after a warm-up of 2.5 min, the median density of the vesicles containing [^{125}I]ASOR and the receptor is about 1.10 g ml^{-1}. If we allow the warm-up to proceed for up to 10 min we still see no degradation, but we find vesicles of lighter density—1.05–1.09 g ml^{-1} in metrizamide. The receptor may not be in that lighter vesicle.

Rothman: Is the binding of asialoglycoprotein needed to cause the receptor to enter the pits, or is the unoccupied receptor normally found there?

Hubbard: We have no way of knowing. We have not done the kinds of experiments that Dr Goldstein mentioned. All we have tried to do is to modulate the conditions by incubating for different times or by washing out at different times, and nothing changes.

Rothman: In your electron microscopy studies do you have evidence for any intermediate vesicles which might be coated vesicles that have lost their coats and not yet fused with anything?

Hubbard: We have not looked very systematically but we see a few endosomes that still have some coat. We don't see anything that is the size of a coated vesicle and that has lost a partial coat, but we do see small uncoated vesicles in the early stages, and these can be the same size as coated vesicles.

Dean: Has sucrose or sucrose-conjugated ASOR been used in experiments similar to your own? Is this diacytosis a direct consequence of the characteristics of the ligand–receptor interaction or could it be more a consequence of the vesicular localization from which the release is occurring? In other words, could you see whether, from the same vesicles, degradation products from sucrose–ASOR or fluid-phase tracer (viz. free sucrose) were also released like that? You could use dual-labelling.

Hubbard: We haven't done those experiments, but it would have to be tested biochemically and kinetically. If they were released with the same kinetics it would imply that they might be in the same vesicle.

Cohn: Has anyone looked at the transport of immunoglobulin A (IgA), to see whether it goes through the same organelle system and is discharged at the bile front?

Schneider: Yes. We are investigating this process by means of morphological and biochemical methods. Rat polymeric IgA–HRP conjugates bind to the sinusoidal membrane of hepatocytes *in vitro* (cultured hepatocytes) or *in vivo* and are mainly associated with coated pits or profiles (Courtoy et al 1981, 1982). Thereafter, the conjugate is found in larger electron-lucent vesicles and finally in smaller vesicles close to the biliary membrane or in the bile canaliculi.

Using cell-fractionation techniques (Limet et al 1982a,b) we found that ^{3}H-labelled polymeric IgA, from 5 to 45 min after intravenous injection, is mainly associated with the particulate (MLP) fraction. After isopycnic centrifugation of this fraction in sucrose gradients, the label equilibrates around densities of 1.13 g ml^{-1}. This suggests the association of IgA with structures distinct from plasma membrane, Golgi and lysosomes. In addition these structures are rich in cholesterol, since they shift to higher equilibration densities in the presence of digitonin.

Hubbard: Do you then find secretory IgA in the bile quantitatively?

Schneider: About two-thirds of the injected dose of ^{3}H-labelled polymeric IgA arrives in the bile as secretory IgA which results from the covalent binding of one molecule of IgA with one molecule of secretory component, its specific receptor. Both *in vivo* and *in vitro* (Limet et al 1980) about one third of polymeric IgA is released from the cells in the serum or the culture medium as degradation products that are soluble in trichloroacetic acid. The degradation process takes place in lysosomes as indicated by cell fractionation experiments as well as by the effect of lysosomotropic drugs (J. N. Limet & Y.-J. Schneider, unpublished results).

Sly: Is there any difference between isolated cells and whole liver in the distribution of the receptors on the cell surface or inside the cell?

Hubbard: When we perfuse the liver with collagenase and prepare isolated hepatocytes, the receptor redistributes around the total cell surface rather than being predominantly around the sinusoidal front as it was *in situ* or in isolated plasma membranes (Zeitlin & Hubbard 1982). Secondly if we perfuse the liver with collagenase but do not mechanically dissociate it subsequently, we see a collagenase-dependent reduction in the number of ASOR-binding sites or receptors. If we incubate those cells for 30–60 min at 25 °C, we see an increased number of receptors appearing on the cell surface, although the total number does not change, suggesting that the internal pool is being mobilized (P. L. Zeitlin and A. L. Hubbard, unpublished observations). We also see a reduced total number of receptors *per cell* when isolated cells are compared with cells in the perfused organ. It seems that receptors are being destroyed and recruited from the inside, which thus reduces the internal pool.

Recycling of the transferrin receptor in HeLa cells

Bretscher: Dr Jeff Bleil and I (Bleil & Bretscher 1982) have taken advantage of the work of Wada et al (1979), Seligman et al (1979), Sutherland et al (1981) and Trowbridge & Omary (1981) who have characterized the transferrin receptor protein. The molecule has a relative molecular mass (M_r) of 180 000, and is a 90 000 S═S bridged dimer. We have worked with HeLa cells grown in suspension. When a membrane preparation from these cells is run on a two-dimensional sodium dodecyl sulphate (SDS) gel, in which the first dimension is non-reducing and the second dimension is reducing, most of the membrane proteins run along the diagonal but the transferrin receptor runs at 180 000 in the first dimension and at 90 000 in the second dimension. It can thus be seen easily with a protein stain, without any other purification. We have labelled cells with lactoperoxidase and iodide at 0 °C, and then run their membranes on a two-dimensional gel. This reveals a lot of radioactivity along the diagonal, and in the transferrin receptor spot.

If cells labelled at 0 °C are treated with trypsin at 0 °C, again most of the membrane-bound labelled polypeptide chains lie on the diagonal in the gel; the transferrin receptor spot is still visible, when stained for protein, but it is no longer labelled. This shows that this receptor on the cell surface can be labelled, but if the cells *are* so-labelled and then treated with trypsin at 0 °C the labelled receptors on the surface are degraded. The remaining unlabelled ones therefore constitute a pool of these molecules inside the cell.

Cells, labelled at 0 °C, were warmed up to 37 °C for various lengths of time and then cooled to 0 °C and treated with trypsin to see whether surface receptors enter the cell and thus become resistant to trypsin. This indeed happens: the transferrin receptor pool inside the cell gradually becomes labelled, with about 75% of the receptors residing inside the cell after a long incubation (90 min). The time taken for half the labelled receptors to enter the cell is 5 min.

These results show that the receptors enter the cells very quickly, and that when equilibrium has been reached between the labelled molecules on the inside and those on the outside, approximately three times as many molecules are inside the cell as on the surface. Presumably, as the cells are warmed up, labelled receptors enter the cell and are replaced by unlabelled ones that were inside the cell at the start of the experiment. From that, one can calculate that the time taken for internalization of a surface-equivalent of receptors (about 2×10^5 molecules) is about 7 min. Since roughly three times as many molecules are inside the cell as are on the surface, the transit time through the cell turns out to be about 21 min. This is similar to the cycling time that Karin & Mintz (1981) estimated for transferrin on teratocarcinoma cells. It seems likely that the transferrin receptor and transferrin enter the cell together, stay together through the cell, and come out together after about 20 min.

Cuatrecasas: How do you know that you are not just labelling some superficial component that has a very small molecular weight, so that when you treat with trypsin you can't see the component on your diagonal map in the gel? Instead of the M_r being 90 000 it could be 88 000 or 87 000. Furthermore, when you do the warm-up, this could redistribute the protein so that it becomes inaccessible to trypsin. You are thus *assuming* it is going inside.

Bretscher: One can label the molecules on the outside surface, and show that only these are labelled because they are all sensitive to trypsin. If you then warm the cells up, the label in the transferrin receptor spot, instead of being trypsin-sensitive, becomes trypsin-resistant. The simple answer is that the receptor has gone inside the cells.

I might add that we have also tried to look at the distribution of transferrin receptor on these cells by using rat anti-receptor antibodies, followed by rabbit anti-rat IgG–ferritin conjugates. Coated pits contain large amounts of ferritin although a lot of the ferritin is not in coated pits. We have not attempted to quantitate this further, as these cells are highly villiated, making it very difficult to do the job properly.

Palade: What could be the rationale for the transferrin receptor–transferrin complex coming out again to the cell surface?

Bretscher: It is believed to be re-utilized many times. The transferrin supposedly goes into the cells, deposits or releases its iron in some lysosomal or low pH compartment, and then the transferrin comes out to exchange, presumably, with other serum transferrin.

Goldstein: Cells have a transferrin receptor that is different from the asialoglycoprotein receptor, but the asialoglycoprotein receptor will apparently recognize asialotransferrin. In the recent report of *diacytosis* (i.e. exocytosis of endocytosed ligand) of asialotransferrin by rat hepatocytes (Tolleshaug et al 1981), which Dr Hubbard mentioned earlier, asialotransferrin was the asialo ligand that showed the most prominent degree of diacytosis. This phenomenon may be related to the fact that internalized transferrin is excreted from cells once it has delivered its iron to lysosomes. There may be something about the structure of the transferrin molecule that allows it to escape lysosomal degradation and to come back out of the cell whether it enters cells through the transferrin receptor or through the asialoglycoprotein receptor.

Bretscher: The main point about the similar time course for the transferrin movement and the transferrin receptor movement is that it is very different from the low density lipoprotein (LDL) system. The transferrin receptor has a long transit time, unlike the LDL receptor.

Cuatrecasas: Also, the transferrin receptor goes in extremely rapidly in the absence of ligand.

Bretscher: I wouldn't say that's in the absence of ligand because the cells were incubated in complete medium, containing serum, and the transferrin is probably there.

Palade: How much of the iodine label is in fatty acids esterified to the polypeptide chain of the transferrin receptor?

Bretscher: We haven't measured that but Omary & Trowbridge (1981) showed that essentially all the iodine label on the protein was released as a 70 000 M_r tryptic fragment when the cells were treated with trypsin. In any case, I think the fatty acid chains may be saturated.

Palade: Well it is [^{3}H]palmitate that is used for labelling, but a certain fraction of the label is recovered as oleate (Schmidt et al 1979).

Goldstein: Is there any receptor other than the transferrin receptor that has palmitate covalently attached to it?

Kornfeld: I don't know about other receptors but one can grow chick embryo fibroblasts in the presence of [^{3}H]palmitic acid and find that many membrane proteins are acylated (Schlesinger et al 1980).

Palade: The G protein of vesicular stomatitis virus is acylated, as are the murine equivalents of human glycophorins (Dolci & Palade 1982). It may turn out that many integral transmembrane proteins have fatty acyl residues.

REFERENCES

Amar-Costesec A, Wibo M, Thines-Sempoux D, Beaufay H, Berthet J 1974 Analytical study of microsomes and isolated subcellular membranes from rat liver. J Cell Biol 62:717-745

Ashwell G, Morell AG 1974 The role of surface carbohydrates in the hepatic recognition and transport of circulating glycoproteins. Adv Enzymol Relat Areas Mol Biol 41:99-129

Bleil JD, Bretscher MS 1982 Transferrin receptor and its recycling in Hela cells. EMBO J 1:351-355

Connolly DT, Townsend RR, Kawaguchi K, Bell WR, Lee YC 1982 Binding and endocytosis of cluster glycosides by rabbit hepatocytes. J Biol Chem 257:939-945

Courtoy PJ, Limet JN, Baudhuin P, Schneider Y-J, Vaerman JP 1981 Receptor-mediated endocytosis of polymeric IgA by cultured rat hepatocytes. Cell Biol Int Rep 5:57

Courtoy PJ, Limet JN, Baudhuin P, Schneider Y-J, Vaerman JP 1982 Ultrastructural aspects of transepithelial transfer of IgA through rat liver. Arch Int Physiol Biochim 90:B11-B12

Dolci ED, Palade GE 1982 Biosynthetic labelling of murine erythroid cell sialoglycoproteins with [^{3}H]palmitic acid. Fed Proc 41:1140

Dunn WA, Hubbard AL, Aronson NN Jr 1980 Low temperature selectively inhibits fusion between pinocytic vesicles and lysosomes during heterophagy of ^{125}I-asialofetuin by the perfused liver. J Biol Chem 255:5971-5978

Hubbard AL, Stukenbrok H 1979 An electron microscopic autoradiographic study of the carbohydrate recognition systems in rat liver. J Cell Biol 83:65-81

Karin M, Mintz B 1981 Receptor-mediated endocytosis of transferrin by developmentally totipotent mouse teratocarcinoma stem cells. J Biol Chem 256:3245-3252

Limet JN, Schneider Y-J, Vaerman JP, Trouet A 1980 Interaction of rat IgA with cultured rat hepatocytes: binding sites, drug effects. Toxicology 18:187-194

Limet JN, Quintart J, Otte-Slachmuylder C, Schneider Y-J 1982a Receptor-mediated endocytosis of hemoglobin–haptoglobin, galactosylated serum albumin and polymeric IgA by the liver Acta Biol Med Ger 41:113-124

Limet JN, Otte-Slachmuylder C, Schneider Y-J 1982b Uptake by hepatocytes, endoctyosis and biliary transfer of polymeric IgA, anti-secretory component IgG, hemoglobin–haptoglobin and horseradish peroxidase. Arch Int Physiol Biochim 90:958-59

Omary MB, Trowbridge IS 1981 Covalent binding of fatty acid to the transferrin receptor in cultured human cells. J Biol Chem 256:4715-4718

Pricer WE Jr, Ashwell G 1976 Subcellular distribution of a mammalian hepatic binding protein specific for asialoglycoproteins. J Biol Chem 251:7539-7544

Schlesinger MJ, Magee AI, Schmidt MFG 1980 Fatty acid acylation of proteins in cultured cells. J Biol Chem 255:10021-10024

Schmidt MFG, Bracha M, Schlesinger MJ 1979 Evidence for covalent attachment of fatty acids to Sindbis virus glycoproteins. Proc Natl Acad Sci USA 76:1687-1691

Seligman PA, Schleicher RB, Allen RH 1979 Isolation and characterization of the transferrin receptor from human placenta. J Biol Chem 254:9943-9946.

Steer CJ, Ashwell G 1980 Studies on a mammalian hepatic binding protein specific for asialoglycoproteins. J Biol Chem 255:3008-3013

Stockert RJ, Gartner U, Morell AG, Wolkoff AW 1980a Effects of receptor-specific antibody on the uptake of desialylated glycoproteins in the isolated perfused rat liver. J Biol Chem 255:3830-3831

Stockert RJ, Howard DJ, Morell AG, Scheinberg IH 1980b Functional segregation of hepatic receptors for asialoglycoproteins during endocytosis. J Biol Chem 255:9028-9029

Sutherland R, Delia D, Schneider C, Newman R, Kemshead J, Greaves M 1981 Ubiquitous cell-surface glycoprotein on tumour cells is proliferation associated receptor for transferrin. Proc Natl Acad Sci USA 78:4515-4519

Tolleshaug H, Chindemi PA, Regoeczi E 1981 Diacytosis of human asialotransferrin type 3 by isolated rat hepatocytes. J Biol Chem 256:6526-6528

Trowbridge IS, Omary MB 1981 Human cell surface glycoprotein related to cell proliferation is the receptor for transferrin. Proc Natl Acad Sci USA 78:3039-3043

Wada HG, Hass PE, Sussman HH 1979 Transferrin receptor in human placental brush border membranes. J Biol Chem 254:12629-12635

Wall DA, Wilson G, Hubbard AL 1980 The galactose-specifc recognition system of mammalian liver: the route of ligand internalization in rat hepatocytes. Cell 21:79-93

Wall DA, Hubbard AL 1981 The galactose-specific recognition system of mammalian liver: receptor distribution on the hepatocyte cell surface. J Cell Biol 90:687-696

Zeitlin PL, Hubbard AL 1982 Cell surface distribution and intracellular fate of asialoglycoproteins: a morphological and biochemical study of isolated rat hepatocytes and monolayer cultures. J Cell Biol 92:634-647

The Golgi apparatus: roles for distinct 'cis' and 'trans' compartments

JAMES E. ROTHMAN

Department of Biochemistry, Stanford University Medical Center, Stanford, California 94305, USA

Abstract The Golgi apparatus seems to consist of distinct *cis* and *trans* compartments that are proposed to act sequentially to refine the protein export of the endoplasmic reticulum by removing escaped endoplasmic reticulum proteins. Refinement may be a multi-stage process that employs a principle akin to fractional distillation; the stack of cisternae comprising the cis Golgi may be the plates in this distillation tower. The trans Golgi, consisting of the last one or two cisternae, may be the receiver that collects from the cis Golgi only its most refined fraction for later distribution to specific locations throughout the cell.

Eukaryotic cells face a widely recognized sorting problem that stems directly from their highly compartmentalized organization. A mixture of proteins that will ultimately reside in such diverse compartments as the surface membrane, secretion granules and lysosomes is segregated, first, away from the cytosol in one compartment, the endoplasmic reticulum (ER). The cell then assumes the formidable task of sorting—physically fractionating—these precursors into distinct sets for delivery to their final and separate destinations, while leaving most of the ER's own proteins behind. Interposed between the ER and multiple final destinations is a mystery: the Golgi apparatus. Why do all eukaryotic cells have a Golgi? Why must the mixture of proteins exported from the ER pass through the Golgi before sorting can be completed? In other words, why can't all the sorting take place equally well in the ER? The importance of these issues is intensified by the current evidence suggesting that the various types of exported proteins may not even be separated from each other during their passage through the Golgi. Thus, glycoproteins destined for plasma membranes, lysosomes and secretion must pass through a distal site in the Golgi for their terminal glycosylation (Banerjee et al 1976,

---

1982 Membrane recycling. Pitman Books Ltd, London (Ciba Foundation symposium 92) p 120-137

Bretz et al 1980, Hino et al 1978a, b, Tartakoff & Vassali 1979) and may have access to every Golgi cisterna (Geuze et al 1979, Papermaster et al 1978, Bergmann et al 1981).

If exported proteins are not sorted out from each other in passing through most of the Golgi, what kind of sorting, if any, does take place? In this paper, I call attention to two new findings, the first from our own work and the second from other laboratories, which suggest a novel but speculative view of the Golgi that simplifies matters and may explain a great deal about the structure and function of this organelle.

The Golgi apparatus: two organelles in tandem

Cellular processes can often be elucidated by the use of simplified experimental systems provided by cells infected with viruses. The limited genetic capacity of many viruses forces them to use pathways provided by their host for the production and maturation of their proteins. In particular, the vesicular stomatitis viral (VSV)-encoded membrane glycoprotein (G protein) follows a pathway to the plasma membrane of infected animal cells that appears to be indistinguishable from that taken by the cell's own surface membrane proteins before infection. G Protein is inserted into the ER membrane, where the precursor oligosaccharide (Li et al 1978) is added. Then, G protein is transported through the Golgi (Bergmann et al 1981) stack (where its oligosaccharides are processed) and on to the plasma membrane before entering virions as they bud from the cell surface. Processing takes place in two major stages. First, the precursor is 'trimmed', as four mannose units (of a total of nine) are removed by a specific α-1, 2-mannosidase (Tabas & Kornfeld 1979) to yield an intermediate containing five mannose units (Li et al 1978). Second, terminal sugars such as galactose are added and two more mannose units are removed in a complex series of steps (Lennarz 1980, Gibson et al 1980, Li et al 1978, Tabas & Kornfeld 1978). Enzymes capable of trimming (α-1, 2-mannosidase) and capable of terminal glycosylation (including *N*-acetylglucosaminyl transferases, galactosyltransferases, sialyltransferases) are highly concentrated in Golgi fractions of liver (Tartakoff 1980, Farquhar 1978, Novikoff 1976), and thus serve as enzymic markers for portions of the Golgi. Experiments (Dunphy et al 1981, Fries & Rothman 1981), based on the apparent reconstitution in a cell-free system (Fries & Rothman 1980, Rothman & Fries 1981) of a part of the intracellular transport pathway of G protein that involves the Golgi, strongly suggest that the Golgi has distinct *cis* and *trans* subcellular compartments. Crude cytoplasmic fractions were first prepared from VSV-infected cells that had been pulse-labelled (usually with [^{35}S]methionine) before homogenization (Fries &

Rothman 1980, Rothman & Fries 1981). A mutant line (Gottlieb et al 1975, Tabas & Kornfeld 1978, Harpaz & Schachter 1980) of Chinese hamster ovary cells (clone 15B), whose Golgi bodies lack an enzyme needed for terminal glycosylation (UDP-*N*-acetylglucosamine:glycoprotein glycosyltransferase I) but which transport G protein normally, was used as the source of this labelled cytoplasmic fraction, the *in vitro* 'donor' of G protein. This fraction was then incubated with a comparable cytoplasmic fraction, but from uninfected wild-type cells. *N*-acetylglucosamine should have been incorporated into the labelled G protein if the G protein had been transferred *in vitro* from the donor membranes of the mutant cell (lacking the enzyme) to the Golgi from the normal cell containing the required terminal glycosyltransferase (Fries & Rothman 1980, Rothman & Fries 1981). In optimal conditions, approximately 40% of the labelled G protein in the donor fraction acquired *N*- acetylglucosamine from the UDP-sugar, as judged by the conversion of G protein to Endo H-resistance. This process also required ATP, wild-type membranes and factors from the high-speed supernatant (Fries & Rothman 1981, Fries & Rothman 1980, Rothman & Fries 1981). Direct confirmation that terminal glycosylation followed the physical transfer of G protein to the added Golgi membranes was obtained by use of a highly purified Golgi fraction from rat liver to substitute for the crude cytoplasmic fraction of wild-type cells (Fries & Rothman 1980, Rothman & Fries 1981).

Which membranes from the infected clone 15B cell serve as donors of G protein? To ascertain this, we prepared cytoplasmic fractions from VSV-infected 15B cells that had been repeatedly 'chased' for increasing periods of time after a brief pulse-label with [^{35}S]methionine. The cells were then tested as donors during prolonged incubations *in vitro* to determine the maximum extent of transfer (Fries & Rothman 1981). G protein present in the cytoplasmic fraction that was prepared immediately after the pulse (0 min chase) was not processed by wild-type Golgi, showing that the ER membranes do not serve as donors in these conditions. The capacity to serve as a donor increased dramatically with chase, and was maximal (approximately 40% of G protein transferred) at 10 min, precisely when the oligosaccharides are being trimmed (Fries & Rothman 1981) and when G protein has been independently shown to have reached Golgi stacks (Bergmann et al 1981). Fractionation on a sucrose density gradient suggested that most of the labelled G protein before incubation, and all the terminally glycosylated G protein after incubation, was present in Golgi membranes (Fries & Rothman 1981). These observations together suggested that freshly trimmed G protein was being transported *in vitro* from the Golgi membranes of the donor 15B cell to those of the wild-type.

Remarkably, the capacity of Golgi membranes in cytoplasmic fraction to donate G protein *in vitro* declined precipitously (half-time ≈ 5 min) with additional chase *in vivo* beyond the time of trimming (Fries & Rothman 1981). This distinguished two pools of G protein: a transferable pool from which G protein can reach the portion of another Golgi that houses terminal glycosyltransferases; and another, non-transferable pool. Both pools appeared to be present in Golgi membranes, partly because little of the G protein would have reached the cell surface (Rothman & Fine 1980) when 80% of G protein had already entered the non-transferable pool (Fries & Rothman 1981), and partly because of the additional evidence described below.

A simple interpretation (Fries & Rothman 1981) of these *in vitro* experiments would be that the two pools represent two distinct compartments of the Golgi through which G protein passes successively and irreversibly. In the light of the cis-to-trans direction of protein transport in the Golgi (Tartakoff 1980), these two postulated compartments have been termed 'cis Golgi' (housing the transferable pool) and 'trans Golgi' (containing the non-transferable pool). In a cell, G protein would generally be transported from the cis to the trans compartment of the same Golgi apparatus. The inter-Golgi transport observed *in vitro* (Fries & Rothman 1981) could then represent transfer from the cis compartment of one Golgi to the trans compartment of another. Because the appearance of G protein in the transferable pool coincides with the trimming of its oligosaccharides, and because transfer *in vitro* to a wild-type Golgi results in terminal glycosylation, it would be expected that the enzymes responsible for trimming would be concentrated in the cis Golgi compartment, and that the terminal glycosyltransferases would be concentrated in the trans Golgi compartment; these enzymes would then serve as markers for the two compartments.

Two portions of Golgi-like membranes with the enzymic properties expected of the postulated cis and trans compartments can be physically resolved in sucrose density gradients (Dunphy et al 1981). These fractionation experiments (Dunphy et al 1981) have taken advantage of the recent discovery (Schmidt et al 1979, Schmidt & Schlesinger 1980) that fatty acid groups are added covalently to G protein during its intracellular transport. Acylation takes place well after the initial glycosylation in the ER but several minutes before terminal glycosylation (Schmidt et al 1979, Schmidt & Schlesinger 1980). This suggested that fatty acid added to G protein might be found mainly within the putative cis Golgi compartment. A short pulse of a ^{3}H-labelled palmitic acid was incorporated into an intermediate form of G protein whose oligosaccharides had already been trimmed (Dunphy et al 1981), as judged by electrophoretic mobility. When the membranes of these cells were fractionated by isopycnic centrifugation in a sucrose density gradient, this freshly acylated, trimmed G protein was found in a peak centred at a density of approximately 1.14 g ml^{-1}. Assay of an α-1,2-mannosidase capable of trimming G protein's high mannose oligosaccharides (Tabas & Kornfeld 1979) revealed that this activity distributed within the gradient in the same way as did the *in vivo* pulse of ^{3}H-labelled fatty acid (Dunphy et al 1981). By contrast, the galactosyltransferase and sialyltransferase activities (Dunphy et al 1981), traditional Golgi markers, were concentrated together in a peak centred at approximately 1.11 g ml^{-1}, distinct from the denser distribution of fatty acid addition and trimming. These distinct denser and lighter patterns of Golgi markers have been termed the cis and trans Golgi distributions, respectively, because of the compartments they appear to represent. Both cis and trans distributions differ from the distribution of ER membranes (Dunphy et al 1981), marked by the position in the gradient of freshly synthesized G protein carrying the precursor oligosaccharide. The sequence of oligosaccharide processing (from precursor to trimmed protein to terminally glycosylated protein) dictates that G protein is transported successively from ER to the cis Golgi distribution and then to the trans Golgi distribution.

As would be expected from the two-compartment model, the enzymically distinguished cis Golgi distribution contains a transferable pool of G protein because most of the freshly acylated G protein in the cis Golgi distribution can be processed by wild-type Golgi *in vitro* (Dunphy et al 1981). Furthermore, the trans Golgi distribution of terminal glycosyltransferases appears to represent the postulated trans compartment because trimmed G protein enters the non-transferable pool with a time course similar to that with which fatty acid-labelled G protein is terminally glycosylated in wild-type cells (Dunphy et al 1981, Schmidt & Schlesinger 1980). Therefore, the cis and trans distributions appear to reflect a compartmental division of the Golgi. The trans Golgi compartment would then act as a sink for an exported protein like G arriving from the cis Golgi compartment.

What portions of the Golgi apparatus constitute the biochemically identified cis and trans compartments? Because G protein enters the Golgi stack at its cis face (Bergmann et al 1981), it is natural to attempt to equate the (earlier) cis compartment with cisternae at the cis end of the stack, and to equate the (later) trans compartment with cisternae at the trans end. Two principal lines of evidence help to establish this link.

First, exported proteins move across the Golgi stack in the cis-to-trans direction; glycoproteins must move from the cis to the trans Golgi compartments.

Secondly, enzymic markers of the trans Golgi compartment are selectively localized in cisternae at the trans end of the stack. Thiamine pyrophosphatase activity can serve as a biochemical marker of the trans compartment because this enzyme fractionates identically to galctosyltransferase in Golgi subfractions of rat liver (Hino et al 1978a,b,c). The histochemical stain used to localize this phosphatase activity results in deposits in only the last one or two cisternae at the trans end of the stack (Friend 1969), suggesting that these comprise the trans compartment. *In vivo* pulse labels of galactose and sialic acid are incorporated almost exclusively into an apparently trans-rich fraction, with very little in the cis-rich fraction of rat liver Golgi (Banerjee et al 1976). Enzyme activities responsible for adding terminal *N*-acetylglucosamine, galactose and sialic acid to glycoproteins are all also somewhat rich in the trans (relative to the cis) Golgi fraction; these three glycosyltransferase activities are always found in the same ratio to each other (Bretz et al 1980).

If these implications are correct, then the trans Golgi consists, in many instances, of only the last one or two cisternae at the trans end, suggesting that the cis Golgi may consist of the entire remainder of the stack, beginning from the cis end. Because the cis and the trans Golgi appear to function as distinct compartments, and because they contain different concentrations of certain enzmes and can be physically resolved, it is plausible to consider them to be different organelles that happen to be physically connected. It remains possible that the Golgi is further divided into more than two distinct compartments; the limited resolution of our experiments cannot rule this out.

The distinct cis and trans compartments of Golgi can help explain a variety of morphological observations. Certain extracellular tracers would be selectively deposited almost exclusively in the cisterna or two closest to the trans end of the stack, owing to plasma membrane recycling (Farquhar 1978), because the trans Golgi compartment would be engaged in transport to the cell surface; the cis compartment would not. The concentrated secretions that can often be visualized (Beams & Kessel 1968) only in the last one or two cisternae at the trans end of the stack would result from the vectorial transport of secreted protein from the cis to the trans compartments. More generally, our results attach a new significance to the previously recognized overall polarity of the Golgi stack. This polarity would not simply be due to a gradient of cisternal properties within a single compartment but would result instead from a division of the stack into functionally distinct compartments.

Does most of the Golgi sort endoplasmic reticulum proteins?

What significance does the division of the Golgi stack into cis and trans compartments hold for its function? An important clue may come from several recent findings (Hino et al 1978a, b, c, Howell et al 1978, Ito & Palade 1978, Borgese & Meldolesi 1980, Meldolesi et al 1980) which suggest that enzymes characteristic of ER also found in large amounts (30–50% or more of their concentration in ER) in cis but not in trans Golgi fractions. It is generally accepted that the Golgi is responsible for the major sorting of proteins exported from ER for separate delivery to multiple destinations (Tartakoff 1980). However, the required concentration in the trans Golgi of the terminal glycosyltransferases that act on glycoproteins destined for plasma membranes, secretion and lysosomes (Paigen 1979, Lalley & Shows 1977, Womak et al 1981) suggests that this aspect of sorting does not take place until the trans Golgi at least is reached. Another form of sorting would be needed to explain the progressive changes in cisternal composition across most of the Golgi stack (Tartakoff 1980).

Therefore, it may be worth considering a different type of sorting in the Golgi, the sorting of ER proteins. The recent findings of ER membrane proteins in Golgi (Hino et al 1978a, b, Howell et al 1978, Ito & Palade 1978, Borgese & Meldolesi 1980, Meldolesi et al 1980) seem to raise the possibility of a high rate of export of ER proteins from ER, and suggest that cells need a Golgi to refine the export by removing the ER membrane proteins that have escaped.

However, the membrane proteins that are intended for export from the ER and for subsequent distribution throughout the cell represent only a trace of the total membrane protein present at any given time in the ER. In fact, a simple calculation based on the relative amounts of ER and plasma membrane, and on the transit and turnover times of their proteins, suggests that in many cells as little as 1 in 10 000 of the total membrane proteins present in ER may be destined for the plasma membrane. These precursors to the plasma membrane would then need to be purified by as much as 10 000-fold during the sorting process, with only traces of ER proteins remaining! It seems unlikely that this degree of selectivity can be achieved in a single export stage in the ER; if so, multiple stages of purification (to remove ER proteins) would be needed, presumably in the Golgi. The qualitative aspects of this analysis make it plausible to consider that essentially every protein that can diffuse freely in the ER membrane will be unavoidably exported to Golgi membranes in the significant amounts that have been observed (Hino et al 1978a, b, Howell et al 1978, Ito & Palade 1978, Paigen 1979, Lalley & Shows 1977, Womak et al 1981).

Most newly synthesized ER proteins are found in rough ER fractions several minutes before they arrive in smooth membrane fractions (Omura & Kuriyama 1971, Fujii-Kuriyama et al 1979, Bar-Nun et al 1980). Therefore, most ER proteins in the Golgi probably arrive there after export from ER rather than by an independent and simultaneous insertion into ER and Golgi membranes (Borgese & Meldolesi 1980, Meldolesi et al 1980). Given the apparent flow of protein through the Golgi stack in the cis-to-trans direction (Tartakoff 1980, Bergmann et al 1981, Bergeron et al 1978), the lower level of ER proteins in trans (as compared to cis) portions (Hino et al 1978a, b,

Howell et al 1978, Ito & Palade 1978, Borgese & Meldolesi 1980, Meldolesi et al 1980) suggests that ER membrane proteins are continuously removed from the export as it passes through the Golgi stack in a *filtration* process or 'refinement'.

The most likely mechanism of refinement would be for ER membrane proteins to be transported back to the ER from the Golgi. The major alternative—that ER membrane proteins are selectively degraded within the Golgi and not returned to ER—cannot be ruled out but is unattractive because of the great disparity between the time scale for turnover of ER membrane proteins (days) and that for the transit of exported proteins through the Golgi stacks (minutes).

The need to improve the fidelity of export from the ER, by refinement, could explain why all eukaryotes need a Golgi. Without one, the sorting of 'exported' proteins into types (plasma membrane, secretion, etc.) at the ER would be difficult, because the selective export to each of these final destinations would be heavily contaminated with ER membrane proteins. If the Golgi apparatus is imposed as a filter between the ER and the final destinations, the contaminating ER membrane proteins would be removed before further sorting takes place. The sorting of proteins in the Golgi may therefore proceed in two sequential stages in attached cis and trans units. The crude export of ER may first be refined by removal of ER membrane proteins during transit through the cis Golgi and then passed on to trans Golgi for further sorting, probably during exit from the Golgi stack.

A stack could improve the overall efficiency with which ER proteins are removed by making this refinement a multi-stage process. ER membrane proteins could be extracted from the export repeatedly at successive cisternae. This would be crudely analogous to the procedure of countercurrent distribution in which the same purification step is applied sequentially in a series of test tubes. Certainly, the operation of a multi-stage process (King 1980) would explain the progressive changes in cisternal properties so characteristic of the Golgi stack.

This view not only explains in a natural way why a stack of cisternae would be vital for the function of the Golgi, but also explains why a distinct and small trans compartment of the stack might be necessary. Only the purest, most fully refined fraction of the export should be utilized for further sorting and delivery to multiple destinations. Of course, the purest fraction in a stack (having had ER proteins extracted the maximum number of times) would be found in the cisterna at the trans end. The trans Golgi compartment, closely applied to this very cisterna and acting as a sink, would continuously receive and trap an optimally refined mixture of exported proteins. If there were no such distinct receiver on the trans side of the stack, to constantly siphon off the cleanest fraction and to provide a unique site for exit from the Golgi, then the cell would probably be forced to take proteins from the cisternae at random. This mixed output of multiple cisternae would of course be much more contaminated by ER proteins than the 'best' product at the trans end of the stack.

The purpose and principle of the Golgi may therefore be essentially that of fractional distillation (King 1980), and the apparatus used may be fundamentally the same. In this hypothesis, the cis Golgi stack is the distillation tower and its membrane-bound cisternae are the plates. The trans Golgi acts as the receiver, bleeding off the best product, often condensing it so that it can be more easily distributed in a concentrated form. The cisternae of the Golgi stack may in fact be no more complex than the plates in a distillation apparatus, simply a series of places where proteins can be delivered and removed.

Acknowledgements

I thank Debra Forseth for preparing the manuscript. Our research has been supported by NIH grants GM25662–03 and AM27044–02.

REFERENCES

Banerjee D, Manning C, Redman C 1976 The *in vivo* effect of colchicine on the addition of galactose and sialic acid to rat hepatic serum glycoproteins. J Biol Chem 251:3887

Bar-Nun S, Kreibich G, Adesnik M, Alterman L, Negishi M, Sabatini D 1980 Synthesis and insertion of cytochrome P-450 into endoplasmic reticulum membranes. Proc Natl Acad Sci USA 77:965

Beams H, Kessel R 1968 The Golgi apparatus: structure and function. Int Rev Cytol 23:209

Bergeron J, Borts D, Cruz J 1978 Passage of serum-destined proteins through the Golgi apparatus of rat liver. J Cell Biol 76:87

Bergmann JE, Tokuyasu KT, Singer SJ 1981 Passage of an integral membrane protein, the Vesicular Stomatitis virus glycoprotein, through the Golgi apparatus en route to the plasma membrane. Proc Natl Acad Sci USA 78:1746

Borgese N, Meldolesi J 1980 Localization and biosynthesis of NADH-cytochrome b_5 reductase, an integral membrane protein in rat liver cells. I: Distribution of the enzyme activity in microsomes, mitochondria, and Golgi complex. J Cell Biol 85:501

Bretz R, Bretz H, Palade G 1980 Distribution of terminal glycosyltransferases in hepatic Golgi fractions. J Cell Biol 84:87

Dunphy WG, Fries E, Urbani LJ, Rothman JE 1981 Early and late functions associated with the Golgi apparatus reside in distinct compartments. Proc Natl Acad Sci USA 78:7453-7457

Farquhar M 1978 Recovery of surface membrane in anterior pituitary cells. J Cell Biol 77:R35

Friend D 1969 Cytochemical staining of multivesicular body and Golgi vesicles. J Cell Biol 41:269

Fries E Rothman J 1980 Transport of Vesicular Stomatitis virus glycoprotein in a cell-free extract. Proc Natl Acad Sci USA 77:3870

Fries E, Rothman J 1981 Transient activity of a Golgi-like membrane as donor of Vesicular Stomatitis virus glycoprotein *in vitro*. J Cell Biol 90:697
Fujii-Kuriyama Y, Negishi M, Mikawa R, Tashiro Y 1979 Biosynthesis of cytochrome P-450 on membrane-bound ribosomes and its subsequent incorporation into rough and smooth microsomes in rat hepatocytes. J Cell Biol 81:510
Geuze J, Slot J, Tokuyasu K, Goedemans W, Griffith J 1979 Immunocytochemical localization of amylase and chymotrypsinogen in the exocrine pancreatic cell with special attention to the Golgi complex. J Cell Biol 82:697
Gibson R, Kornfeld S, Schlesinger S 1980 A role for oligosaccharides in glycoprotein biosynthesis. Trends Biochem Sci 5:290
Gottlieb C, Baenzinger J, Kornfeld S 1975 Deficient uridine diphosphate-N-acetylglucosamine: glycoprotein N-acetylglucosaminyltransferase activity in a clone of Chinese hamster ovary cells with altered surface glycoproteins. J Biol Chem 250:3303
Harpaz N, Schachter H 1980 Control of glycoprotein synthesis. J Biol Chem 255:4894
Hino Y, Asano A, Sato R, Shimizu S 1978a Biochemical studies of rat liver Golgi apparatus. J Biochem (Tokyo) 83:909
Hino Y, Asano A, Sato R 1978b Biochemical studies on rat liver Golgi apparatus II. J Biochem (Tokyo) 83:925
Hino Y, Asano A, Sato R 1978c Biochemical studies on rat liver Golgi apparatus III. J Biochem (Tokyo) 83:935
Howell K, Ito A, Palade G 1978 Endoplasmic reticulum marker enzymes in Golgi fractions—what does this mean? J Cell Biol 79:581
Ito A, Palade G 1978 Presence of NADPH-cytochrome P-450 reductase in rat liver Golgi membranes. J Cell Biol 79:590
King CJ 1980 Separation processes, 2nd edn. McGraw Hill, New York
Lalley P, Shows T 1977 Lysosomal acid phosphatase deficiency: liver specific variant in the mouse. Genetics 87:305
Lennarz WJ (ed) 1980 The biochemistry of glycoproteins and proteoglycans. Plenum Press, New York
Li E, Tabas I, Kornfeld S 1978 The synthesis of complex-type oligosaccharides. J Biol Chem 253:7762
Meldolesi J, Corte G, Pietrini G, Borgese N 1980 Localization and biosynthesis of NADH-cytochrome b_5 reductase, an integral membrane protein, in rat liver cells. II: Evidence that a single enzyme accounts for the activity in its various subcellular locations. J Cell Biol 85:516
Novikoff AB 1976 The endoplasmic reticulum: a cytochemist's view (a review). Proc Natl Acad Sci USA 73:2781
Omura T, Kuriyama Y 1971 Role of rough and smooth microsomes in the biosynthesis of microsomal membranes. J Biochem (Tokyo) 69:651
Paigen K 1979 Acid hydrolases as models of genetic control. Annu Rev Genet 13:417
Papermaster D, Schneider B, Zorn M, Kraehenbuhl J 1978 Immunocytochemical localization of opsin in outer segments and Golgi zones of frog photoreceptor cells. J Cell Biol 77:196
Rothman JE, Fine RE 1980 Coated vesicles transport newly synthesized membrane glycoproteins from endoplasmic reticulum to plasma membrane in two successive stages. Proc Natl Acad Sci USA 77:780
Rothman J, Fries E 1981 Transport of newly synthesized Vesicular Stomatitis viral glycoprotein to purified Golgi membranes. J Cell Biol 89:162
Schmidt M, Schlesinger M 1980 Relation of fatty acid attachment to the translation and maturation of vesicular stomatitis and Sindbis virus membrane glycoproteins. J Biol Chem 255:3334

Schmidt M, Bracha, M, Schlesinger M 1979 Evidence for covalent attachment of fatty acids to Sindbis virus glycoproteins. Proc Natl Acad Sci USA 76:1687
Tabas I, Kornfeld S 1978 The synthesis of complex-type oligosaccharides. J Biol Chem 253:7779
Tabas I, Kornfeld S 1979 Purification and characterization of a rat liver Golgi α-mannosidase capable of processing asparagine-linked oligosaccharides. J Biol Chem 254:11655
Tartakoff A 1980 The Golgi complex: crossroads for vesicular traffic. Int Rev Exp Pathol 22:228
Tartakoff A, Vassali P 1979 Plasma cell immunoglobulin M molecules: their biosynthesis, assembly, and intracellular transport. J Cell Biol 83:284
Womak J, Yan D, Potier M 1981 Gene for neuraminidase activity on mouse chromosome 17 near H2: pleiotropic effects on multiple hydrolases. Science (Wash DC) 212:63

DISCUSSION

Sabatini: For your cell fractionation studies with Chinese hamster ovary (CHO) cells, what was the endoplasmic reticulum (ER) marker that you used?

Rothman: We use the one that is most pertinent to the pathway via which vesicular stomatitis viral-encoded membrane glycoprotein (G protein) is transported: our ER marker is the distribution of G protein, pulse-labelled *in vivo* for two minutes with [^{35}S]methionine, representing the earliest biosynthetic compartment.

Sabatini: Have you shown conclusively that the activity of α-1, 2-mannosidase is found in the Golgi? Is it true that some of the sugars are trimmed from the oligosaccharides possibly before the protein reaches the Golgi?

Rothman: The evidence that the mannosidase-containing compartment represents the Golgi is mainly by analogy to the rat liver, where the α-1, 2-mannosidase and the galactosyltransferase are co-purified by over 100-fold, as Tabas & Kornfeld (1979) have shown. We therefore term this activity 'Golgi-associated' because it is associated with the Golgi in other instances. The significance of our experiments is that they demonstrate two sets of Golgi-associated functions in two separate compartments, each of which is distinct from the biosynthetic compartment (the rough ER).

Sabatini: Can we therefore discard the idea that α-1, 2-mannosidase, as well as the fatty acid activity, is present in the smooth portions of ER?

Rothman: There may, of course, be some present in smooth ER in tissues that, like liver, have an extensive network of smooth ER. The bulk of mannosidase is not found in smooth ER of liver, although the acyltransferase may well be. The next stage will be to localize these enzymes *in situ*.

Sabatini: If you postulate that the exit from the *trans* face of the Golgi is *selective*, i.e. only for proteins that leave the Golgi and not for proteins that belong there, could one not, equally, postulate that the exit from the transitional element constitutes the real filter, and that there isn't such a crude passage of material from the ER to the Golgi?

Rothman: Yes, but that would not account for the reported prevalence of the ER proteins in the Golgi, as I discussed in my paper. Essential to this is the issue of how good a filter is the transitional element in relation to the degree of filtration ($>10^6$-fold) that is required.

Sabatini: Well, it is not a prevalence; it is the presence of *some* activities (Jarasch et al 1979, Borgese & Meldolesi 1980). For example, NADH cytochrome b_5 reductase and cytochrome b_5 are incorporated into the membranes post-translationally (Borgese & Gaetani 1980, Rachubinski et al 1980, Okada et al 1982) and are widely distributed in different organelles where they could have a function. With Drs G. Kreibich and M. Rosenfeld we have studied ribophorin glycoproteins, which we believe are associated with the ribosomes and are exclusive markers for the rough ER. The fact that they are glycoproteins is quite exceptional, because many other proteins that we have studied from the ER are not. Our preliminary analysis, with Dr P. Atkinson, of the oligosaccharides in the purified ribophorins shows that they lack the terminal sugars, which suggests that they don't ever go to the *trans* parts of the Golgi. We have measured between six and eight mannoses in the oligosaccharides of the ribophorins. Could the removal of the other mannoses, as well as the glucoses, take place in the ER?

Rothman: If you are seeing about six mannoses, this is less than the nine mannoses that one would expect without α-1,2-mannosidase action. Dr Martin Snider and I have similar results (unpublished) from preliminary experiments in which we analysed the bulk oligosaccharides that are found in rough microsome fractions. One interpretation of these and your results is that the glycoprotein has passed into and back from the cis Golgi. The other interpretation is that there is a small amount of α-1, 2-mannosidase in the ER that acts over a long time. This experiment does not distinguish between the alternative explanations. Pulse–chase experiments are needed.

Sabatini: We are unable to detect any bulk amount of ribophorins in purified Golgi fractions.

Rothman: You would not expect to if most of the ribophorin is bound to ribosomes at any given time.

Palade: Some of the points so far discussed may reflect imperfections in our cell fractionation procedures. Even for generally accepted Golgi markers, such as galactosyltransferase and sialyltransferase, recovery is not better than 60% in the best Golgi fractions now available. The remaining 40% is located in a residual microsomal fraction and in other fractions that are discarded during the procedure. All those fractions (including the residual microsomes) still have elements that, on morphological grounds, appear to be Golgi elements.

Rothman: This points out the need for immunocytochemistry to get those problems.

Sabatini: Some immunocytochemical studies and the immunoabsorption experiments from Dr Palade (Ito & Palade 1978) have produced results that do not completely agree with those from other laboratories. For example, Matsuura & Tashiro (1979), who studied the distribution of cytochrome *P*–450 by immunocytochemistry, have found no evidence for its presence in the Golgi. How can we be sure that the activity that transfers fatty acids and the activity of the α-1,2-mannosidase are absent from the ER? In kinetic experiments you yourself indicated that the fatty acid is introduced somewhere between the ER and the Golgi.

Rothman: The important point is not whether the mannosidase is entirely absent from ER but, instead, that the bulk of it separates from both ER and traditional Golgi markers (such as galactosyltransferase).

I don't think that fatty acid *is* introduced in the same compartment as α-1, 2-mannosidase. I think it arrives there very rapidly after incorporation at an earlier site.

Sabatini: So the activity may be an ER marker?

Rothman: I would not be surprised if the acyltransferase were a bulk ER marker but I would certainly be surprised if it were a Golgi marker.

Kornfeld: In rat liver preparations the bulk of the α-1,2-mannosidase activity co-purifies with galactosyltransferase, and can be purified about 100-fold by subcellular fractionation (Tabas & Kornfeld 1979). Therefore one would conclude that most of the membrane-associated α-1,2-mannosidase activity represents the Golgi enzyme. On the other hand, Godelaine et al (1981) observed that the addition of carbonyl cyanide *m*-chlorophenylhydrazone (CCCP) to thyroid slices causes processing to stop after the removal of one mannose residue to form the Man_8 species. This indicates that there may be another α-mannosidase which is localized in the rough ER.

Sabatini: How many mannoses are present in the pulse-labelled G-protein found in the ER?

Rothman: It is certainly at least nine mannoses.

Sabatini: Is there no mannose 5 or 6 at all?

Rothman: No. We know this is true from the size of the G protein relative to a series of markers. The oligosaccharide attached to G protein in the ER fractions has nine mannoses and probably also contains glucose. This is also known in much greater detail from the work of Dr Kornfeld, who could perhaps comment on this.

Kornfeld: We know from the kinetics of mannose processing of G protein in CHO cells that mannose removal starts at about 10–20 minutes after the protein has been synthesized, although glucose is removed almost instantly. This result would be most consistent with the protein being in the Golgi when the mannose removal occurs.

Palade: You said that the two enzyme activities (galactosyltransferase and α-1,2-mannosidase) co-purify, but what are their recoveries?

Kornfeld: Our best recovery is about 40–50%, and the recovery of the galactosyltransferase and the α-1,2-mannosidase is equal.

Hubbard: Bergmann et al (1981) used immunoelectron microscopy to visualize G protein through the stacks of the Golgi. What did their pictures show? Would you predict a concentration of the G protein on the trans side, and was there any?

Rothman: I'm not sure what to predict; I wouldn't be surprised if it did concentrate, but I also wouldn't be surprised if the immunoelectron microscopy technique were not capable of showing it.

Sabatini: We have shown (see p 184-208) that G protein is not concentrated in the trans face of the Golgi, but that it is equally distributed throughout the Golgi cisternae.

Meldolesi: I don't know whether the work of Nica Borgese and myself (1980), which you have mentioned, can be used to support your interpretation, because the Golgi subfractionation procedure we used does not result in the complete separation of the cis and the trans Golgi. However, one piece of information that emerges clearly from our data is that the cis compartment of the Golgi (identified by the presence of NADH-cytochrome b_5 reductase and low cholesterol) is probably smaller than the trans compartment, whereas in your scheme it is the other way around.

Rothman: Would the cis compartment be smaller on the basis that it contains less total protein?

Meldolesi: Yes.

Rothman: Our model proposes that the protein becomes concentrated in the trans compartment, and so this might not be the right comparison.

Meldolesi: Instead of protein, one could consider the amount of membrane in the trans compartment, which we have found to be larger than that in the cis compartment. However, one should acknowledge that Golgi recoveries are generally incomplete, so that no definite conclusion can be reached on this issue at present.

Rothman: In your case we cannot tell how sensitive the shift made by digitonin is to the amount of cholesterol contained in the membrane. For example, in the trans compartment there could be one mole of cholesterol per mole of phospholipid and in the cis compartment there might be only 0.1 moles per mole, and yet both could shift via digitonin and be lumped together into your 'trans' fraction. The amounts of protein or lipid in each fraction might be misleading.

Sabatini: You mentioned the specific features of proteins in different compartments. We have studied a number of ER membrane proteins that we have synthesized (Okada et al 1982) *in vitro*, and we have some partial sequence information for them. None of them, except the two ribophorins (Rosenfeld et al 1981) is a glycoprotein or contains amino-terminal signals that are removed proteolytically. This is true for cytochrome *P*–450 and its reductase, and for cytochrome *P*–448, epoxide hydrolase (Okada et al 1982), and the Ca^{2+} ATPase of the sarcoplasmic reticulum (Chyn et al 1979). The two ribophorins, however, are the only ER glycoproteins so far characterized that contain transient amino-terminal segments that are removed proteolytically.

Pearse: You referred to the ATP and cytosol-dependent release of clathrin from cages. Is there any change in the isoelectric point of clathrin when it is released enzymically from coated vesicles?

Rothman: We don't know. We looked initially for a phosphorylation or an addition of the adenine part of ATP, which would be among the most obvious possibilities, but we haven't seen anything yet, probably because the cytosol fraction is too crude. Once we have purified the cytosolic factor we should be able to do some more refined studies of mechanisms.

Goldstein: You mentioned that ATP-released clathrin will not reassemble. Perhaps the simplest model is that ATP has caused a covalent modification of clathrin—either phosphorylation or adenylation—and that the phosphate or adenyl residue must be removed before reassembly can occur.

Rothman: Yes. Or, a non-covalent modification could take place (for example, a bound nucleotide). However, there may have been some trivial modification that is irrelevant to the release, and we have not yet ruled this out.

Goldstein: Have you tested the effects of purified phophatases or phosphodiesterases to determine whether the ATP effect is causing a phophorylation or an adenylation?

Rothman: Such studies are indirect though valuable. We feel that the best route towards studying mechanism is via purification of the cytosolic factor.

Hubbard: What is your evidence that the cytosolic factor is a protein?

Rothman: Only in that it behaves as if it were a protein during several steps of partial purification.

Goldstein: Does GTP replace ATP?

Rothman: Yes, but only partially. However, the cytosol is so crude that we do not feel it appropriate yet to take tests of nucleotide specificity seriously. This must await purification.

Baker: You said that this dissociation works with isolated clathrin cages as well as with intact vesicles, so is it correct to say that the vesicle within its coat is immaterial to all this?

Rothman: We have measured only the *extent* of release, and we are just beginning to make comparisons of rates between empty cages and coated vesicles as substrates. We have not yet ruled out the possibility of differences in *rate* between the two.

Baker: If ATP is present, you could be driving a proton pump in the membrane of the vesicle and this may alter the stability of the cage.

Rothman: To investigate this, we tested whether proton ionophores will prevent the uncoating of coated vesicles, and they do not.

Baker: The protein ionophores won't block the pump either.

Rothman: That is correct, but ionophores will eliminate the *effect* of such a pump by preventing the accumulation of a proton gradient.

Branton: You mentioned that when you spray the supernatant to look at these coats that have become dissociated, they still seem to be aggregated into larger polygonal forms. When we spray intact associated coats they always fall apart into triskelions unless they are covalently cross-linked. The implication is therefore that the coats you are looking at *are* covalently cross-linked.

Rothman: I am certain that the ATP-released clathrin is not covalently cross-linked (except possibly via disulphide bridges) because all the released clathrin is recovered in the 180 000 M_r clathrin band after sodium dodecyl sulphate polyacrylamide gel electrophoresis in the presence of 2-mercapto-ethanol.

Geisow: What calcium levels are present in your association/dissociation system for the clathrin–membrane interaction?

Rothman: One can equally well do the ATP-dependent uncoating in the presence of calcium, in the presence of EGTA, or with the addition of calmodulin. If one tries to substitue calmodulin for the cytosol it doesn't replace it.

Branton: Have you ruled out proteases in the decoating reaction?

Rothman: Yes, as best as we can at the moment. We see no detectable shift in the M_r for clathrin; the light chains (known to be especially sensitive to proteolysis) are not affected in amount or in M_r. The recovery of intact clathrin in the supernatant plus pellet is 99%. If coats are purposely treated with a variety of proteases so as to cut both clathrin and the light chains, these fragments remain in coats and will even re-assemble. So if proteolysis is going on, it would have to be ATP-dependent, very selective for clathrin, and cut uniquely at a site very near the N- or C-terminus.

Dean: To return to the original subject of discussion, the retrieval of ER proteins from the Golgi to the ER was also suggested by John Mayer but for quite different reasons, at a previous Ciba Foundation symposium (Mayer et al 1980). He has found practically no correlations between the molecular characteristics of the proteins in many organelles (including ER) and their rates of turnover, whereas many people have found the converse. As a consequence he proposed the same recycling that you have proposed. The idea still requires complete verification but it might be relevant to your work.

Bretscher: Is the point of having a stack of membrane with countercurrent distribution that the proportion of plasma membrane or other proteins in the ER is very low?

Rothman: The assumption is that one would need multiple steps of purification, and therefore a stack, if a single step doesn't suffice.

Bretscher: In the passage of newly synthesized proteins from the Golgi to a variety of other compartments, there is presumably a high degree of separation of proteins (i.e. basolateral from apical plasma membrane proteins, and so on). If one can only obtain a 10- or 20-fold purification in one step, then in order to separate proteins for different destinations beyond the trans side of the Golgi, one would presumably need several 'Golgi stacks', one for each of these other compartments?

Rothman: I don't think so. The collection of proteins that needs to be exported to a variety of membranes is much more concentrated in the trans cisterna of the Golgi than in the ER. In other words the full purification needed from that point onwards might be small (say 10-fold) in relation to what has already been accomplished (e.g. 1000-fold). This might well be obtained with only one more step—in other words, during exit from a single trans cisterna of the Golgi.

Bretscher: With one of the viral spike proteins that goes to the apical membrane of an epithelial cell, how many go to the basolateral side? How great a degree of purification is there? It may be enormous.

Rothman: I don't know if reliable data are available on this. But we need numbers for this and for earlier transport steps.

Sabatini: We have been unable to show conclusively that the envelope glycoproteins come out sorted from the Golgi.

Rothman: That confirms the suggestion that such sorting occurs at the level of exit from the Golgi or beyond.

Bretscher: My own suggestion would be that the function of the Golgi is to separate *cholesterol*, and not proteins, countercurrently away from the ER.

Palade: The compartmentation of the Golgi complex is probably more extensive than we yet realize. Phenomena comparable to those described by Dr Rothman occur *in situ* when cells are deprived of ATP; the loss of clathrin cages affects at least a certain fraction of the total population of coated vesicles (see p 287-291

REFERENCES

Bergmann JE, Tokuyasu KT, Singer SJ 1981 Passage of an integral membrane protein, the Vesicular Stomatitis virus glycoprotein, through the Golgi apparatus en route to the plasma membrane. Proc Natl Acad Sci USA 78:1746-1750

Borgese N, Gaetani S 1980 Site of synthesis of rat liver NADH-cytochrome b_5 reductase, an integral membrane protein. FEBS (Fed Eur Biochem Soc) Lett 112:216-220

Borgese N, Meldolesi J 1980 Localization and biosynthesis of NADH-cytochrome b_5 reductase, an integral membrane protein in rat liver cells. I: Distribution of the enzyme activity in microsomes, mitochondria, and Golgi complex. J Cell Biol 85:501-515

Chyn T, Martonosi A, Morimoto T, Sabatini DD 1979 in vitro synthesis of the Ca^{2+} transport ATPase by ribosomes bound to sarcoplasmic reticulum membranes. Proc Natl Acad Sci USA 76:1241-1245

Godelaine D, Spiro MJ, Spiro RG 1981 Processing of the carbohydrate units of thyroglobulin. J Biol Chem 256:10161-10168

Ito A, Palade GE 1978 Presence of NADPH-cytochrome P-450 reductase in rat liver Golgi membranes. J Cell Biol 79:590-597

Jarasch E-D, Kartenbeck J, Bruder G, Fink A, Morre JD, Franke WW 1979 B-Type cytochrome in plasma membranes isolated from rat liver in comparison with those of endomembranes. J Cell Biol 80:37-52

Matsuura S, Tashiro Y 1979 Immunoelectron microscopic studies of endoplasmic reticulum—Golgi relationships in the intracellular transport process of lipoprotein particles in rat hepatocytes. J Cell Sci 39:273-290

Mayer RJ, Russell SM, Burgess RJ, Wilde CJ, Paskin N 1980 Coordination of protein synthesis and degradation. In: Protein degradation in health and disease. Excerpta Medica, Amsterdam (Ciba Found Symp 75) p 253-272

Okada Y, Frey AB, Guenthner TM, Oesch F, Sabatini DD, Kreibich G 1982 Studies on the biosynthesis of microsomal membrane proteins—site of synthesis and mode of insertion of cytochrome b_5, cytochrome b_5 reductase, cytochrome P-450 reductase and epoxide hydrolase. Eur J Biochem 122:393-402

Rachubinski RA, Verma DPS, Bergeron JJM 1980 Synthesis of rat liver microsomal cytochrome b_5 by free ribosomes. J Cell Biol 84:705-716

Rosenfeld MG, Marcantonio EE, Harnik VM, Sabatini DD, Kreibich G 1981 Synthesis and cotranslational processing of ribophorins. J Cell Biol 91:404a

Tabas I, Kornfeld S 1979 Purification and characterization of a rat liver Golgi α-mannosidase capable of processing asparagine-linked oligosaccharides. J Biol Chem 254:11655-11663

Steps in the phosphorylation of the high mannose oligosaccharides of lysosomal enzymes

STUART KORNFELD, MARC L. REITMAN, AJIT VARKI, DANIEL GOLDBERG and CHRISTOPHER A. GABEL

Division of Hematology-Oncology, Departments of Internal Medicine and Biological Chemistry, Washington University School of Medicine, 660 South Euclid, St Louis, Missouri 63110, USA

Abstract The phosphomannosyl recognition marker of acid hydrolases, which mediates their translocation to lysosomes, has been shown to be synthesized in two steps. First, *N*-acetylglucosamine 1-phosphate is transferred to an acceptor mannose by UDP-*N*-acetylglucosamine:lysosomal enzyme *N*-acetylglucosamine-1-phosphotransferase, resulting in a phosphate group in diester linkage between the outer *N*-acetylglucosamine and the inner mannose. Next, an α-*N*-acetylglucosaminyl phosphodiesterase removes the *N*-acetylglucosamine, leaving the phosphate in monoester linkage with the underlying mannose residue. This exposed phosphomannosyl residue serves as the essential component of a recognition marker which leads to binding to high-affinity receptors and subsequent translocation to lysosomes. We propose that the first enzyme in this scheme, *N*-acetylglucosaminylphosphotransferase, catalyses the initial, determining step by which newly synthesized acid hydrolases are distinguished from other newly synthesized glycoproteins and thus are eventually targeted to lysosomes. The absence of this enzyme activity, as in inclusion-cell (I-cell) disease and pseudo-Hurler polydystrophy, precludes the receptor-mediated targeting of newly synthesized acid hydrolases to lysosomes. As a consequence, the enzymes are secreted into the extracellular milieu.

It is now well established that adsorptive pinocytosis of lysosomal enzymes is mediated by phosphomannosyl residues on these enzymes (Kaplan et al 1977, Sando & Neufeld 1977, Kaplan et al 1978, Ullrich et al 1978, Natowicz et al 1979, Distler et al 1979, von Figura & Klein 1979, Bach et al 1979, Fischer et al 1980a, b, Hasilik & Neufeld 1980a, b, Gonzalez-Noriega et al 1980). These residues are the essential components of a recognition marker that is necessary for binding to high-affinity receptors on the cell surface and for

1982 Membrane recycling. Pitman Books Ltd, London (Ciba Foundation symposium 92) p 138-156

subsequent translocation to lysosomes (Hickman et al 1974). Initially adsorptive pinocytosis was believed to be the major route by which newly synthesized acid hydrolases were delivered to lysosomes (Neufeld et al 1977). Recently Sly and co-workers have demonstrated that most of the receptor is located intracellularly rather than on the cell surface (Fischer et al 1980a). From this finding and other evidence, these investigators have proposed that the primary function of this receptor system is to mediate the intracellular transport of acid hydrolases from the Golgi to lysosomes (Sly & Stahl 1978, Sly et al 1981). According to this model adsorptive pinocytosis serves as an alternative route for enzyme delivery to lysosomes.

Our work has been concerned with two basic aspects of this pathway. First, we have investigated the biochemical steps in the phosphorylation of the mannose residues of lysosomal enzymes and, second, we have attempted to determine the basis by which the cell selectively phosphorylates the mannose residues of lysosomal enzymes. Our initial goal was to elucidate the structure of the phosphorylated oligosaccharide units present on lysosomal enzymes because these structures might provide clues about the biosynthetic steps in phosphorylation. We labelled cells in tissue culture with [2-^{3}H]mannose and then we isolated the lysosomal enzyme β-D-glucuronidase (EC 3.2.1.31) by immunoprecipitation. The [2-^{3}H]mannose-labelled oligosaccharide units of the enzyme were then purified and structurally analysed. This approach offered several advantages: it enabled us to determine the structure of the units by using extremely small amounts of material; and it allowed us to define the kinetics of phosphorylation in the intact cell. Previous workers had established that the initial glycosylation of asparagine-linked oligosaccharides occurs by the *en bloc* transfer of a high mannose-type oligosaccharide to the nascent polypeptide chain (Kiely et al 1976, Parodi & Leloir 1979). With the [2-^{3}H]mannose label one could therefore follow the fate of the oligosaccharide units of newly synthesized lysosomal enzymes as they proceeded from their site of synthesis in the rough endoplasmic reticulum to their final destination in the lysosomes.

When we analysed the phosphorylated oligosaccharide units of newly synthesized β-D-glucuronidase, we found that most of the phosphate groups were present in diester linkage between the sixth OH group of mannose residues in the underlying oligosaccharide and the first carbon of outer, α-linked *N*-acetylglucosamine residues (Tabas & Kornfeld 1980). A similar finding was also made by Hasilik et al 1980. Our studies of the structure of these unusual oligosaccharides (Varki & Kornfeld 1980b) showed that the phosphates are linked to five different mannose residues on the oligosaccharide, and that individual molecules may contain one or two phosphates, thus generating many isomers. A composite picture is shown in Fig. 1, where the asterisks identify the mannose residues that can be phosphorylated in the

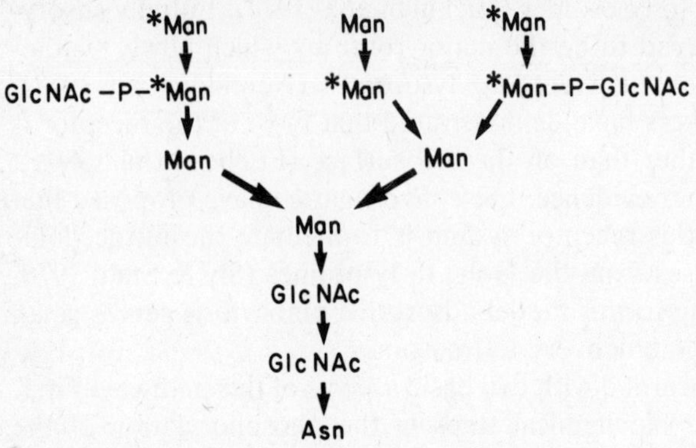

FIG. 1. Structure of a phosphorylated high mannose oligosaccharide that contains two phosphodiester units. The asterisks identify the mannose residues that are phosphorylated in various isomers (after Varki & Kornfeld 1980b).

various isomers. The most common sites for phosphorylation of oligosaccharides that have two diesters are also shown.

The finding of phosphodiester-linked moieties on the oligosaccharides of newly synthesized β-D-glucuronidase suggested the existence of a novel biosynthetic pathway, which is outlined in Fig. 2. In this scheme the

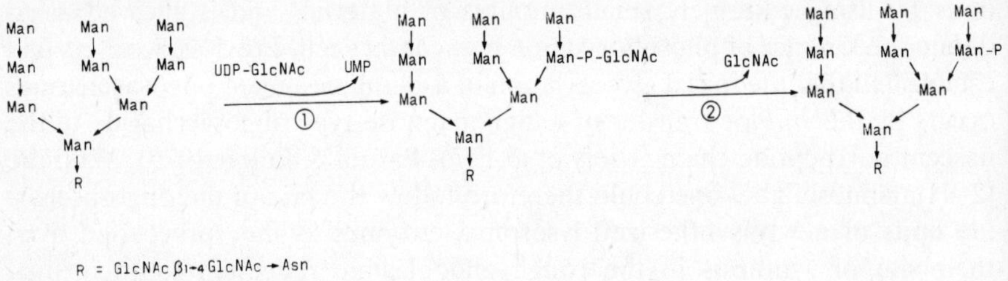

FIG. 2. Proposed mechanism for the phosphorylation of the high mannose units of acid hydrolases. The enzymes involved are (1) UDP-*N*-acetylglucosamine:lysosomal enzyme *N*-acetylglucosaminyl-1-phosphotransferase and (2) α-*N*-acetylglucosaminyl phosphodiesterase (R = *N*-acetylglucosamine β1 → 4 *N*-acetylglucosamine → asparginine).

phosphomannosyl residues are synthesized in two steps. First, *N*-acetylglucosamine l-phosphate is transferred to an acceptor mannose to generate a phosphate group in diester linkage between the outer 'blocking' *N*-acetylglucosamine and the inner mannose. Next, the *N*-acetylglucosamine is removed, leaving the phosphate in monoester linkage to the underlying mannose residue. This mechanism is compatible with the finding that alkaline phosphatase-susceptible phosphomonesters are essential for high-affinity binding to the receptors (Kaplan et al 1977).

In order to prove that the diesters are actually converted to monoesters *in vivo* (i.e. by 'uncovering'), we examined the kinetics of the phosphorylation pathway in the murine macrophage line P388D$_1$ (Goldberg & Kornfeld 1981). Cells were incubated with [2-^{3}H]mannose for 15–20 min and then chased in unlabelled media for various times up to 5 h. β-D-Glucuronidase was immunoprecipitated and its oligosaccharide units were examined (for extent of phosphorylation and uncovering) by chromatography of the oligosaccharides on quaternary aminoethyl-Sephadex using an elution system that separates neutral oligosaccharides from oligosaccharides with one or two phosphomonoesters or phosphodiesters. The results are shown in Fig. 3. At the end of the 20-min labelling period most of the β-D-glucuronidase oligosaccharides were neutral; the only detectable phosphorylated species was a small amount of oligosaccharide with one covered phosphate. During the first 40 min of the chase, the amount of oligosaccharide with one covered phosphate greatly increased, and material having two covered phosphates became detectable, as did material with uncovered phosphates. The total extent of phosphorylation increased from 5% at the end of the pulse period to 20% after 40 min of chase and to 25% after 80 min of chase. As the chase proceeded, most of the phosphodiesters were converted to phosphomonoesters, as predicted by the scheme shown in Fig. 2. Since the total amount of labelled mannose in newly synthesized β-D-glucuronidase increased only slightly during the first 40 min of the chase period, the increase in the amount of phosphorylated species during this time must have resulted from phosphorylation of neutral high-mannose-type oligosaccharides transferred to the enzyme during the 20 min pulse. This finding demonstrates that phosphorylation is a post-translational event which takes place on protein-bound oligosaccharide and not on lipid-linked oligosaccharide precursors.

In our most recent experiments we have attempted to define the state of the oligosaccharides on the lysosomal enzymes when these molecules first bind to the high-affinity intracellular receptors. To approach this question we did a pulse-chase experiment similar to the one described above, but in this case we isolated a total cell-membrane fraction at each time point. This fraction, which contained the high-affinity receptor with its bound ligand, was treated with mannose 6-phosphate to release the bound lysosomal enzymes. The oligosaccharide units of the released glycoproteins were then analysed and compared to the phosphorylated oligosaccharides that remained in the soluble fraction of the cell. The results were striking. There was a tremendous enrichment in oligosaccharides with one and two phosphomonoesters in the material released from the receptor with mannose 6-phosphate. In contrast, most of the enzyme in the soluble fraction contained oligosaccharides with phosphodiester units. These data indicate that conversion of phosphodiesters to phosphomonoesters precedes binding to the intracellular receptors.

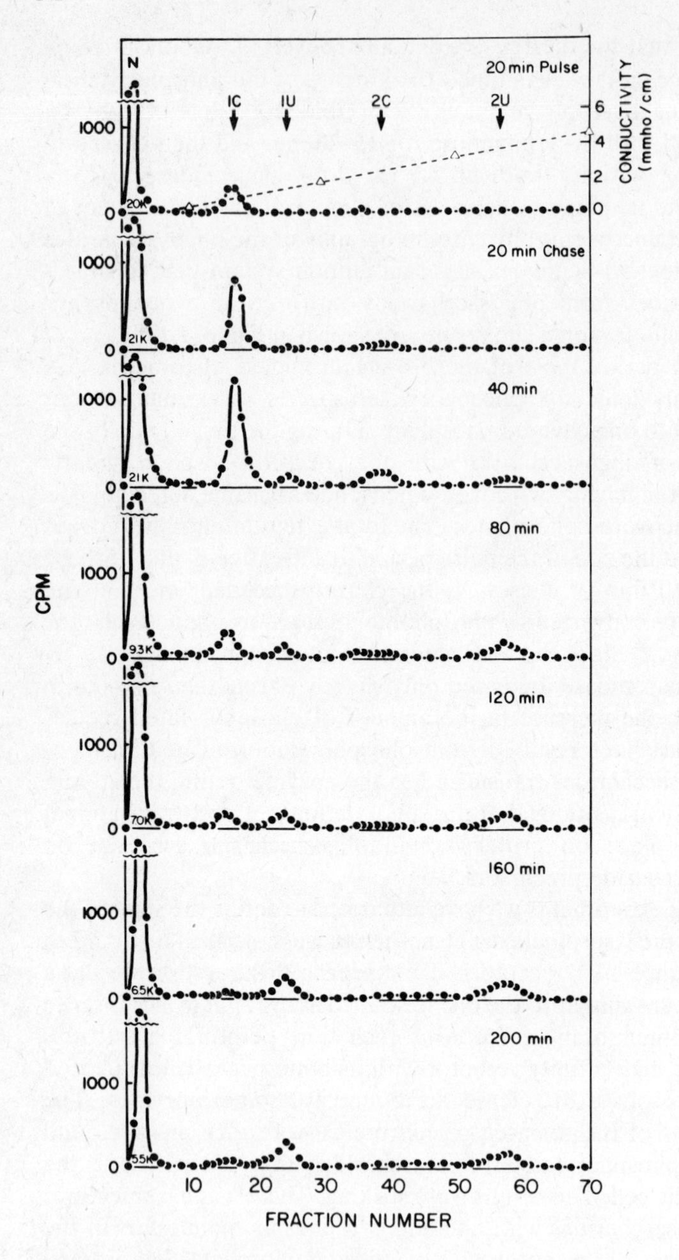

FIG. 3. Kinetics of phosphorylation of oligosaccharides in β-D-glucuronidase. Mouse P388D$_1$ cells were incubated with [2-^{3}H]mannose for 20 min and then chased in unlabelled media. At the indicated times β-D-glucuronidase was immunoprecipitated and its oligosaccharide units were isolated and fractionated on quaternary aminoethyl-Sephadex. The column was eluted with a

UDP-*N*-Acetylglucosamine:lysosomal enzyme *N*-acetylglucosamine-1-phosphotransferase

As shown in Fig. 2, the proposed biosynthetic pathway requires two enzymes: an *N*-acetylglucosaminylphosphotransferase and an α-*N*-acetylglucosaminyl phosphodiesterase. Assays for each of these enzymes have been developed and the enzyme activities have been detected in a number of tissues. UDP-*N*-acetylglucosamine:lysosomal enzyme *N*-acetylglucosamine-1-phosphotransferase activity has been detected in homogenates of Chinese hamster ovary cells, human diploid fibroblasts and rat liver (Reitman & Kornfeld 1981a, Hasilik et al 1981). Using [β-^{32}P]UDP-[^{3}H]*N*-acetylglucosamine as donor, we showed that the enzyme transfers *N*-acetylglucosamine 1-phosphate to the sixth OH group of mannose residues in high mannose-type oligosaccharides of lysosomal enzymes. The enzyme is not inhibited by tunicamycin or stimulated by dolichol phosphate, indicating that the reaction does not proceed via a dolichol-pyrophosphoryl-*N*-acetylglucosamine intermediate. Fibroblasts from patients with inclusion-cell (I-cell) disease (mucolipidosis II) and pseudo-Hurler polydystrophy (mucolipidosis III) are severely deficient in this enzyme activity (Hasilik et al 1981, Reitman et al 1981, Varki et al 1981). These autosomal recessive storage diseases are characterized by a general failure to target acid hydrolases to lysosomes in spite of normal rates of acid hydrolase synthesis (Hickman & Neufeld 1972) and by the failure to incorporate [^{32}P]phosphate into newly synthesized acid hydrolases (Hasilik & Neufeld 1980b, Bach et al 1979). The absence of the *N*-acetylglucosaminylphosphotransferase explains the lack of phosphorylation of the newly synthesized acid hydrolases. This failure to generate the phosphomannosyl recognition signal precludes the receptor-mediated targeting of the newly synthesized acid hydrolases to lysosomes, and consequently the enzymes are secreted into the extracellular milieu.

One of the most intriguing questions about this system concerns the basis for the specificity of the selective phosphorylation of the oligosaccharide units of lysosomal enzymes. If we assume, as the data indicate, that lysosomal enzymes and non-lysosomal glycoproteins are each glycosylated with the identical oligosaccharide precursor, then we must explain why only the lysosomal enzymes are phosphorylated. The simplest explanation is that all

---

linear gradient of pyridinium acetate, pH 5.3, from 2 mM to 500 mM. The neutral oligosaccharide peaks are cut off at the top; total counts per min (cpm) are given under each peak. The abbreviations are: N, neutral oligosaccharide standard; 1C, oligosaccharide standard containing one covered phosphate; 1U, standard containing one uncovered phosphate; 2C and 2U, standards containing two covered and two uncovered phosphates, respectively (from Goldberg & Kornfeld 1981).

lysosomal enzymes share a common protein recognition site or marker and that this marker causes one of two possible effects. First, it could lead to the segregation of newly synthesized lysosomal enzymes into a specialized subcellular compartment which contains the *N*-acetylglucosaminylphosphotransferase. In this case the transferase would not have to recognize lysosomal enzymes specifically but, rather, it would phosphorylate the oligosaccharide units of any glycoprotein in the compartment. The other possibility is that the *N*-acetylglucosaminylphosphotransferase is able to distinguish between lysosomal and non-lysosomal enzymes on the basis of this common protein

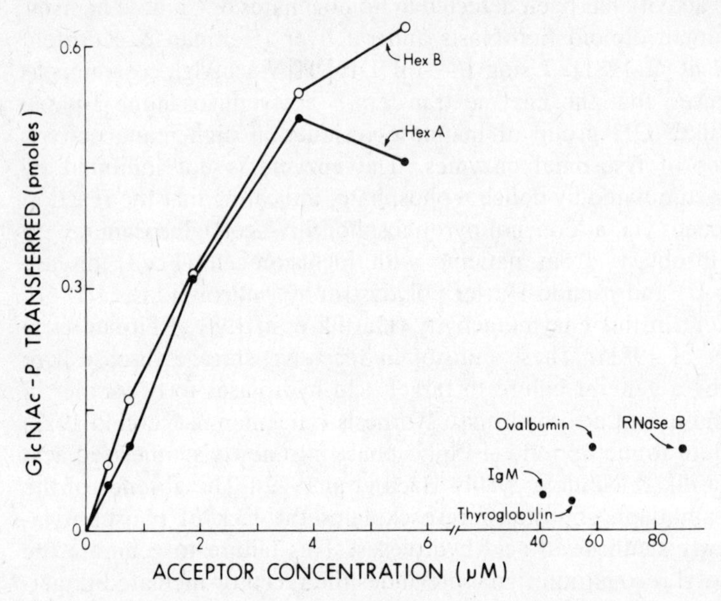

FIG. 4. Activity of *N*-acetylglucosaminylphosphotransferase towards various glycoproteins containing high mannose-type oligosaccharides. The two lysosomal enzyme acceptors were human placental β-hexosaminidase A and B (after Reitman & Kornfeld 1981b).

recognition marker. To evaluate this, we partially purified the *N*-acetylglucosaminylphosphotransferase from rat liver and determined its ability to phosphorylate the oligosaccharide units of a series of lysosomal and non-lysosomal glycoproteins (Reitman & Kornfeld 1981b). Fig. 4 shows the results of a typical experiment using glycoprotein acceptors that have in common the presence of high mannose-type oligosaccharide units. It is evident that the lysosomal enzymes β-hexosaminidase A and B are the best acceptors, being active at very low protein concentrations. In contrast, none of the four non-lysosomal glycoproteins is a good acceptor. The data in Table 1 show some of the kinetic parameters of the enzyme towards various acceptors. The apparent K_m values for the three lysosomal enzymes are in the low micromolar range whereas the K_m for ribonuclease B is approximately 100-fold greater. The acceptor activity of the other non-lysosomal glycoproteins was so low that it was not possible to determine their apparent K_m and V_{max} values. Table 1 also presents the kinetic parameters for two carbohydrate acceptors, α-methylmannoside and Man_{5-8}*N*-acetylglucosamine oligosaccharide. These molecules have very high apparent K_m values, being 10^3–10^4-fold greater than those of lysosomal enzymes. Their V_{max} values are 10–100-fold greater than those of the glycoprotein acceptors. The relative catalytic efficiency (V_{max} divided by apparent K_m) of the transferase towards the various acceptors is also shown in Table 1. The three lysosomal enzymes are phosphorylated at least 100-fold more efficiently than either ribonuclease B, Man_{5-8}*N*-acetylglucosamine oligosaccharide or α-methylmannoside. This preference of the *N*-acetylglucosaminylphosphotransferase for lysosomal enzymes in these *in vitro* assays demonstrates the remarkable specificity of a glycosyltransferase for the class of glycoproteins on which it acts *in vivo*.

TABLE 1 Kinetic parameters of *N*-acetylglucosaminylphosphotransferase activity towards various acceptors

	Apparent K_m *mM*	V_{max} [a]	*Relative acceptor activity* [b]
Human placental β-hexosaminidase A	0.020	251	163
Human placental β-hexosaminidase B	0.006	270	610
Porcine hepatic α-N-acetylglucosaminidase	0.009	181	258
Bovine pancreatic ribonuclease B	0.9	116	1.6
Man_{5-8}*N*-acetylglucosamine oligosaccharide	32	3103	1.2
α-Methylmannoside	113	8940	1.0

[a]pmol *N*-acetylglucosamine 1-phosphate transferred $h^{-1}mg^{-1}$ enzyme protein. [b]The relative acceptor activity is defined as the V_{max} divided by the apparent K_m, normalized to α-methylmannoside (after Reitman & Kornfeld 1981b).

To probe further the substrate specificity of the *N*-acetylglucosaminylphosphotransferase, we denatured the lysosomal enzymes by heat and then used them as acceptors. In all instances the denatured enzyme lost its ability to serve as an acceptor of *N*-acetylglucosamine 1-phosphate even though the protein remained in solution. Although the loss or modification of some heat-labile moiety cannot be ruled out, we believe these results strongly suggest that the conformation of the acceptor protein is recognized by

N-acetylglucosaminylphosphotransferase. In other words, the *N*-acetylglucosaminylphosphotransferase appears to be specific for a particular protein conformation that is unique to lysosomal enzymes. The nature of this conformation requirement is obscure at present. Since both ribonuclease B (with an exposed oligosaccharide unit [Baynes & Wold 1976, Tarentino et al 1974]) and free high mannose-type oligosaccharides are poor acceptors, it is unlikely that accessibility of the carbohydrate chain to the transferase is the sole conformational requirement for recognition.

The specificity of the *N*-acetylglucosaminylphosphotransferase towards lysosomal enzymes makes it unnecessary to postulate a mechanism for the specific segregation of newly synthesized acid hydrolases which would precede exposure to the *N*-acetylglucosaminylphosphotransferase. From these data we propose that the *N*-acetylglucosaminylphosphotransferase is the initial and determining enzyme for the pathway that eventually results in the segregation of acid hydrolases into lysosomes.

α-*N*-Acetylglucosaminylphosphodiesterase

An *α*-*N*-acetylglucosaminylphosphodiesterase capable of removing the 'covering' *N*-acetylglucosamine residues has been purified from rat liver (Varki & Kornfeld 1981) and human placenta (Waheed et al 1981a). The enzyme is distinct from a previously described lysosomal *α*-*N*-acetylglucosaminidase, which can also remove covering *N*-acetylglucosamine residues (Varki & Kornfeld 1980a, Waheed et al 1981b), and it is active against *N*-acetylglucosamine residues that are *α*-linked to an underlying phosphate, although it is inactive against *p*-nitrophenyl *α*-*N*-acetylglucosaminide. Furthermore, the activity is inhibited by PO_4 ions. Recently we have obtained evidence that the enzymic reaction may proceed by a glycosidase mechanism rather than by a true phosphodiesterase mechanism. We are pursuing this further to determine if a change in the nomenclature of the enzyme is appropriate.

Subcellular localization of *N*-acetylglucosaminylphosphotransferase and *α*-*N*-acetylglucosaminylphosphodiesterase

Both the *N*-acetylglucosaminylphosphotransferase and *α*-*N*-acetylglucosaminylphosphodiesterase have been localized to Golgi-enriched smooth membranes (Waheed et al 1981b, Varki & Kornfeld 1980a). We have succeeded in separating both activities from the trans Golgi marker galactosyltransferase by subjecting mouse lymphoma and $P388D_1$ macrophage membranes to centrifugation on a continuous sucrose density gradient. The precise location of these two enzyme activities is not known.

REFERENCES

Bach G, Bargal R, Cantz M 1979 I-cell disease: deficiency of extracellular hydrolase phosphorylation. Biochem Biophys Res Commun 91:976-981

Baynes JW, Wold F 1976 Effect of glycosylation on the *in vivo* circulating half-life of ribonuclease. J Biol Chem 251:6016-6024

Distler J, Hieber V, Sahagian G, Schmickel R, Jourdian GW 1979 Identification of mannose 6-phosphate in glycoproteins that inhibit the assimilation of β-glucuronidase by fibroblasts. Proc Natl Acad Sci USA 76:4235-4239

von Figura K, Klein U 1979 Isolation and characterization of phosphorylated oligosaccharides from α-N-acetylglucosaminidase that are recognized by cell-surface receptors. Eur J Biochem 94:347-354

Fischer HD, Gonzalez-Noriega A, Sly WS 1980a β-Glucuronidase binding to human fibroblast membrane receptors. J Biol Chem 255:5069-5074

Fischer HD, Natowicz M, Sly WS, Bretthauer R 1980b Fibroblast receptor for lysosomal enzymes mediates pinocytosis of multivalent phosphomannan fragment. J Cell Biol 84:77-86

Goldberg D, Kornfeld S 1981 The phosphorylation of β-glucuronidase oligosaccharides in mouse P388D$_1$ cells. J Biol Chem 256:13060-13067

Gonzalez-Noriega A, Grubb JH, Talkad V, Sly WS 1980 Chloroquine inhibits lysosomal enzyme pinocytosis and enhances lysosomal enzyme secretion by impairing receptor recycling. J Cell Biol 85:839-852

Hasilik A, Neufeld EF 1980a Biosynthesis of lysosomal enzymes in fibroblasts (synthesis as precursors of higher molecular weight). J Biol Chem 255:4937-4945

Hasilik A, Neufeld EF 1980b Biosynthesis of lysosomal enzymes in fibroblasts (phosphorylation of mannose residues). J Biol Chem 255:4946-4950

Hasilik, A, Klein U, Waheed A, Strecker G, von Figura K 1980 Phosphorylated oligosaccharides in lysosomal enzymes: identification of α-N-acetylglucosamine(1)phospho(6)mannose diester groups. Proc Natl Acad Sci USA 77:7074-7078

Hasilik A, Waheed A, von Figura K 1981 Enzymatic phosphorylation of lysosomal enzymes in the presence of UDP-*N*-acetylglucosamine. Absence of the activity in I-cell fibroblasts. Biochem Biophys Res Commun 98:761-767

Hickman S, Neufeld EF 1972 A hypothesis for I-cell disease: defective hydrolases that do not enter lysosomes. Biochem Biophys Res Commun 49:992-999

Hickman S, Shapiro LJ, Neufeld EF 1974 A recognition marker required for uptake of a lysosomal enzyme by cultured fibroblasts. Biochem Biophys Res Commun 57:55-61

Kaplan A, Achord DT, Sly WS 1977 Phosphohexosyl components of a lysosomal enzyme are recognized by pinocytosis receptors on human fibroblasts. Proc Natl Acad Sci USA 74:2026-2030

Kaplan A, Fischer D, Sly WS 1978 Correlation of structural features of phosphomannans with their ability to inhibit pinocytosis of human β-glucuronidase by human fibroblasts. J Biol Chem 253:647-650

Kiely ML, McKnight GS, Schmike RT 1976 Studies on the attachment of carbohydrate to ovalbumin nascent chains in hen oviduct. J Biol Chem 251:5490-5495

Natowicz MR, Chi MM-Y, Lowry OH, Sly WS 1979 Enzymatic identification of mannose

6-phosphate on the recognition marker for receptor-mediated pinocytosis of β-glucuronidase by human fibroblasts. Proc Natl Acad Sci USA 76:4322-4326
Neufeld EF, Sando GN, Garvin AJ, Rome LH 1977 The transport of lysosomal enzymes. J Supramol Struct 6:95-101
Parodi AJ, Leloir L 1979 The role of lipid intermediates in the glycosylation of proteins in the eucaryotic cell. Biochim Biophys Acta 559:1-37
Reitman ML, Kornfeld S 1981a UDP-*N*-Acetylglucosamine: glycoprotein *N*-acetylglucosamine-1-phosphotransferase. J Biol Chem 256:4275-4281
Reitman ML, Kornfeld S 1981b Lysosomal enzyme targeting. *N*-acetylglucosaminylphosphotransferase selectively phosphorylates native lysosomal enzymes. J Biol Chem 256:11977-11980
Reitman ML, Varki A, Kornfeld S 1981 Fibroblasts from patients with I-cell disease and pseudo-Hurler polydystrophy are deficient in uridine 5′-diphosphate-*N*-acetylglucosamine: glycoprotein *N*-acetylglucosaminylphosphotransferase activity. J Clin Invest 62:1574-1579
Sando GN, Neufeld EF 1977 Recognition and receptor-mediated uptake of a lysosomal enzyme, α-L-iduronidase, by cultured human fibroblasts. Cell 12:619-627
Sly WS, Stahl P 1978 Receptor mediated uptake of lysosomal enzymes. In: Silverstein SC (ed) Transport of macromolecules in cellular systems. Dahlem Konferenzen, Berlin (Life Sci Res Rep 11) p 229-244
Sly WS, Natowicz M, Gonzalez-Noriega A, Grubb JH, Fischer HD 1981 The role of the mannose-6-phosphate recognition marker and its receptor in the uptake and intracellular transport of lysosomal enzymes. In: Callahan JW, Lowden JA (eds) Lysosomes and lysosomal storage disease. Raven Press, New York, p 131-145
Tabas I, Kornfeld S 1980 Biosynthetic intermediates of β-glucuronidase contain high mannose oligosaccharides with blocked phosphate residues. J Biol Chem 255:6633-6639
Tarentino A, Plummer TH, Maley F 1974 The release of intact oligosaccharides from specific glycoproteins by Endo-β-*N*-acetylglucosaminidase H. J Biol Chem 249:818-824
Ullrich K, Mersmann G, Weber E, von Figura K 1978 Evidence for lysosomal enzyme recognition by human fibroblasts via a phosphorylated carbohydrate moiety. Biochem J 170:643-650
Varki A, Kornfeld S 1980a Identification of a rat liver α-*N*-acetylglucosaminyl phosphodiesterase capable of removing 'blocking' α-*N*-acetylglucosamine residues from phosphorylated high mannose oligosaccharides of lysosomal enzymes. J Biol Chem 255:8398-8401
Varki A, Kornfeld S 1980b Structural studies of phosphorylated high mannose-type oligosaccharides. J Biol Chem 255:10847-10858
Varki A, Kornfeld S 1981 Purification and characterization of rat liver α-*N*-acetylglucosaminyl phosphodiesterase. J Biol Chem 256:9937-9943
Varki A, Reitman ML, Kornfeld S 1981 Identification of a variant of mucolipidosis III (pseudo-Hurler polydystrophy): a catalytically active *N*-acetylglucosaminyl-phosphotransferase that fails to phosphorylate lysosomal enzymes. Proc. Natl Acad Sci USA 78:7773-7777
Waheed A, Hasilik A, von Figura K 1981a Processing of the phosphorylated recognition marker in lysosomal enzymes. J Biol Chem 256:5717-5721
Waheed A, Pohlmann R, Hasilik A, von Figura K 1981b Subcellular location of two enzymes involved in the synthesis of phosphorylated recognition markers in lysosomal enzymes. J Biol Chem 256:4150-4152

DISCUSSION

Sabatini: Are the substrates that you use enzymes that are recovered from secretory fluids and not from lysosomes?

Kornfeld: They are from various sources: some are from tissues; some are from secretions.

Sabatini: Would the enzyme that has reached the lysosomes, lost the phosphate and, in most cases, also lost a peptide piece, be a substrate?

Kornfeld: Yes. The enzymes that we use as acceptors have lost their pro-piece, so this piece evidently does not contain the information necessary for oligosaccharide phosphorylation. It is possible, however, that if the pro-piece were still present, the enzymes might be even better substrates.

Sabatini: Although the enzymes that have lost the peptide can be phosphorylated because they have a recognition segment for phosphorylation, is it possible that they cannot be taken to the lysosomes because they have lost the piece for address to the lysosomes?

Kornfeld: We have not phosphorylated a molecule *in vitro* and shown that it is taken up by cells and targeted to lysosomes. However, all the studies of Dr Sly and others in this field, on acid hydrolase uptake and targeting, utilize enzymes that have lost their pro-piece. Yet these enzymes go into lysosomes. Furthermore, in I-cell disease, the only known defect is the lack of phosphorylation, and in fibroblasts, at least, targeting of newly synthesized acid hydrolases to lysosomes does not occur.

Sabatini: But in the liver cells is the requirement for phosphomannose an open question?

Kornfeld: Yes.

Sly: We have some data on isolated oligosaccharides which agree with your findings on the binding to the receptor (K. Creek & W. Sly, unpublished observations). The diphosphorylated oligosaccharide itself is subject to pinocytosis and is delivered to lysosomes. In fact, a single Endo-H-released oligosaccharide with two phosphates as monoesters is endocytosed very efficiently by fibroblasts and is delivered to lysosomes. The signal for targeting is thus present in a single oligosaccharide.

Sabatini: So the signal is not only necessary for uptake but is also sufficient for delivery to lysosomes?

Sly: I don't know if we can say that. Although the phosphorylated oligosaccharide by itself is delivered to lysosomes this doesn't rule out the possibility that a portion of the polypeptide could also contribute to binding.

Sabatini: The observation shows only that the mature enzyme could reach the liver lysosome by pinocytosis; it doesn't show that it would reach the liver lysosomes by the intracellular route, which may require the peptide.

Dean: Have you looked at a variety of other hydrolytic enzymes as possible substrates? And do you have evidence for any late processing which is specific to secretion in the $P388D_1$ cells?

Kornfeld: We haven't studied any non-lysosomal hydrolases except ribonuclease B. We are finding some unusual oligosaccharides with hybrid-type structures on the secreted lysosomal enzymes. I don't think that there are any unique steps; it is just a matter of the phosphate changing the normal sequence of processing by the usual processing enzymes.

Jourdian: What is the nature of the hybrid molecules?

Kornfeld: The phosphorylated hybrid oligosaccharides have features of both high mannose and complex-type molecules. The phosphate is present on the high mannose portion and presumably prevents mannose processing. On the other side of the molecule we find the sequence: sialic acid to galactose to *N*-acetylglucosamine, which is typical of a complex-type molecule.

Jourdian: Are the hybrid oligosaccharides susceptible to the action of endohexosaminidase H?

Kornfeld: Some are.

Jourdian: We find that the greatest affinity or avidity of binding of β-galactosidase fractions to the phosphomannosyl receptor is for those that contain hybrid-type oligosaccharides.

Baker: In general terms, to what extent is protein synthesis, followed by post-translational modification, relevant to making membranes recycle? To what extent are these processes generating any part of the 'motor' for recycling? Or is membrane recycling driven totally independently? Is it correct to say that there is a membrane 'conveyor belt' operating independently of the nature and quantity of molecules loaded onto it?

Kornfeld: I don't think that glycosylation is at all concerned with membrane recycling. In certain circumstances one can block glycosylation with tunicamycin and show that normal cell growth occurs or, with certain viruses, that normal virus assembly takes place. I would consider that the glycosylation reactions serve to put recognition tags on proteins for targeting and for influencing their physical properties, but these reactions do not directly affect membrane recycling.

Baker: Is protein synthesis driving the system at all?

Palade: The present evidence indicates that a conveyor belt operates continuously, and probably at a constant rate, between the endoplasmic reticulum and the Golgi complex (Jamieson & Palade 1967) and between the Golgi complex and the lysosomes (Friend & Farquhar 1967); you also mentioned this in your paper, Dr Kornfeld. The rate of movement from the Golgi complex to the cell surface and from the surface to lysosome, however, can be regulated in many cells. To the extent so far tested, these activities do not depend on sustained protein synthesis (Jamieson & Palade 1968).

Kornfeld: There is plenty of evidence that receptors migrate from the surface internally and back out again in the presence of inhibitors of protein synthesis (see also p 109-119).

Palade: The conveyor belts operate continuously and some of them can be regulated over very large ranges, from a high level, as in secretagogue-stimulated exocytosis, to a basic level, as for miniature potentials recorded at resting neuromuscular junctions (Katz 1969).

Bretscher: If β-glucuronidase has three high mannose oligosaccharide chains, and if any one of them can get phosphorylated, does that imply that the protein has three recognition sites for the enzyme that puts on the phospho-*N*-acetylglucosamine?

Kornfeld: Perhaps it does. All three of the oligosaccharide units of β-glucuronidase may be phosphorylated. It could be that there is a recognition marker for each glycosylation site.

Bretscher: Perhaps this would make it a very restrictive sequence.

Rothman: Have you analysed the products of the enzyme incubations to find out which oligosaccharides are present, and the rates at which they are glycosylated?

Kornfeld: We have not tried that, but hope to do so.

Goldstein: Assuming that you found a mutation that destroys the activity of the phosphodiesterase or uncovering enzyme, what cellular fate would you predict for the uncovered lysosomal enzymes? Would they be secreted extracellularly?

Kornfeld: Oligosaccharides with the uncovered phosphates clearly bind with much higher affinity to the receptor than do species with the covered phosphates, but that doesn't mean that molecules with covered phosphates have no affinity for the receptor. Most of the lysosomal enzymes are tetramers: they have an average of 12 oligosaccharides per molecule. This means that there could be a very large number of diesters on one molecule which might produce a high enough affinity to allow some of the enzyme molecules to bind to the receptor. This could explain why the lack of the uncovering enzyme might not be associated with a clinically apparent disease. In assays of about 50 fibroblast cell lines from patients with I-cell disease and pseudo-Hurler polydystrophy, we have not found any line that is deficient in uncovering activity. Every line we have tested has had a defect in the phosphorylating enzyme.

Sly: We have done some uptake experiments on oligosaccharides with covered phosphate to see if they are taken up (K. Creek & W. Sly, unpublished observations). The results agree with what you say. Although they are taken up, the uptake is very low, with about 30-fold less efficiency than the uptake for those with two uncovered phosphates. However, if one had a multivalent ligand, containing several groups that bind with low affinity, it might still be segregated. This is certainly true for low density lipoprotein (LDL) substituted with many pentamannosyl monophosphate groups. LDL with one such group has low affinity, but LDL with 40 such groups binds 1000-fold better, and is taken up very efficiently.

Goldstein: Dr Sly, you proposed several years ago that the function of lysosomal acid phosphatase was to remove the mannose 6-phosphate from lysosomal enzymes, thereby trapping the enzymes within lysosomes. Do you still believe that?

Sly: Yes. I believe that lysosomal acid phosphatase cleaves the phosphate from the recognition marker, and this forms a trap for delivered enzyme in the lysosome.

Sabatini: In I-cell disease do the non-phosphorylated enzymes become terminally glycosylated?

Kornfeld: Yes.

Sabatini: In that case, terminal glycosylation is probably normally prevented by the phosphate. Do you agree?

Kornfeld: I think that is true.

Sabatini: This then further supports the idea that the phosphate must be added at a site that precedes the one where terminal glycosylation takes place.

Kornfeld: Yes it does.

Cohn: What are the rates of synthesis of some of these enzymes in I-cell disease compared with normal cells?

Kornfeld: The studies of Vladutiu & Rattazzi (1979) would indicate that the rate of lysosomal enzyme synthesis in I-cell disease is normal.

Cohn: Has this been done carefully with specific antibodies against the enzyme?

Kornfeld: No; the studies have not been done using antibodies. These investigators followed the increase in total enzyme activity of growing fibroblasts and found that I-cell fibroblasts appear to make the same amount of enzyme as normal fibroblasts but secrete most of it into the medium.

Geisow: Do you know whether tunicamycin causes secretion of newly synthesized lysosomal enzymes? Since tunicamycin blocks addition of carbohydrate, the enzymes might be released.

Kornfeld: We haven't done those experiments with lysosomal enzymes but one would predict that tunicamycin should cause secretion.

Sabatini: We have recently done that and found that tunicamycin causes massive secretion of completely unprocessed lysosomal polypeptides (Rosenfeld et al 1982).

Geisow: It has been found in many systems, although not exclusively, that blocking glycosylation by tunicamycin or 2-deoxyglucose causes proteolysis to occur in non-lysosomal secretory proteins. Could this be because failure of a recognition step causes the newly synthesized acid hydrolases to enter a secretory pathway, rather than to be picked up and taken to the lysosome?

Kornfeld: Possibly, although the pH in the secretory vesicles would not be optimal for lysosomal proteases. The susceptibility of non-glycosylated glycoproteins to proteolysis appears to be mainly a function of the particular protein involved. The results obtained with tunicamycin are extremely variable depending on the system studied. Even within the same system the results may be variable. For instance, in one strain of vesicular stomatitis virus (VSV), tunicamycin totally blocks the migration of the G protein to the surface because the protein aggregates, whereas with another variant of VSV the G protein is expressed normally on the cell surface in the presence of tunicamycin (Gibson et al 1979). It appears that each individual glycoprotein has its own requirement for glycosylation in terms of its ability to maintain its normal physical properties.

Geisow: In some cases, however, the oligosaccharide chain is added to a large molecule, so one would imagine that there was minimal perturbation of conformation of that molecule.

Sly: What is your tentative assignment of the processing steps and the binding to receptors? Do you believe that the phosphotransferase is a late endoplasmic reticulum (ER) enzyme?

Kornfeld: Our studies of enzyme localization utilize membranes separated on the basis of density, and we cannot extrapolate from these experiments to the morphological compartment without other evidence. The *N*-acetylglucosaminylphosphotransferase could be in the transitional zone of the ER or in the cis portion of the Golgi. I don't think it is in the trans Golgi.

Sly: Where do you think the receptors are occupied?

Kornfeld: We don't know for sure, but our kinetic data are consistent with the molecules going onto the receptor when they reach the trans Golgi.

Hubbard: How does the receptor distribute in that gradient?

Kornfeld: We are just starting to do those experiments.

Sabatini: I would interpret the findings of Dr Sly to mean that the receptor is mainly in microsomes derived from the ER. If that were true then, teleologically, I would expect the phosphate to be uncovered in the ER.

Sly: We have some fractionation data which suggested that the bulk of the receptor was in the ER, or at least sedimented with the ER, and that that receptor was occupied (Fischer et al 1980). If uncovering of the phosphate doesn't occur until the Golgi, and uncovering is required to bind to receptors, I believe that our interpretation of those experiments may be wrong, and perhaps lysosomes contaminated the fraction that we called ER. I hope that Dr Kornfeld's fractionation work will provide an answer to this question, but it is not clear yet.

Kornfeld: Phosphate uncovering doesn't occur until 30–40 min after the protein has been made, so if the receptors are in the ER the protein has to be sitting in the ER all this time. In addition it appears that mannose trimming also occurs before the molecules go onto the receptor. Since our data indicate that the 'uncovering' enzyme and the processing α-mannosidase are localized in the Golgi elements, it seems unlikely that the newly synthesized lysosomal enzymes go onto the receptor in the ER.

Sabatini: Have you looked at the effect of carbonyl cyanide *m*-chlorophenylhydrazone (CCCP) on these processes?

Kornfeld: We are just starting to do that.

Jourdian: Binding of oligosaccharides, or glycoproteins containing mannose 6-phosphate blocked with *N*-acetylglucosamine, to receptors is extremely poor. If one chemically synthesizes D-mannopyranose-6-(2-acetamido-2-deoxy-α-D-glucopyranosyl phosphate) one finds that it is a very poor inhibitor of binding. These results suggest that removal of the *N*-acetylglucosamine residue is required for binding.

Rothman: Dr Kornfeld said that he isolated 'hybrid' molecules from secretions and it has just been suggested that hybrids are better substrates for binding. Are such hybrid oligosaccharides found in lysosomes; and are they found in larger amounts in secretions than other phosphorylated forms that lack terminal sugars?

Kornfeld: We have looked carefully for these species only in secretions. We are now starting to look in lysosomes, but that is more difficult because as soon as the molecules reach the lysosome the phosphate is removed. One therefore cannot accumulate these species in lysosomes and then purify them.

Sly: Then why is the hybrid binding better?

Jourdian: It seems to bind better than the isozymic forms that carry mannose-rich chains which contain exposed mannose 6-phosphate residues.

Sly: Is it better than the binding of oligosaccharides that contain two uncovered phosphates?

Jourdian: I can't answer that.

Kornfeld: What are you comparing it with?

Jourdian: We have eluted highly purified enzyme from diethylaminoethyl cellulose (DEAE) with increasing concentrations of sodium chloride, and we have arbitrarily divided the eluted enzyme into fractions that exhibit low, intermediate, high and very high uptake when added exogenously to cultures of skin fibroblasts (Sahagian et al 1981). We compared the carbohydrate content of the low and intermediate fractions to that of the very high-uptake fractions and found that the oligosaccharide portion of the very high-uptake fractions released with endohexosaminidase H contained increased concentrations of sialic acid and galactose (J. Distler et al, unpublished work).

Kornfeld: But you are not determining which oligosaccharide on the molecules is responsible for the high-uptake behaviour. Those same molecules might also have species with two uncovered phosphates.

Jourdian: That is possible.

Cohn: Does the secreted molecule often have phosphate associated with it?

Kornfeld: Yes.

Cohn: Do you think, then, that the secreted molecule never 'sees' the lysosome and that it goes by a separate intracellular pathway?

Kornfeld: That probably varies with the cell line being studied. We have recently made an observation that seems to complicate the issue considerably. We have found that the mouse $P388D_1$ cells do not have detectable mannose 6-phosphate binding protein, and while they secrete half the enzyme that they make, they retain the other half, presumably in lysosomes, although we haven't isolated lysosomes from these cells. In other cell types that *do* have detectable receptor, very little enzyme is secreted. In addition, in I-cell disease several cell types seem to have normal levels of lysosomal enzymes even though the phosphorylating enzyme is absent. So there may be alternative ways of getting acid hydrolases into lysosomes, including pathways that are independent of the phosphomannosyl recognition marker. We are starting to screen a variety of cell types to determine which of them have mannose 6-phosphate receptor and how that correlates with secretion.

Sabatini: Randy Schekman has worked with yeast cells, which contain a vacuole that may be equivalent to a lysosome, because it contains acid hydrolases which I believe bear phosphate groups linked to mannoses. He has characterized many interesting mutants that are unable to secrete proteins at the cell surface (Novick et al 1981, Esmon et al 1981). He has found that secretory mutants are still capable of segregating carboxypeptidase Y to the vacuole. This suggests that passage of this hydrolase to the cell surface is not required for it to reach lysosomes. In addition, I understand that there are mutant yeasts which are unable to add the phosphate and yet the enzyme still reaches the vacuoles.

Kornfeld: Do you know if those vacuoles contain only lysosomal enzymes?

Sabatini: I don't know, and one could question whether they can be regarded as equivalent to lysosomes.

Palade: How long is the interval between the time of enzyme synthesis and the time of arrival of the same enzyme in the lysosomes?

Kornfeld: We have not directly traced the enzymes into lysosomes. The best marker we have of arrival in lysosomes is the loss of phosphate from the oligosaccharides. If one accepts that the loss occurs in the lysosomes, then it takes approximately one hour for newly synthesized enzyme to reach the lysosomes. This is consistent with the studies of Skudlarek & Swank (1981) who showed that the loss of the pro-piece from newly synthesized enzyme begins at about one hour.

Sabatini: But that evidence applies only to β-glucuronidase (Rosenfeld et al 1982). For other enzymes it takes hours and hours to completely remove the pro-pieces.

Palade: The evidence for rapid transit along the secretory pathway is obtained from certain cell types that produce secretory proteins and discharge them continuously into the extracellular medium. Such proteins begin to appear within the medium after considerably shorter times (15–20 min). Therefore, lysosomal enzymes must be contained and delayed in some compartment (perhaps the Golgi) which is different from the compartments involved in the secretory traffic. When exactly is the G protein acylated?

Kornfeld: I don't know for this cell line.

Rothman: In Chinese hamster ovary cells fatty acid is added almost precisely 10 min after synthesis. Our studies (reported in my paper, p 120-137) indicate that at least two glucose residues are removed before the fatty acid is added, but that the fatty acid is added before any mannoses are removed.

REFERENCES

Esmon B, Novick P, Schekman R 1981 Compartmentalized assembly of oligosaccharides on exported glycoproteins. Cell 25:451-460

Fischer HD, Gonzalez-Noriega A, Sly WS, Morré DJ 1980 Phosphomannosyl-enzyme receptors in rat liver. J Biol Chem 255:9608-9615

Friend DS, Farquhar MG 1967 Function of coated vesicles during protein adsorption in the rat vas deferens. J Cell Biol 35:357-376

Gibson R, Schlesinger S, Kornfeld S 1979 The nonglycosylated glycoprotein of vesicular stomatitis virus is temperature-sensitive and undergoes intracellular aggregation at elevated temperatures. J Biol Chem 254:3600-3607

Jamieson JD, Palade GE 1967 Intracellular transport of secretory proteins in the pancreatic exocrine cell. I: Role of the peripheral elements of the Golgi complex. J Cell Biol 34:577-596

Jamieson JD, Palade GE 1968 Intracellular transport of secretory proteins in the pancreatic exocrine cell. III: Dissociation of intracellular transport from protein synthesis. J Cell Biol 38:580-588

Katz B 1969 The release of neurotransmitter substances. Liverpool University Press, Liverpool

Novick P, Ferro S, Schekman R 1981 Order of events in the yeast secretory pathway. Cell 25:461-470

Rosenfeld MG, Kreibich G, Popov D, Kato K, Sabatini DD 1982 Biosynthesis of lysosomal hydrolases: their synthesis in bound polysomes and the role of co- and posttranslational processing in determining their subcellular distribution. J Cell Biol 92:135-143

Sahagian GG, Distler J, Jourdian GW 1981 Characterization of a membrane-associated receptor from bovine liver that binds phosphomannosyl residues of bovine testicular β-galactosidase. Proc Natl Acad Sci USA 78:4289-4293

Skudlarek MD, Swank RT 1981 Turnover of two lysosomal enzymes in macrophages. J Biol Chem 256:10137-10144

Vladutiu GD, Rattazzi M 1979 Excretion–reuptake route of β-hexosaminidase in normal and I-cell disease cultured fibroblasts. J Clin Invest 63:595-601

Membrane recycling in secretory cells: pathway to the Golgi complex

MARILYN GIST FARQUHAR*

Section of Cell Biology, Yale University School of Medicine, 333 Cedar Street, P.O. Box 3333, New Haven, Connecticut 06510, USA

Abstract The pathway taken by membrane that is recovered by endocytosis from the surface of secretory cells was investigated with electron-dense tracers (dextrans and cationized ferritin). The cell types examined included exocrine cells of the parotid and lacrimal glands, endocrine cells of the anterior pituitary gland, and immunoglobulin-secreting cells from lymph nodes or myeloma cell lines. In all cases, when the cells were incubated at 37 °C the tracers were initially taken up by endocytosis and they later appeared in the stacked Golgi cisternae, in immature secretion granules or vacuoles and in lysosomes. Similar results were obtained after covalent labelling of surface membrane constituents when myeloma cells were radioiodinated and the fate of the labelled components was followed by autoradiography. Initially only the cell surface was labelled, and the autoradiographic grains were concentrated over the plasmalemma. After incubation at 37 °C some of the labelled components were internalized (by endocytosis), and the majority of the internal autoradiographic grains were found over Golgi cisternae and over associated secretory vacuoles, which were the only organelles significantly labelled. The findings indicate the existence of considerable membrane traffic from the plasmalemma to the stacked Golgi cisternae and forming secretion granules or vacuoles in all these cell types. Membrane is thus continually recovered from the cell surface of secretory cells and funnelled through the Golgi complex; moreover, the plasmalemma-to-Golgi traffic appears to represent a major route of membrane traffic in secretory cells. A large portion of this traffic appears to be associated with the recycling of the membrane containers used in the packaging of secretory products.

Multiple pathways of membrane traffic exist for exocytosis, endocytosis and membrane recycling in various cells, and specific pathways are amplified in individual cell types according to their functions. Our interest has centred on

---

*Unable to attend the symposium because of illness. Paper read by Dr G. E. Palade, who also led the discussion that follows.

1982 Membrane recycling. Pitman Books Ltd, London (Ciba Foundation symposium 92) p 157-183

the traffic of membrane and secretory products between the Golgi complex and the plasmalemma. This traffic occurs most extensively in secretory cells and is believed to be related—at least in part—to recycling of secretory granule membranes. This paper summarizes our work on this topic, and will serve as an introduction to interested readers. Those who wish to delve further into the topic can consult a recent review (Farquhar 1981) or the original references for more detailed information.

Background information

For the past few years, we have been studying the pathways taken by endocytosed surface membrane in secretory cells. We were interested in determining the fate of the granule membrane inserted into the plasmalemma at the time of exocytosis. Release of secretory products by exocytosis involves the continual insertion into the plasmalemma of a considerable amount of membrane derived from secretory granules or vacuoles. At the time that exocytosis was first discovered by Palade (1959), it was recognized that membrane must be continually removed from the cell surface to compensate for that added during exocytosis in order to maintain a constant cell size. To explain the mechanism involved it has been variously proposed that membrane is recycled, or dismantled to macromolecular components that are subsequently reassembled, or degraded and synthesized anew. The evidence has been contradictory, but recycling had generally been ruled out until a few years ago because of results obtained using electron-dense tracers, usually horseradish peroxidase and native ferritin (which we now know to be primarily content markers rather than membrane markers). When exocytosis was stimulated in the presence of tracers, in various secretory cells, the tracer was found initially in endocytic vesicles and later in lysosomes. The results of such experiments established that, after exocytosis, surface membrane is recovered intact, in the form of small vesicles. However, because the tracers eventually became concentrated in lysosomes the majority of workers concluded that the recovered surface membrane is degraded in lysosomes. These tracer data appeared to correlate with the early turnover data, indicating that the proteins of the granule membranes and their contents turn over at the same rate. Hence it became widely accepted until recently (see Holtzman et al 1977) that granule membranes are recovered intact and degraded rather than being reutilized.

Several years ago we (Farquhar et al 1975) and others obtained results using horseradish peroxidase in pituitary and other cells which indicated that, in addition to its presence in lysosomes, a small amount of the tracer became concentrated in elements of the Golgi complex. Subsequently we began to study membrane traffic systematically in secretory cells, and turned to tracers other than native ferritin or horseradish peroxidase. We have since obtained evidence on several cell types for the existence of considerable membrane traffic from the cell surface to the Golgi complex in secretory cells.

Results obtained with dextrans on exocrine cells of parotid and lacrimal glands

Our first informative results came from experiments with Volker Herzog on exocrine cells from the parotid and lacrimal glands, in which we used dextrans as tracers (Herzog & Farquhar 1977). Dextrans were selected because in contrast to protein tracers they are uncharged and relatively inert polysaccharide molecules. When we incubated isolated acini from the parotid or lacrimal glands with dextrans *in vitro*, or when we infused dextrans up the parotid duct *in vivo* after stimulating granule discharge (by using isoproterenol for the parotid and carbamylcholine for the lacrimal gland cells) we found, as in previous work, that exocytosis was followed by endocytosis. Particles of the tracer were taken up via coated pits (Fig. 2) into smooth-surfaced apical vesicles (Fig. 1). However, the fate of the tracer differed from that detected previously: the tracer was found not only in lysosomes, but also in multiple stacked Golgi cisternae (Fig. 3), concentrated in their dilated rims, and in condensing vacuoles in the Golgi region, suggesting that the incoming vesicles had fused with all these compartments.

The novel findings include, first, the demonstration that the tracer can reach—directly or indirectly—the Golgi stack and, second, the demonstration of the rapidity with which movement of vesicles and fusion with Golgi cisternae occurs—i.e. within 5 min in the *in vivo* experiments (Fig. 3).

Results obtained with cationic ferritin on anterior pituitary cells

The next observations were made on the mammotroph or prolactin-producing cells of the rat anterior pituitary gland by using cationized ferritin as tracer (Farquhar 1978a). We chose this tracer because its net positive charge (isoelectric pt$>$7.8) causes it to bind avidly to the cell membrane and it therefore might be expected to act as a membrane marker under appropriate conditions. We chose to study the pituitary mammotroph because for some time we had been using it as a model for investigating secretory processes in endocrine cells and we and others had consequently accumulated a good deal of information on mechanisms and pathways of secretion in this cell type (see Farquhar 1977). From morphological observations and autoradiographic findings (Farquhar et al 1978) we know that the secretory product, prolactin,

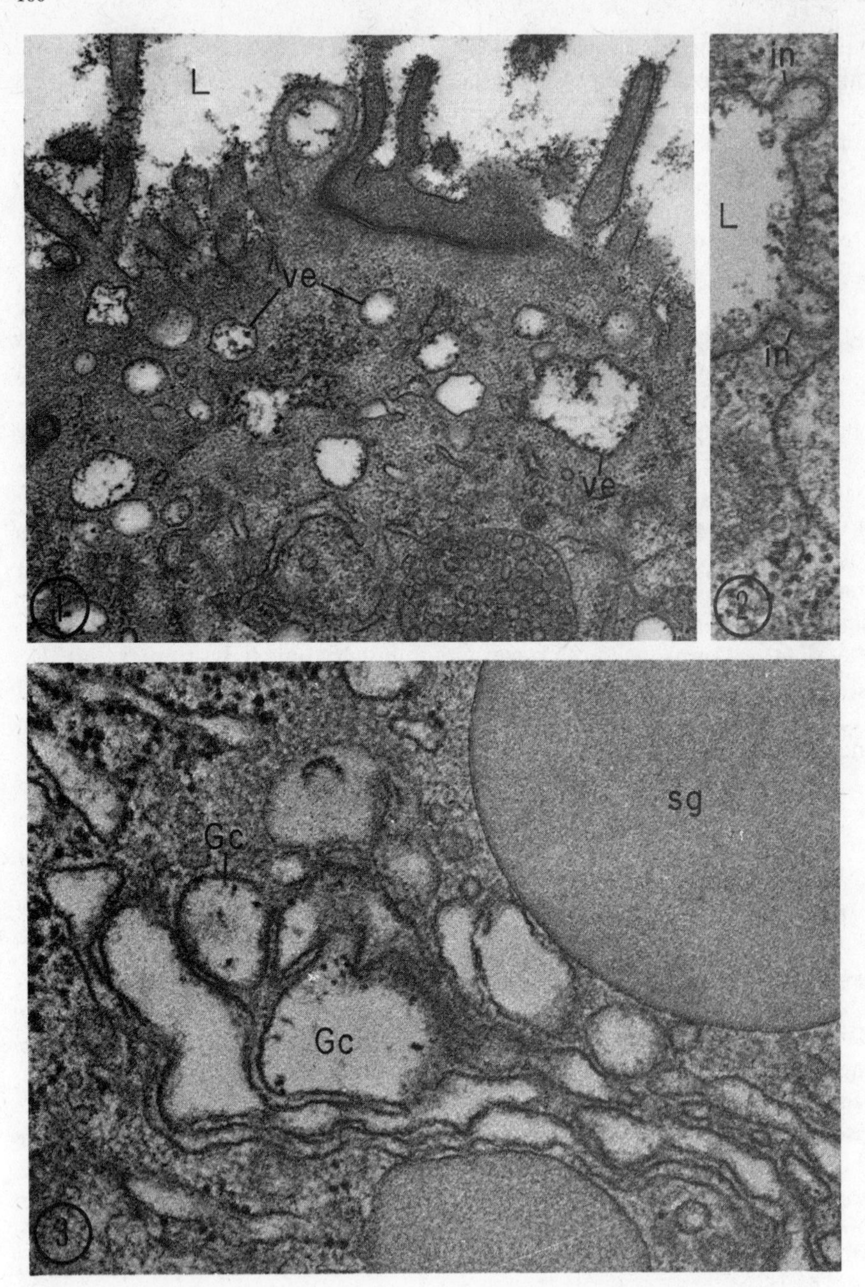

FIGS. 1–3. Results obtained using dextrans to trace the fate of surface membrane in cells from the lacrimal (Fig. 1) and parotid (Figs. 2 and 3) glands. The illustrations show that after exocytosis membrane is recovered by endocytosis, and the resultant endocytic vesicles subse-

is concentrated into granules in the dilated rims of the cisternae closest to the trans side of the Golgi stack. These granules are assembled into mature granules over a period of 1–3 h. The mature secretion granule is eventually discharged by exocytosis or, when secretion is shut off, it can fuse with, and be destroyed in, lysosomes by crinophagy (Smith & Farquhar 1966, Farquhar 1969, 1977).

Mammotrophs can be stimulated by pretreatment of the animals *in vivo* with oestrogen, which leads to increased prolactin secretion (see Farquhar et al 1978). When we incubated a suspension of dissociated pituitary cells, containing stimulated mammotrophs, with cationized ferritin at 0 or 25 °C, we found that the tracer immediately bound to the cell surface in a continuous layer 1–2 molecules deep (Fig. 4). If the cells were subsequently incubated at 37 °C, the findings were very similar to those obtained on parotid and lacrimal cells: cationized ferritin was taken up by endocytosis and appeared in the stacked Golgi cisternae, in forming granules and in lysosomes (Figs. 5–7). Molecules of the tracer were often seen around secretion granules that were condensing within the cisternae closest to the trans side of the Golgi stack (Fig. 6), indicating that vesicles containing the tracer had fused with the dilated rims of the trans-most cisternae. Although cationized ferritin was found in all cisternae of the stack, often within multiple cisternae (Fig. 7), it was most abundant in the trans-most cisternae. Some of the cationized ferritin was located close to the membranes of the transporting vesicles and Golgi cisternae, whereas some of it appeared to form part of the contents (Figs. 7–8).

In the granules, cationized ferritin was found initially (i.e. before 30 min) only in early, forming granules located within or close to the trans-most cisterna. Only later (after 30 min) was it found in polymorphous aggregating granules or in mature granules (after 2–3 h). Typically the tracer was seen in multiple granules and was located in a peripheral rim between the membrane and the dense contents (Fig. 5).

Cationized ferritin was seen in larger amounts in the Golgi cisternae and in granules of mammotrophs that had been obtained from oestrogen-treated females, which are known to be producing high levels of prolactin (Farquhar et al 1978). These cells were rapidly synthesizing and discharging prolactin, as

---

quently fuse with Golgi cisternae. Fig. 1 is from an acinus isolated from the rat lacrimal gland and incubated *in vitro* with dextran (T-10) for 10 min. Multiple endocytic vesicles (ve) containing dextran are seen in the apical cytoplasm. Figs. 2 and 3 are from parotid acinar cells of rats given isoproterenol (to stimulate exocytosis), followed by an infusion of dextran T-40 into the duct, and fixed 5 min after the infusion. Dextran is seen within coated invaginations (in) of the luminal (L) cell membrane and within several of the stacked Golgi cisternae (Gc). sg, secretion granule. Magnification: Fig. 1, × 17 000; Fig. 2, × 48 000; Fig. 3, × 57 000. From Herzog & Farquhar (1977).

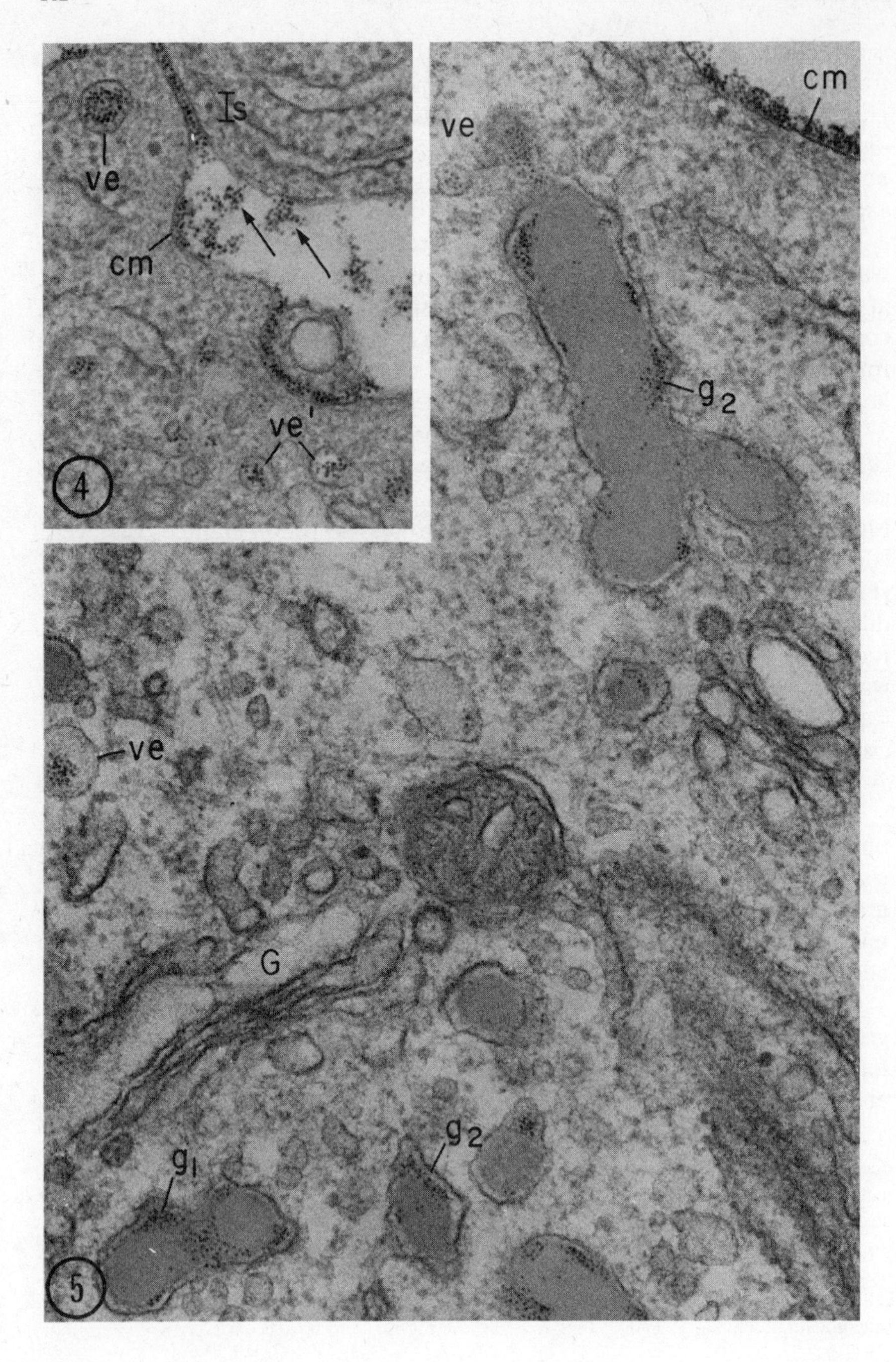
Is
ve
cm
ve'
4
ve
cm
g2
ve
G
g1
g2
5

indicated by the presence of many forming secretion granules and few stored mature granules. Thus, the uptake of ferritin into compartments along the secretory pathway was greater in stimulated mammotrophs than in mammotrophs from normal cycling females.

The same findings pertained to other cell types. Fig. 8 shows a somatotroph, or growth-hormone-producing cell, incubated with the tracer for 1 h. Cationized ferritin is present throughout the stacked Golgi cisternae, in numerous vesicles, and around all the many granules formed during the incubation period. Such a distribution indicates that endocytic vesicles carrying the tracer have fused with Golgi cisternae during the formation of all these granules.

These results led to the following conclusions (summarized in Fig. 11, p 170): (1) after exocytosis membrane is recovered by endocytosis and the resultant endocytic vesicles fuse with multiple stacked Golgi cisternae, lysosomes, and with forming granules; (2) although they can fuse with all the stacked Golgi cisternae, the vesicles preferentially fuse with the trans-most cisternae; (3) fusion with mature granules is infrequent; and (4) fusion with compartments other than Golgi, lysosomes or forming granules (e.g. endoplasmic reticulum, mitochondria, nuclear envelope) is non-existent. The fact that the incoming vesicular traffic is directed primarily to the trans-most cisternae—the very same cisternae from which the granules arise—strongly suggests that a major portion of this traffic is connected with the recycling of granule membranes.

Results obtained with native ferritin and cationized ferritin (pI <7.8) on anterior pituitary cells

When native or anionic ferritin (isoelectric pt, pI, approx. 4.8) was used as the tracer under similar conditions and at concentrations up to 10 times that

FIGS. 4–5. Mammotrophs, from oestrogen-treated female rats, incubated with cationized ferritin (CF) (0.1 mg ml^{-1}) to trace the surface membrane. Fig. 4 shows that initially (after 15 min of incubation at 37 °C) CF binds to the cell membrane (cm) and is taken up by endocytosis. Numerous endocytic vesicles (ve) containing CF are seen in the cytoplasm near the plasmalemma. The CF is aggregated on the free cell surface (arrows) but forms a regular layer one to two molecules deep in the intercellular spaces (Is). Inside the vesicles (ve′) the CF is also aggregated; consequently, some CF molecules are closely associated with the vesicle membrane, whereas others are not. Fig. 5 shows that after longer periods of incubation (one hour at 37 °C) CF is seen within immature secretory granules (g_1 and g_2) as well as in endocytic vesicles (ve). The incoming CF-labelled vesicles apparently fuse preferentially with the dilated rims of the trans-most Golgi cisternae (see Fig. 6), and the CF becomes trapped within the forming granules, where it is typically located at the periphery of the dense content. cm, cell membrane; G, Golgi cisternae. Magnification: Fig. 4, × 70 000; Fig. 5, × 76 000. From Farquhar (1978a).

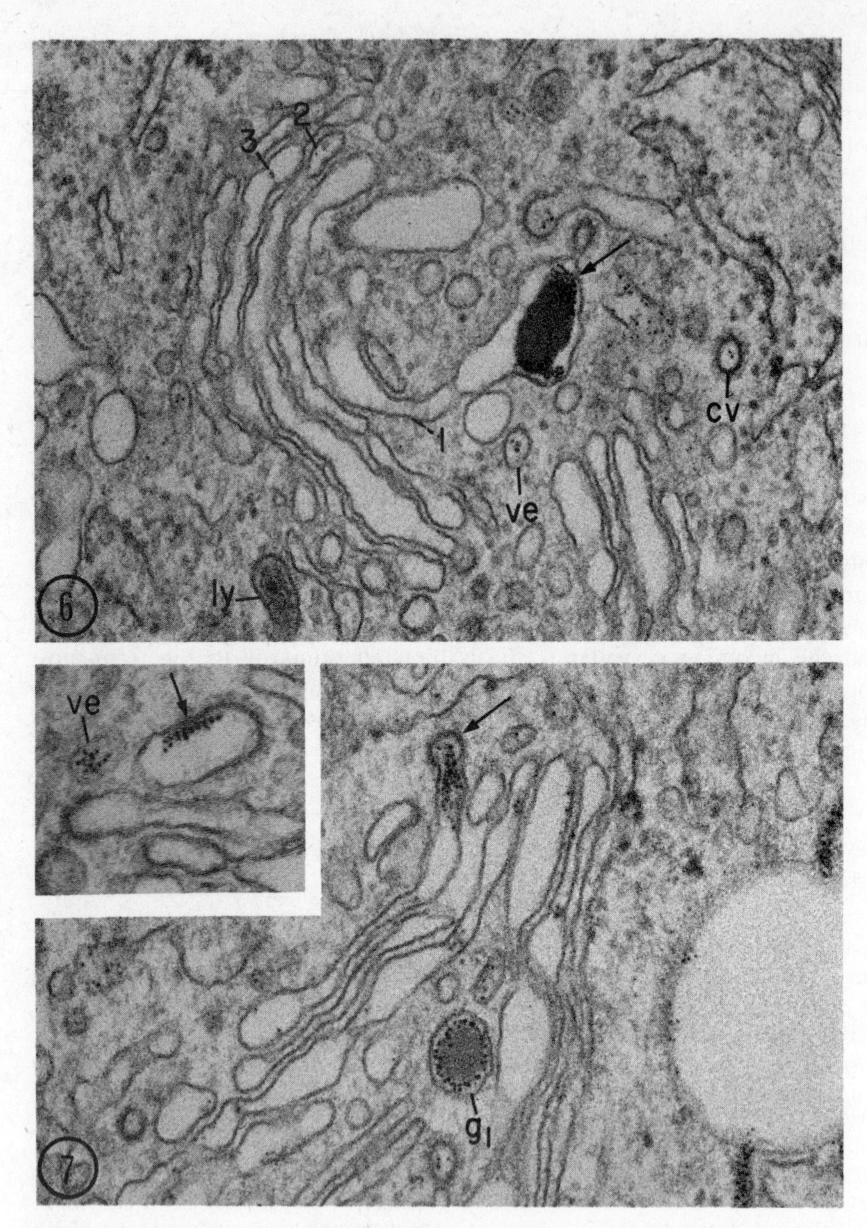

FIGS. 6 and 7. Fields from pituitary mammotrophs incubated with cationized ferritin (CF) for 60 min, showing that the tracer is taken up by endocytosis and appears within multiple Golgi cisternae. Fig. 6 shows CF within three Golgi cisternae (1–3). The tracer is particularly concentrated around a secretory granule that is forming within the trans-most Golgi cisterna (arrow), suggesting that the incoming vesicles carrying CF fuse preferentially with these Golgi of cationized ferritin, it did not bind to the cell membrane of pituitary cells, it was taken up in very small amounts by fluid-phase pinocytosis, and it was found only in lysosomes. It never appeared in the stacked Golgi cisternae or in forming granules of any cell type.

Experiments with cationized ferritins of different pI's clearly showed that the key factor in determining how the tracer is routed intracellularly is its ability to bind to the cell membrane: fractions with pI's of 4.8–7.3 do not bind to the cell membranes under the conditions tested, they act as fluid-phase or content markers, and they appear only in lysosomes. On the other hand, fractions with pI's greater than 7.8 bind to the cell membrane, act at least initially as membrane markers, and appear in Golgi cisternae and in forming secretory granules as well as in lysosomes.

The results obtained with differently charged ferritins raised the following question: are there two recovery routes (plasmalemma-to-Golgi, and plasmalemma-to-lysosomes), or one route with two stops (plasmalemma-to-lysosomes-to-Golgi)? The latter would require that the incoming vesicle should fuse with lysosomes where it loses its content, and the membrane carrying the marker should then move on to fuse with the Golgi cisternae. Our results did not allow us to choose between these two possibilities.

Further questions concerning the kinetics of internalization, and whether there are two recovery routes or one, are not so easily answered by work on a dissociated pituitary cell system, which is composed of multiple cell types and is limited in the amount of tissue available. Hence we searched for a system of secretory cells that would be more suitable for investigation of these problems, and we turned to myeloma cells in culture for this purpose.

Results obtained with cationized ferritin on myeloma cells

There is now abundant evidence that the production of immunoglobulins in plasma cells or myeloma cells involves the same operations and uses the same pathways as in exocrine and endocrine cells except that plasma cells or myeloma cells do not concentrate their secretory product into granules, but

---

elements. Note that the CF is concentrated at the periphery of the forming granule, adhering to its dense contents. CF is also seen within several vesicles (ve) one of which is coated (cv) in the Golgi region and within a lysosome (ly). Fig. 7 shows CF within multiple stacked Golgi cisternae and within a forming granule (g_1). One cisterna loaded with CF has a coated region at its tip (arrow), suggesting that a CF-loaded vesicle has just fused with the Golgi cisterna. The centre inset depicts another Golgi cisterna with a row of CF molecules attached to its membrane, which appears coated on part of its surface (arrow). Magnification: Fig. 6, × 70 000; Fig. 7, × 87 000; inset, × 100 000. From Farquhar (1978a).

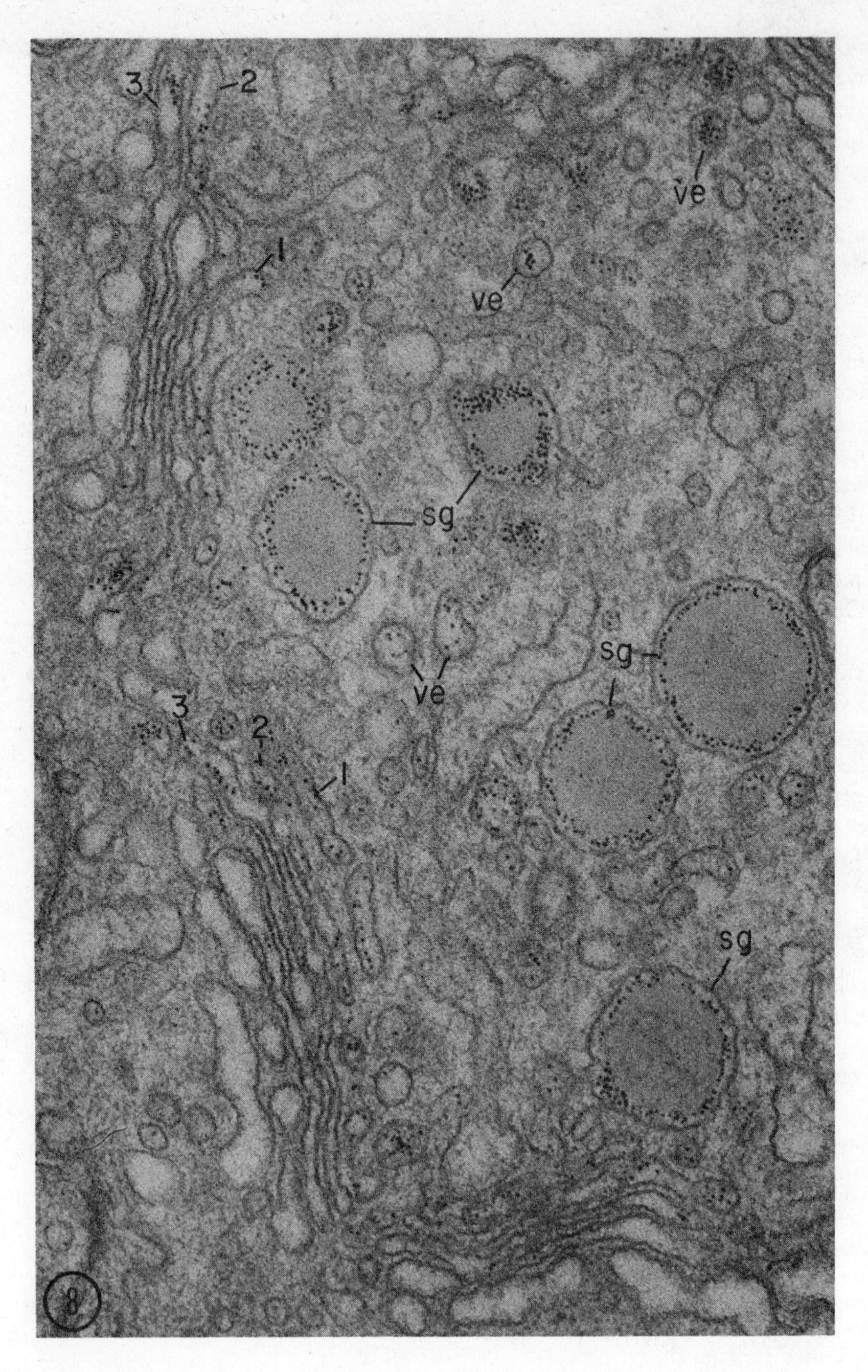
3
2
ve
1
ve
sg
sg
ve
sg
3
2
1
sg
8

instead package them in dilute solution in small vesicles or larger vacuoles (Palade 1975, Tartakoff & Vassalli 1977). Results obtained by Tartakoff and Vassalli indicate that cultured myeloma cells more or less continuously secrete immunoglobulins into the incubation medium. Therefore, it seemed likely that there might be extensive membrane traffic to the cell surface and back, connected with the recycling of containers for secretory products in these cells. If so, myeloma cells should represent a highly favourable system for study of membrane recycling because numerous established cell lines are available, and the cells can be easily maintained and grown in large quantities in culture. We selected two mouse cell lines to study—RPC 5.4 and X63 Ag 8—because they have conspicuous Golgi complexes and high rates of secretion of immunoglobulins (5–10% newly synthesized protein per hour).

When we did experiments with cationized ferritin on plasma cells from lymph nodes or on cultured myeloma cells under the same conditions as for anterior pituitary cells, we obtained essentially the same results (Ottosen et al 1980): cationized ferritin bound to the cell membrane and was taken up by endocytosis, primarily in coated vesicles (Fig. 9); the endocytic vesicles carrying the tracer subsequently fused with lysosomes, multiple stacked Golgi cisternae (Fig. 10), and with secretory vacuoles present in the Golgi region.

In myeloma cells in which the secretory compartments were identified by immunocytochemistry, the secretory product could be seen throughout the rough endoplasmic reticulum, Golgi cisternae and secretory vacuoles; moreover, the secretory product and the tracer could be identified within the same Golgi cisternae (Ottosen et al 1980).

These findings indicated that the myeloma cells have a plasmalemma-to-Golgi traffic similar to that found in exocrine and endocrine glandular cells. This traffic is presumably connected, at least in part, with the reutilization of containers for secretory products. Because the myeloma cells behave like glandular cells, they appear to present a suitable and promising system for further studies of mechanisms and pathways of membrane retrieval and recycling in secretory cells. Because cationized ferritin was found in both lysosomes and Golgi cisternae, as in pituitary cells, we could not choose, from these results, between the possibility of either one or two recovery routes.

FIG. 8. Somatotroph or growth hormone-secreting cell from a male rat incubated for 60 min in cationized ferritin (CF) (0.05 mg ml^{-1}), showing uptake of CF into multiple Golgi cisternae and secretion granules (sg). CF molecules are present within several stacked Golgi cisternae (1–3), multiple smooth vesicles (ve) in the Golgi region, and numerous secretion granules (sg) of varying size. Note that the CF molecules are located exclusively at the periphery of the dense granule contents. Magnification: × 85 000. From Farquhar (1978a).

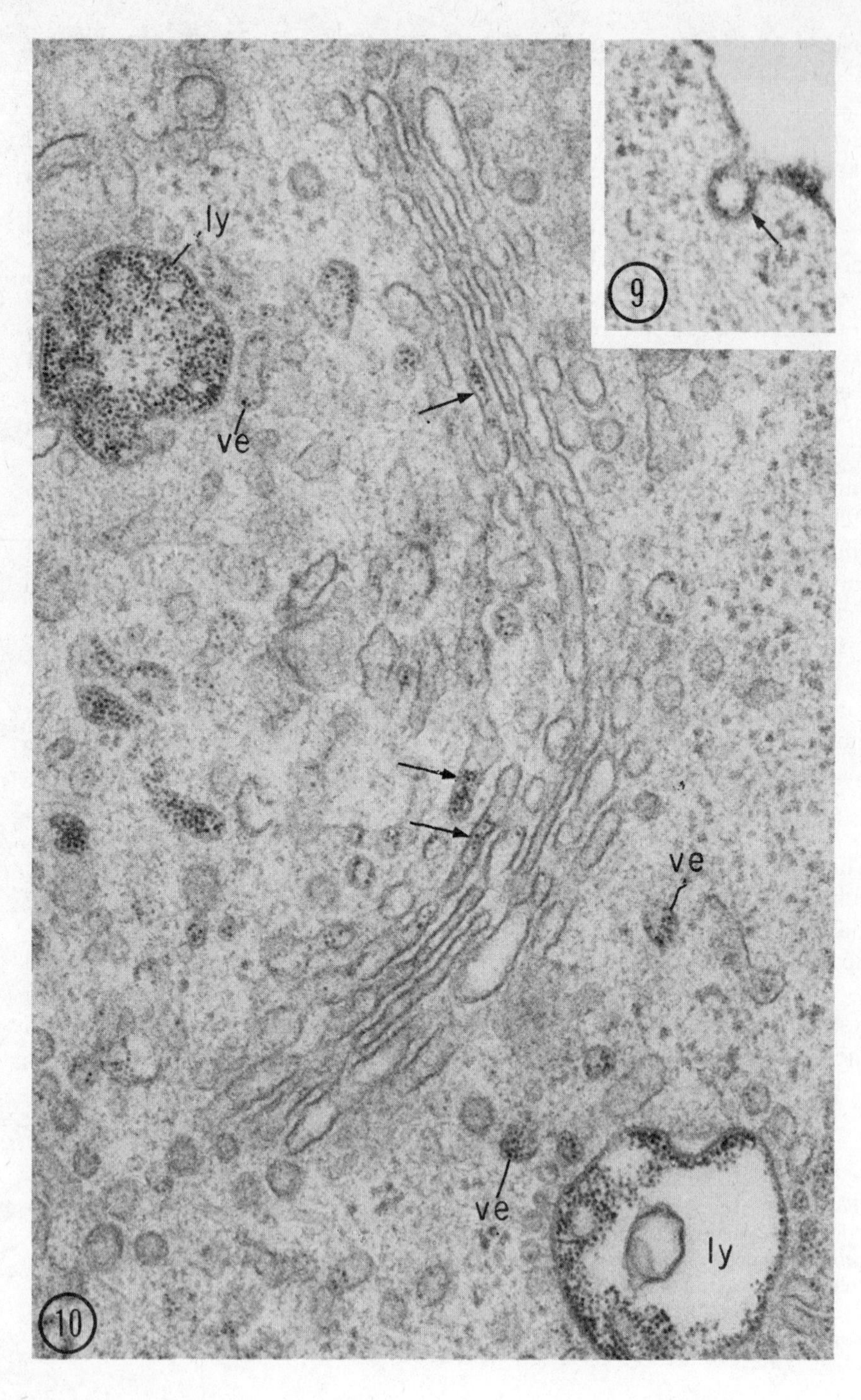
ly
ve
9
ve
ve
ly
10

Results obtained on radioiodinated myeloma cells

Cationized ferritin has been a useful tracer because it has allowed us to detect a previously unidentified route of membrane traffic directed to the Golgi complex in pituitary and immunoglobulin-secreting cells. However, it is not an ideal tracer because its binding to membranes depends on charge interaction, and it is susceptible to displacement or relocation if the vesicle fuses with a compartment in which there is competition for binding with acidic groups of higher charge density than those of the vesicle membrane. In addition, it appears to lead to increased numbers of lysosomes in myeloma cells and pituitary cells.

To avoid these difficulties and to check our results by a different approach, we decided to undertake alternative methods of membrane labelling by using covalently bound labels. In collaboration with Paul Wilson (Wilson et al 1981), we have recently radio-iodinated cell membrane constituents of myeloma cells (using the lactoperoxidase–glucose oxidase procedure) and have followed the fate of the labelled components by electron microscopic autoradiography. We found that immediately after iodination at 0 °C, the majority of the autoradiographic grains were associated with the plasmalemma. When the cells were subsequently incubated at 37 °C for periods up to 2 h, some of the grains were internalized, presumably by endocytosis, and many of these were associated with Golgi complex. Quantitative autoradiographic analysis revealed that, depending on the cell line studied, 25–50% of the internalized grains were associated with the Golgi complex, and were found either directly over Golgi cisternae and secretory vacuoles or within two half-distances (1600 Å) of them.

Calculations of the grain density, or the grains per unit area, which is a measure of the relative specific activity of labelled compartments, indicated that the Golgi complex was the only organelle that was significantly labelled. There was never any significant labelling of lysosomes. Interestingly enough, in radioiodinated cells, as in control cells, lysosomes are few in number (0.2% of the total cell area). They are much more numerous in cells treated with cationized ferritin which, as already mentioned, appears to lead to increased numbers of lysosomes.

FIGS. 9 and 10. Portions of mouse myeloma cells (RPC 5.4 cell line), fixed after incubation in cationized ferritin (CF) (0.05 mg ml^{-1}) for 10 min (Fig. 9) or 60 min (Fig. 10). Fig. 9 shows that CF binds to the cell surface and is taken up by endocytosis, often in coated invaginations (arrow), which subsequently fuse with multiple Golgi cisternae and with lysosomes. Fig. 10 shows CF in several stacked Golgi cisternae (arrows) and in small vesicles (ve) located near the Golgi cisternae or near lysosomes (ly). The two large lysosomes in the field also contain a considerable amount of CF. Magnification: Fig. 9, × 76 000; Fig. 10, × 57 000. From Ottosen et al (1980).

Thus, the results of our autoradiographic analysis of radioiodinated myeloma cells also demonstrate that cell membrane is recovered and funnelled through the Golgi complex in these cells. Because lysosomes are few in number and because so few grains are ever associated with them, it is unlikely that the majority of the endocytic vesicles make a lysosomal stop-over en route to the Golgi complex, thereby strongly suggesting the existence of a direct plasmalemma-to-Golgi route in these cells (Fig. 11).

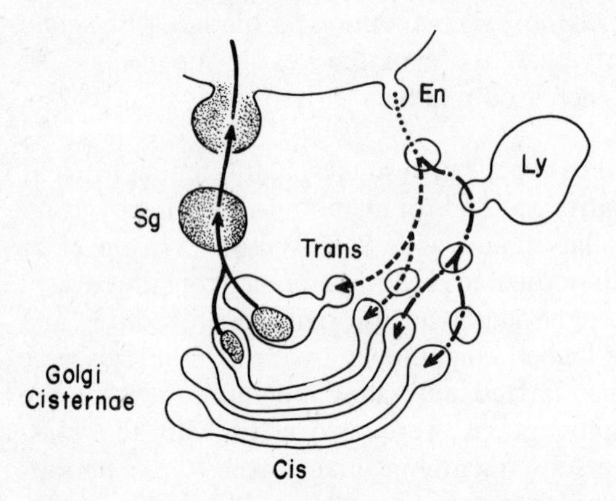

FIG. 11. Diagram showing two possible routes that could be taken by surface membrane to reach the stacked Golgi cisternae. Following exocytosis of secretory granules (solid line), patches of surface membrane are recovered by endocytosis and fuse with the dilated rims of multiple stacked Golgi cisternae. Membrane recovered by endocytosis can either fuse directly with the Golgi cisternae (dashed line) or fuse first with lysosomes and then with the Golgi cisternae (dotted and dashed line). The available evidence suggests that both routes may be used in different cell types.

Summary and conclusions

We have obtained data on six different secretory cell types, including exocrine, endocrine and immunoglobulin-secreting cells, by using electron-dense tracers which indicate that, after exocytosis, surface membrane is recovered and fuses with multiple stacked Golgi cisternae. We have also used alternative, covalent labelling procedures on myeloma cells, which lead to the same conclusion. In addition, we have found that Golgi traffic is most prevalent in secretory cells that are actively packaging secretory products, and the traffic is especially heavy to those portions of the Golgi (condensing vacuoles and dilated rims of trans-most cisternae) in which secretory products are normally concentrated and packaged. Similar data have been obtained by others on three other secretory cell types—i.e. the thyroid epithelial cell

(Herzog & Miller 1979) the exocrine pancreatic cell (Herzog & Reggio 1980) and the pancreatic β-cell (Orci et al 1978).

We have demonstrated the existence of considerable membrane traffic from the plasmalemma to the stacked Golgi cisternae and condensing granules in secretory cells. The most likely explanation for at least part of this traffic is that it represents the recovery of granule membrane that is subsequently reutilized in the packaging of secretory granules—i.e. that it represents recycling of these membranes. A number of findings support this assumption: (1) we have detected this pathway primarily in cells producing proteins for export, where such traffic could be expected to be heaviest; (2) traffic is most abundant in secretory cells actively concentrating and packaging secretory granules and is minimal or absent in secretory cells that are quiescent or not actively packaging; and (3) the traffic is heaviest to the trans Golgi components, where packaging of secretory products takes place. Moreover, the newer turnover data indicate that membrane proteins of the secretory granules turn over at a much slower rate than content proteins (Meldolesi 1974, Castle et al 1975, Wallach et al 1975). Therefore, although the evidence is still indirect, there seems to be little reason to doubt at present that the membranes of discharged secretory granules are recovered and reutilized in the packaging of successive generations of secretory granules.

Implications of the plasmalemma-to-Golgi traffic

The demonstration of considerable membrane traffic from the plasmalemma to the Golgi in secretory cells has a number of implications. First, the findings lead us to question the usual flow diagrams indicating the route taken by membranes and secretory products to and through the Golgi complex. According to the prevailing dogma, both membranes and secretory products are believed to move unidirectionally across the Golgi stack, being funnelled into the stack on one face (variously called the cis, entry or immature face), and emerging from it on the opposite face (trans, exit or mature). The cisternae are believed to be progressively displaced one by one, with the membranes being 'used up' as containers for secretory products on the trans side. New cisternae are believed to be assembled by coalescence of vesicles on the cis face. The fact that the main flow of the plasmalemma-to-Golgi traffic is to the dilated rims of the trans cisternae rather than to the cis side (as would be expected if a cis-to-trans flow pattern pertained) suggests that the main flow, both of secretory products from the rough endoplasmic reticulum and of membrane recycling from the cell surface, is to the dilated rims of multiple Golgi cisternae. We have discussed this problem in detail elsewhere (Farquhar 1978b, 1981, Farquhar & Palade 1981).

Secondly, the existence of pathway from the plasmalemma to the Golgi complex is potentially important in itself. A great deal of interest and excitement has been generated by the work of Goldstein et al (1982) on the role of traffic directed to lysosomes in the regulation of intracellular metabolic events through receptor-mediated endocytosis and intralysosomal digestion. The demonstration of a pathway from the cell surface to the Golgi is also exciting, and has broad implications, because it provides a route for molecules from the cell surface to reach a biosynthetic compartment with unique functions. The established functions of the Golgi complex that appear to belong exclusively to this organelle are: (1) packaging of secretory products; (2) terminal glycosylation of glycoproteins and glycolipids; (3) sulphation of glycosaminoglycans and glycoproteins; (4) proteolytic processing of proproteins; and (5) phosphorylation (see review by Farquhar & Palade 1981).

The pathway of membrane traffic from the plasmalemma to the Golgi complex provides a mechanism for various informational molecules, such as hormones and catecholamines, from the extracellular environment to reach the Golgi apparatus and to influence intracellular events. Indeed, the work of Jofesberg et al (1979) and Posner et al (1981) has already shown that insulin and prolactin reach the Golgi complex in hepatocytes, although the biological consequences of the uptake of these hormones into Golgi elements is not yet clear.

The pathway of recycling membrane traffic to the Golgi complex also provides a transit pathway whereby surface membrane components (receptors, enzymes and other proteins) could be modified or 'repaired' (e.g. reglycosylated, sulphated or phosphorylated) in passage. No examples of this phenomenon are yet available, but it could potentially take place if the molecules were brought into contact with the proper compartments or subcompartments of the Golgi complex.

These conjectures need to be experimentally tested, and to do so should keep many of us busy for some time to come!

Acknowledgements

The research summarized here was supported by research grant AM 17780 from the National Institutes of Health, Bethesda, Maryland (USA).

REFERENCES

Castle JD, Jamieson JD, Palade GE 1975 Secretion of granules of the rabbit parotid gland. Isolation, subfractionation, and characterization of the membrane and content subfractions. J Cell Biol 64:182-210

Farquhar MG 1969 Lysosome function in regulating secretion: disposal of secretory granules in cells of the anterior pituitary gland. In: Dingle JT, Fell HB (eds) Lysosomes in biology and pathology. North-Holland, Amsterdam, vol II, p 462-482

Farquhar MG 1977 Secretion and crinophagy in prolactin cells. Adv Exp Med Biol 80:37-94

Farquhar MG 1978a Recover of surface membrane in anterior pituitary cells. Variations in traffic detected with anionic and cationic ferritin. J Cell Biol 77:R35-R42

Farquhar MG 1978b Traffic of products and membranes through the Golgi complex. In: Silverstein SC (ed) Transport of macromolecules in cellular systems. Dahlem Konferenzen, Berlin, (Life Sci Res Rep 11) p 341-362

Farquhar MG 1981 Membrane recycling in secretory cells: implications for traffic of products and specialized membranes within the Golgi complex. Methods Cell Biol 23:399-427

Farquhar MG, Palade GE 1981 The Golgi apparatus (complex)—(1954–1981)—from artifact to center stage. J Cell Biol 91:77s-103s

Farquhar MG, Reid JA, Daniell L 1978 Intracellular transport and packaging of prolactin. A quantitative electron microscope autoradiographic study of mammotrophs dissociated from rat pituitaries. Endocrinology 102:296-311

Farquhar MG, Skutelsky E, Hopkins CR 1975 Structure and function of the anterior pituitary and dispersed pituitary cells. In: Tixier-Vidal A, Farquhar MG (eds) The anterior pituitary gland. Academic Press, New York, p 83-135

Goldstein JL, Anderson RGW, Brown MS 1982 Receptor-mediated endocytosis and the cellular uptake of low density lipoprotein. In: Membrane recycling. Pitman Books, London (Ciba Found Symp 92) p 77-95

Herzog V, Farquhar MG 1977 Luminal membrane retrieval after exocytosis reaches most Golgi cisternae in secretory cells. Proc Natl Acad Sci USA 74:5073-5077

Herzog V, Miller F 1979 Membrane retrieval in epithelial cells of isolated thyroid follicles. Eur J Cell Biol 19:203-215

Herzog V, Reggio H 1980 Pathways of endocytosis from luminal plasma membrane in rat exocrine pancreas. Eur J Cell Biol 21:141-150

Holtzman E, Schacher S, Evans J, Teichberg S 1977 Origin and fate of membranes of secretion granules and synaptic vesicles: membrane circulation in neurons, gland cells and retinal photoreceptors. Cell Surf Rev 4:166-246

Jofesberg Z, Posner BI, Patel B, Bergeron JJM 1979 The uptake of prolactin into female rat liver. Concentration of intact hormone in the Golgi apparatus. J Biol Chem 254:209-214

Meldolesi J 1974 Dynamics of cytoplasmic membranes in guinea pig pancreatic acinar cells. I: Synthesis and turnover of membrane proteins. J Cell Biol 61:1-13

Orci L, Perrelet A, Gorden P 1978 Less-understood aspects of the morphology of insulin secretion and binding. Rec Prog Horm Res 34:95-117

Ottosen PD, Courtoy PJ, Farquhar MG 1980 Pathways followed by membrane recovered from the surface of plasma cells and myeloma cells. J Exp Med 152:1-19

Palade GE 1959 Functional changes in the structure of cell components. In: Hayashi T (ed) Subcellular particles. Ronald Press, New York, p 64-80

Palade GE 1975 Intracellular aspects of the process of protein secretion. Science (Wash, DC) 189:347-358

Posner BI, Bergeron JJM, Josefsberg Z et al 1981 Polypeptide hormones: intracellular receptors and internalization. Rec Prog Horm Res 37:539-579
Smith RE, Farquhar MG 1966 Lysosome function in the regulation of the secretory process in cells of the anterior pituitary gland. J Cell Biol 31:319-347
Tartakoff AM, Vassalli P 1977 Plasma cell immunoglobulin secretion. Arrest is accompanied by alterations of the Golgi complex. J Exp Med 146:1332-1345
Wallach D, Kirshner N, Schramm M 1975 Non-parallel transport of membrane proteins and content proteins during assembly of the secretory granule in rat parotid gland. Biochim Biophys Acta 375:87-105
Wilson P, Sharkey D, Haynes N, Courtoy P, Farquhar MG 1981 Iodinated cell membrane components are transported to the Golgi complex. J Cell Biol 91:417a (abstr)

DISCUSSION

Rothman: The movement of tracers from the cell surface to the Golgi (into its trans cisternae, preferentially) suggests a route of entry (and exit) from the trans face. What are the implications for the proposal that movement through the Golgi occurs only by progression of whole cisternae from the cis to the trans face?

Palade: Until recently, the prevailing assumption was that membrane is added continuously to elements on the cis side of Golgi stacks, while being continuously removed from the trans side. This concept implied a vectorial movement of membranes in the cis-to-trans direction; membrane movement to and from the dilated rims of the Golgi cisternae was not considered significant. At present we have only one reliable example of cisternae moving progressively in the cis-to-trans direction. It comes from observations made on algae that produce decorative body scales. Brown & Willison (1977) and others have shown that these scales become progressively more complex as the cisternae approach the trans side of the stacks. Extensive recent work on a wide variety of cell types has not come up with other examples of vectorial cis-to-trans cisternal movement but has emphasized, instead, the active traffic of vesicular carriers to the side of Golgi stacks, i.e. to the dilated rims of Golgi cisternae (Farquhar & Palade 1981). Vesicular traffic from the plasmalemma appears to favour Golgi cisternae located in the trans part of the stacks, and traffic from the endoplasmic reticulum (ER) to the Golgi complex is supposed to favour the cisternae closest to the cis side of the stacks. But there is no evidence at present about the location within the stack of the boundary between the two vesicular circuits.

Sabatini: Was tracer observed also throughout the Golgi stacks?

Palade: Yes, in the work of Brown & Willison (1977).

Rothman: So if cisternae are formed from vesicles at the cis face of the stack, and then they leave the Golgi when they reach the trans end, one would not expect recycling from the plasma membrane into the trans face.

Palade: We can try to reconcile these apparently conflicting pieces of evidence while keeping in mind that our information is still not extensive. We may assume that we are dealing with two main types of movement that differ in rate and direction within the Golgi complex: a slow movement of cisternae along the cis-to-trans axis (in keeping with the findings on scale-producing algae) and a concomitant movement, at a considerably faster rate, by which membrane containers come to the sides of the piled Golgi cisternae from different sources, and depart for different destinations from the same (or different) Golgi locations. This second movement is connected with vesicular transport and membrane recycling. The first movement may represent an exception, rather than the rule. If it exists, it would have to be accompanied by extensive and sudden changes in membrane chemistry, since Golgi cisternae differ quite sharply from one another along the cis-to-trans axis within each stack.

Geisow: The results of Dr Farquhar do not necessarily prove that membrane is fed to the trans Golgi cisterna. The rate-determining step could be exit of membrane from the trans face, where a preferential accumulation would then occur. Unless very careful kinetic analyses were done (which is probably difficult on an electron microscopic time scale) one might just see a pile-up of material in the trans Golgi stack.

Palade: The advantage of electron microscopy for this type of study is that the compartments of interest can be directly and reliably identified. If the tracer is reasonably good, as is cationized ferritin, it will be detected in all compartments to which it has been brought by recycling vesicles. It will be present also in compartments that have acquired it from the original receiving compartment. This is probably true for secretion granules. The presence of the tracer in a given compartment does not necessarily mean that the membrane that brought it there is still there. It means, though, that the recycling membrane was there at some time.

Hopkins: What is the evidence from the experiments on cationized ferritin that secretion is necessary and that it is driving the system?

Palade: The mammotrophs in Dr Farquhar's (1978) experiments were isolated from the pituitaries of animals in which secretion of prolactin was stimulated by sustained oestrogen treatment. Lacrimal and parotid acinar cells (Herzog & Farquhar 1977) were stimulated by appropriate secretagogues (carbamylcholine and isoproterenol, respectively). The pancreatic exocrine cells in the experiments of Herzog & Reggio (1980) were also stimulated by secretagogues. The labelling of Golgi elements was very rapid in parotid and pancreatic exocine cells (<5 min) when exocytosis was induced by the appropriate secretagogues.

Meldolesi: Why should we stick to the idea that the Golgi cisternae move as a whole in the cis-to-trans direction? The evidence accumulated over the last few years demonstrates that membrane traffic in the cytoplasm of eukaryotic cells is much more rapid and extensive than previously envisaged. Such a traffic, and the ensuing movement of molecules and membrane patches, could easily account also for the apparently steady and extra-slow translocation of large structures through adjacent Golgi cisternae.

Palade: I agree. We still speak of a progressive movement of Golgi cisternae along the cis-to-trans direction because of inherited prejudices clearly expressed in the terminology that recognizes a forming face and a mature face to the Golgi stacks. Yet we have to keep in mind the evidence obtained by Brown & Willison (1977) and others on scale-producing algae. We have to rationalize it, instead of disregarding it.

Rothman: But one must also mention the recent evidence on G protein from J. E. Bergmann & S. J. Singer (unpublished results) showing by electron microscopic immunocytochemistry that G protein always enters a unique face of the Golgi from the ER. Moreover, the evidence from Dr Farquhar's work that we just heard strongly suggests that the exit from the Golgi stacks is preferentially at the trans face, where the recycling occurs. Those observations suggest that entry is at what is morphologically called the 'cis' face, and that exit is primarily from the 'trans' face of the Golgi stack.

Sabatini: I agree, but I believe Dr Palade was referring to the whole cisterna moving as a unit, and I find that concept difficult to reconcile with the idea that there are specific components of each cisterna that must be permanent if they confer any identity on the Golgi.

Rothman: Could there be a slow cis-to-trans movement of whole cisternae that is related to the way that the Golgi replicates and, superimposed on this, a fast cis-to-trans movement of individual molecules, which provides the rapid transport necessary for secretion, membrane synthesis, and so on?

Palade: If Golgi cisternae were indeed continuously added to the cis side while being removed from the trans side, it should be possible—in principle—to label biosynthetically the proteins of an individual Golgi cisterna and to follow its change in position within the stack as a function of time. The approach is inherently difficult since autoradiographic resolution may prove limiting, and since synthesis of membrane protein may be too slow to give a detectable signal. The labelling of content proteins would require extremely short, well-defined pulses, and membrane recycling may complicate the picture from the beginning. In any case, autoradiographic experiments provide no evidence yet for an initially preferential labelling of Golgi cisternae in a cis location.

Cohn: The places in the cell where the label accumulates could indicate areas in which the marker is being dissociated from the membrane, e.g. in the large accumulations in the lysosome. Perhaps the same thing happens in the trans segments of the Golgi apparatus. Autoradiographic information suggests a rapid circulation through the lysosomal plasma membrane compartment, which reaches equilibrium by the time of these autographs. Does this mean a *less* rapid movement from the Golgi element, since that is heavily labelled? In other words, are the two markers telling us something different about accumulation and rates?

Palade: The tracer may be detected primarily in compartments from which the rate of transport is slowed down, but it is also possible that it accumulates in a given compartment because it binds preferentially to another set of slow-moving anionic sites. This does happen in the secretory granules of the pituitary mammotrophs.

Another point deserving comment is that Dr Farquhar's results do not exclude transit of membrane carriers through the lysosomes in other cell systems; they simply indicate that in the myeloma cells used in these autoradiographic experiments recycling vesicles come preferentially to the Golgi complex. I should stress, however, that in Dr Farquhar's interpretation the evidence from these experiments is compatible with direct vesicular traffic from the plasmalemma to the Golgi complex (Wilson et al 1981). A two-station recovery pathway, with soluble tracers that are left behind in lysosomes, and membrane markers (and therefore membranes) that are brought back to the Golgi complex, is suggested by experiments on other cell types (Herzog & Miller 1979). But the evidence we have at present does not exclude a direct recovery pathway from the plasmalemma to the Golgi complex.

Cohn: Are there differences in the hydrogen ion concentrations in different faces of the Golgi?

Palade: That is anybody's guess at the moment!

Hopkins: Do the experiments insist that this label goes *through* the Golgi? For example, could the label in the granules arrive there directly from vesicles derived from the surface? Are the labelling kinetics such that the Golgi must lie between the surface and the granules?

Palade: The results obtained so far (Farquhar 1978, Herzog & Reggio 1980) indicate that on recovery the label appears first in Golgi elements and then in secretion granules.

Sabatini: You stated that the grains were found throughout the stacks of the cisternae, but some of the autoradiographs indicate that they were perhaps on one side of the stack of cisternae. Does the iodinated membrane in any way parallel the distribution of the content?

Palade: Autoradiographic resolution for ^{125}I is not high enough to provide a clear answer to such questions. In addition, it is often difficult to distinguish the cis from the trans side of Golgi stacks in myeloma cells. The same kind of difficulty was noticed by Bergmann et al (1981) in cultured cells.

Bretscher: Would it be useful to place a label in the lysosomes and to see whether it goes from there to the Golgi? Cationized ferritin could be derivatized with something like citraconic anhydride, which would make it negatively charged. This could be delivered to cells that are in chloroquine, when it would accumulate in lysosomes. Then, if one were to take away the chloroquine, the pH in lysosomes would drop and the citraconyl groups would come off and generate cationized ferritin. One could then see where that goes, and whether it goes to the Golgi.

Palade: This or an even better scheme could well be used in the future. We should recognize that at present the pathway from lysosomes to the Golgi complex is less clearly established and that the evidence is mostly indirect.

Geisow: We have said nothing about the other side, the cytoplasmic side, of all these membranes, which are fusing (presumably specifically). Dr Palade inferred that there must be markers to tell the membranes were to go next. Perhaps it it is really the cell architecture—physical barriers—that determines where the membranes go next. This really could apply both to the endocytic leg and the exocytic leg of the route of membrane through cells. An experimental approach to this idea would be the disassembly of the cytoskeleton of the cell, which we know rearranges all the internal apparatus.

Palade: Experimental disassembly of the cytoskeleton has been attempted, but the results obtained were varied and partly contradictory (see Redman et al 1975). I do not believe that the locomotor apparatus of the cell can impart sufficient specificity to vesicular traffic, although it may facilitate traffic by guiding vesicular carriers over long distances. Previous attempts to relegate traffic control to microtubules or microfilaments—which are supposed to direct secretion vacuoles to their proper destination for discharge, for instance—have produced questionable results. We should realize that specificity requirements in vesicular traffic are much more demanding than could be provided by the cytoskeletal system. Microfilaments, microtubules and their associated proteins may guide traffic over long distances, as in neurons and axons (Grafstein & Forman 1980), but it is unlikely that they can control delivery of vesicular carriers to their correct addresses.

Cohn: There has to be a motive force generated so that an endocytic vesicle 'flows' in a saltatory fashion through the cytoplasm and is directed towards this perinuclear station or *hof*. We don't understand what this represents in terms of associations between endocytic vesicles and contractile elements in the cell.

Meldolesi: Herzog & Reggio (1980) used two tracers—dextran and horseradish peroxidase (HRP)—in exactly the same conditions, and dextran reached the Golgi while HRP reached the lysosomes. Apparently the size of the carrier vesicles for the two tracers was different from the very beginning: one (dextran) was classically coated, fairly large, and addressed to the Golgi, and the other (HRP) was smaller, apparently uncoated and went to lysosomes. The tracer may therefore determine the features and final destiny of vesicles originating from exactly the same place in cells. Dr Farquhar has shown that, in omega-shaped plasmalemma infoldings that appear like discharge granules, one sometimes sees coated vesicles pinching off. The big issue is whether what is recycled is exactly the same membrane that had been added during exocytosis, or whether that membrane has somehow changed by intermixing with the rest of the plasmalemma before recycling. If this is so, where is the sorting step by which the recycled membrane regains the molecular features of granule membrane? The observations of coated pits in direct continuity with omega-shaped or flask-shaped images, which we consider to be discharge granules, is common in secretory cells. In particular, in the mouse exocrine pancreas, where coated pits of the luminal plasmalemma are extraordinarily numerous, over 80% of these coated pits are continuous with the membrane of discharged zymogen granules (Koike et al 1980). These observations suggest that the granule membrane can be recycled directly, and that the sorting step, which assures the fidelity of the recycled membrane with respect to the granule membrane, could be located at the cell surface and could rely on the molecular filtration ability of coated vesicles.

Sabatini: Dr Rothman, I believe, demonstrated specificity of transfer even in an *in vitro* system between different membrane compartments.

Rothman: Yes. We have a cell-free system (see this volume, p 120-137) in which G protein, donated specifically by the early compartment of the Golgi apparatus, is rapidly and efficiently processed by glycosyltransferases present in an exogenously supplied Golgi. This is almost surely a specific fusion or transport process. For example, even though the glycosyltransferases in the added Golgi apparatus are concentrated in a very small fraction of the crude membrane fraction added, the delivery is efficient and selective, as shown by the speed and efficiency of processing. If this reflects the specificity in the cell, it would demonstrate that pre-existing cellular structure is not required because membranes from one cell would have provided G protein, and Golgi from another cell would have processed it.

Geisow: Have you looked at this association morphologically with electron microscopy?

Rothman: We are trying to do that now.

Sly: What are the effects of cytoskeletal drugs like colchicine or cytochalasin B on vesicular stomatitis virus (VSV) maturation and secretion?

Rothman: Those agents have no effect, whether present separately or in combination (Rothman et al, unpublished results).

Sly: Does that suggest that the traffic is independent of the cytoskeleton?

Rothman: Of course, if it can be assumed that those drugs are acting on the cytoskeleton in the way that one supposes.

Sly: Does the disruption of the Golgi alter the traffic?

Rothman: In hepatic cells colchicine blocks secretion from Golgi to plasma membrane (Banerjee et al 1976). In most cases there is no positive evidence.

Hubbard: Haven't Dr Farquhar and Paul Wilson looked at the composition of the polypeptides that are iodinated in the myeloma system? Looking at grain distributions may be misleading, since we don't know whether a subset of iodinated membrane molecules is internalized. The question is how specific or general is the movement of membrane molecules into the Golgi from the plasma membrane.

Mellman: I have labelled closely related cells myself in other contexts. Many proteins are iodinated; and there doesn't seem to be any preferential labelling of surface immunoglobulin G (IgG), even if it exists on these cells. If only a highly restricted sub-set of proteins were participating in endocytosis and recycling (which I would not expect), it would seem that a net redistribution of grains to the Golgi region should not have been detected. In other words, if a few cell-surface polypeptides (representing less than 5% of the grains initially present on the plasma membrane) were relocated completely to the Golgi, there would have been an insignificant increase in grain density in that compartment. This might indicate that a surprisingly wide variety of plasma membrane proteins can go through that route.

Hubbard: The evidence says that the route exists but not that it represents membrane retrieval. Tartakoff et al (1981) have information suggesting that, in a myeloma cell that was secreting immunoglobulin, the extent of membrane movement, detected by content markers like cationized ferritin, was unchanged when secretion stopped. The question therefore is not about whether the Golgi is involved but about what, specifically, are the molecules that are moving into that region, and whether they represent retrieved molecules.

Palade: Your comment is essentially correct. To know exactly what is going on, we'll have to recover from the Golgi complex (and identify) membrane proteins labelled at the cell surface at the beginning of such experiments.

Hopkins: What about other cells that are clearly retrieving membrane from the surface, such as macrophages? How often will label like cationized ferritin find its way to the trans face of the Golgi?

Mellman: Thyberg (1980) obtained some limited evidence for the movement of cationized ferritin into Golgi stacks in macrophages, but the work requires verification.

Sabatini: Dr Farquhar and yourself (Farquhar & Palade 1981) have recently stressed that it is unclear what pathway within the Golgi apparatus the secretory products follow. This pathway may be different in different cells, because in some cases the granules appear at the rim of the cisterna and in others they are formed by condensing vacuoles. Dr Farquhar's work on HRP coupled to the antibody in immunoglobulin-secreting cells shows clearly that immunoglobulin goes through all the stages of the Golgi, because all cisternae are equally labelled.

Palade: But Ottosen et al (1980) were looking at IgG present in all the compartments of the secretory pathway of the cell. They were not looking at a well-defined cohort of newly synthesized IgG molecules moving in time along that pathway, as can be done (by autoradiography) in pulse–chase experiments. A limited amount of backflow is not excluded. In fact morphological evidence suggests that recovered membranes from secretion granules carry a small residue of secretory proteins back into the cell. We may assume, however, that cytochemical tests primarily reveal newly synthesized proteins from the endoplasmic reticulum because these proteins are expected to predominate quantitatively.

Widnell: Do you have any evidence for specific proteins in the membrane which are involved in the transport of the cationized ferritin? For example, changing the charge of certain proteins by adding polylysine stimulates their uptake into lysosomes (Shen & Ryser 1978).

Palade: No. At present there is no evidence for 'specific receptors' for cationized ferritin on the plasmalemma. We should not go away believing that the Golgi complex is as simple as it has been assumed during our discussions. Associated with the two or three cis-most cisternae is a strong reducing agent that reduces OsO_4 to lower oxides of osmium and to metallic osmium (Friend & Murray 1965). This reaction was used extensively to 'impregnate', and thereby to visualize, the Golgi complex in light microscopy. The thiamine pyrophosphatase (TPPase) activity mentioned by Dr Rothman is detected only in the trans-most cisterna and occasionally in its immediate neighbours (Novikoff & Goldfischer 1961). Usually a few cisternae (unstained by either procedure) are interposed between those impregnated by OsO_4 and those giving a positive reaction for TPPase (Friend 1969).

Another subcompartment with special cytochemical characteristics (positive reactions for acid phosphatase and aryl sulphatase) and special morphological features (a 'rigid lamella') is quite often found on the trans side of Golgi stacks at some distance from the TPPase-positive cisternae. This subcompartment has been described by Novikoff (1964) and Novikoff et al (1971) as a connection between the Golgi complex, the ER and the lysosomes; hence the acronym GERL (Golgi endoplasmic reticulum lysosome) often used to describe it. Further information about GERL and its assumed functions can be found in Novikoff (1976) and Farquhar & Palade (1981).

In our discussions, we have ignored this complex level of subcompartmentation and have superimposed on it the simpler level (the two subcompartments) proposed and discussed by Dr Rothman (1981). We still have to understand how the two types of subcompartmentation are related to one another. Moreover, when a single Golgi cisterna is considered in orthogonal view, certain components or enzymic activities are found to be differentially distributed between its flat centre-piece and its dilated rims (Farquhar et al 1974).

Coated vesicles are found on the cis as well as on the trans side of the stacks of Golgi cisternae. Those on the cis side are slightly different from the usual coated vesicles, in the sense that they appear to have a less well organized (or more difficult to preserve) coat. If ATP synthesis is arrested (by incubating pancreatic lobules under N_2 or in the presence of an uncoupler of oxidative phosphorylation), transport of secretory proteins from the ER to the Golgi complex is blocked, and a series of striking structural modifications appears concomitantly in the Golgi zone. Practically all peripheral vesicles, as well as the protrusions of the transitional ER elements, disappear from the cis side of the Golgi stacks; large deposits of fibrillar material appear in their place; many of these deposits consist of (or contain) recognizable, vesicle-free geodetic cages which look like clathrin cages; less frequently, similar but smaller deposits are found on the trans side of the Golgi stacks; and the usual coated vesicles appear in increased numbers around the trans-most Golgi elements, including the GERL (Palade & Fletcher 1977). If ATP synthesis is resumed, everything goes back to the previous state in about 10 to 15 min. In other words, the fibrillar masses disappear, the peripheral Golgi vesicles reappear, and so on. The changes described could be related to the events described by Dr Rothman in his paper (p 120-137). Clearly comparable events can be induced experimentally in the intact pancreatic acinar cells of the guinea-pig.

REFERENCES

Banjerjee D, Manning C, Redman C 1976 The in vivo effect of colchicine on the addition of galactose and sialic acid to rat hepatic serum glycoproteins. J Biol Chem 251:3887-3892

Bergmann JE, Tokuyasu KT, Singer SJ 1981 Passage of an integral membrane protein, the vesicular stomatitis virus glycoprotein, through the Golgi apparatus en route to the plasma membrane. Proc Natl Acad Sci USA 78:1746-1750

Brown RM Jr, Willison JHM 1977 The Golgi apparatus and plasma membrane involvement in secretion and cell surface deposition with special emphasis on cellulose biosynthesis. In: Brinkley BR, Porter KR (eds) International Cell Biology 1976-1977. The Rockefeller University Press, New York

Farquhar MG 1978 Recovery of surface membrane in anterior pituitary cells. Variations in traffic detected with anionic and cationic ferritin. J Cell Biol 77:R35-R42

Farquhar MG, Palade GE 1981 The Golgi apparatus (complex)—(1954–1981)—from artifact to center stage. J Cell Biol 91:77s-103s

Farquhar MG, Bergeron JJM, Palade GE 1974 Cytochemistry of Golgi fractions prepared from rat liver. J Cell Biol 60:8-25

Friend DS 1969 Cytochemical staining of multivesicular body and Golgi vesicles. J Cell Biol 41:269-279
Friend DS, Murray MJ 1965 Osmium inpregnation of the Golgi apparatus. Am J Anat 117:135-149
Grafstein B, Forman DS 1980 Intracellular transport in neurons. Physiol Rev 60:1167-1283
Herzog V, Farquhar MG 1977 Luminal membrane retrieval after exocytosis reaches most Golgi cisternae in secretory cells. Proc Natl Acad Sci USA 74:5073-5077
Herzog V, Miller F 1979 Membrane retrieval in epithelial cells of isolated thyroid follicles. Eur J Cell Biol 19:203-215
Herzog V, Reggio H 1980 Pathways of endocytosis from luminal plasma membrane in rat exocrine pancreas. Eur J Cell Biol 21:141-150
Koike H, Tanaka Y, Rez G, Meldolesi J 1980 Membrane interactions in pancreatic acinar-cells—exocytosis, recycling and autophagocytosis. In: Ribet A et al (eds) Biology of normal and cancerous exocrine pancreatic cells. Elsevier/North-Holland, Amsterdam (Inserm Symp Ser 15) p 215-228
Novikoff AB 1964 GERL, its form and function in neurons of rat spinal ganglia. Biol Bull 127:358
Novikoff AB 1976 The endoplasmic reticulum: a cytochemist's view (a review). Proc Natl Acad Sci USA 73:2781-2787
Novikoff AB, Goldfischer S 1961 Nucleosidediphosphatase activity in the Golgi apparatus and its usefulness for cytological studies. Proc Natl Acad Sci USA 47:802-810
Novikoff PM, Novikoff AB, Quintana N, Haun Y 1971 Golgi apparatus, GERL, and lysosomes of neurons in rat dorsal root ganglia studied by thick sections and thin section cytochemistry. J Cell Biol 50:859-886
Ottosen PD, Courtoy PJ, Farquhar MG 1980 Pathways followed by membrane recovered from the surface of plasma cells and myeloma cells. J Exp Med 152:1-19
Palade GE, Fletcher M 1977 Reversible changes in the morphology of the Golgi complex induced by arrest of secretory transport. J Cell Biol 75:371a (abstr)
Redman CM, Banerjee D, Howell K, Palade GE 1975 Colchicine inhibition of plasma protein release from rat hepatocytes. J Cell Biol 66:42-59
Rothman JE 1981 The Golgi apparatus: two organelles in tandem. Science (Wash DC) 213:1212-1219
Shen W-C, Ryser HJ-P 1978 Conjugation of poly-L-lysine to albumin and horseradish peroxidase: a novel method of enhancing the cellular uptake of proteins. Proc Natl Acad Sci USA 75:1872-1876
Tartakoff A, Montesano R, Vassalli P 1981 Is immunoglobulin secretion coupled to endocytosis? 21st Annu Meet Am Soc Cell Biol: Abstr no 24016
Thyberg J 1980 Internalization of cationized ferritin into the Golgi complex of cultured mouse peritoneal macrophages. Effects of colchicine and cytochalasin B. Eur J Cell Biol 23:95-103
Wilson P, Sharkey D, Haynes N, Courtoy P, Farquhar MG 1981 Iodinated cell membrane components are transported to the Golgi complex. J Cell Biol 91:417a (abstr)

Biogenesis of epithelial cell plasma membranes

MICHAEL J. RINDLER, IVAN EMANUILOV IVANOV, ENRIQUE J. RODRIGUEZ-BOULAN* AND DAVID D. SABATINI

*Department of Cell Biology, New York University Medical Center, 550 First Avenue, New York, NY 10016, and *Department of Pathology, Downstate Medical Center, 450 Clarkson Avenue, Brooklyn, NY 11203, USA*

Abstract Polarized monolayers of cultured epithelial cells, such as the kidney-derived MDCK cell line, when infected with enveloped viruses, provide a convenient model system for study of the intracellular routes followed by newly synthesized glycoproteins to reach specific domains of the plasma membrane. The polarized nature of the monolayers is reflected in the asymmetric assembly of enveloped viruses, some of which, such as influenza and simian virus 5 (SV5), bud from the apical surfaces of the cells, while others, such as vesicular stomatitis virus (VSV), emerge from the basolateral surfaces. MDCK cells can sustain double infection with viruses of different budding polarity, and within such cells the envelope glycoproteins of the two viruses are synthesized simultaneously and assembled into virions at different sites. Immunoelectron microscopic observations of doubly infected cells show that glycoproteins of influenza and VSV traverse the same Golgi apparatus. This indicates that critical sorting steps must take place during or after passage of the glycoproteins through the organelle. Following passage through the Golgi, the HA glycoprotein accumulates almost exclusively at the apical surface, where the influenza virions assemble. Significant amounts of the G protein, however, are detected on both plasma membranes in singly and doubly infected cells, although VSV virion assembly is limited to basolateral domains. These observations indicate that the site of VSV budding is not exclusively determined by the presence of G polypeptides on a given cell-surface domain. It is possible that other cellular or viral components are responsible for the selection of the appropriate budding domain or that the G protein found on the apical surface must be transferred to the basolateral domain before it becomes competent for assembly.

Cells in epithelia with a transport function form monolayers which serve as selective permeability barriers between different compartments. Such cells are endowed with a striking degree of functional polarization which is reflected in the asymmetric distribution of enzymic and transport activities

1982 Membrane recycling. Pitman Books Ltd, London (Ciba Foundation symposium 92) p 184-208

(George & Kenny 1973, Lojda 1974, Kinne 1976) associated with the plasma membrane. Two distinct domains, which are separated by intercellular junctions and characterized by specific structural differentiations, can be recognized in the plasma membrane of these cells. Typically, apical (or luminal) regions are associated with activities involved in the uptake of substances from the external medium. These activities include disaccharidases (Sigrist et al 1975), leucine aminopeptidase (Desnuelle 1979), alkaline phosphatase (Hugon & Borgers 1966), and systems of Na^+-dependent sugar transport (Chesney et al 1974) and amino acid transport (Kinne 1976). On the other hand, Na^+, K^+ ATPase (Kyte 1976, Ernst & Mills 1977), adenylate cyclase activity and several hormone receptors (Schwartz et al 1974, Shlatz et al 1975) have been identified as components of the basolateral domains, as have histocompatibility antigens (Kirby & Parr 1979).

There is considerable interest in understanding the biogenetic mechanisms which ensure that newly synthesized proteins are incorporated into cellular membranes with the specificity of distribution necessary for their function. The biogenesis of epithelial plasma membranes is particularly interesting because it must involve not only mechanisms that govern the transport of specific polypeptides to the cell surface but also sorting-out processes which account for the segregation of polypeptides into distinct surface domains. In the light of current models for organelle and membrane biogenesis (see Sabatini et al 1982), it is reasonable to propose that newly synthesized integral plasma membrane proteins of epithelial cells possess addressing signals which determine their intracellular itinerary and their ultimate destination to one or other plasma membrane domain. The capacity of the mature proteins to return to their correct site of function in the course of membrane recycling (Louvard 1980) indicates that addressing signals are permanent structural features of the polypeptides.

The potential of cultured cells for *in vitro* studies of epithelial cell polarity has only recently been recognized, and several cell lines of kidney or bladder origin have been adopted as useful model systems (Mills et al 1979, Rindler et al 1979, Perkins & Handler 1981, Handler et al 1979). One of the best characterized is the MDCK cell line derived in 1958 from dog kidney by Madin & Darby (see ATCC, 1981). These cells grow rapidly in culture to form confluent monolayers which exhibit some of the properties of distal convoluted tubules, including the ability to transport fluid and electrolytes vectorially in an apical-to-basolateral direction (Misfeldt et al 1976, Cereijido et al 1978).

A polarized distribution of plasma membrane proteins in MDCK cells has been demonstrated by the accessibility of the proteins to selective iodination on different faces of cultured monolayers (Richardson & Simmons 1979). In addition, some characteristic plasma membrane activities—leucine aminopeptidase and Na^+, K^+ ATPase (Louvard 1980, Cereijido et al 1980)—have been localized to the same domains in which they are found in natural epithelia. It is likely that other ecto-enzymic activities that have been detected in MDCK cells, such as alkaline phosphatase and γ-glutamyl transpeptidase (Rindler et al 1979), will also be found to be specifically segregated.

Asymmetric budding of enveloped viruses from epithelial cells

Some of the most pertinent studies (c.f. Katz et al 1977) on the biosynthesis of plasma membrane proteins have used enveloped viruses (rhabdo-, paramyxo-, myxo-, toga- and retroviruses) as models. These viruses contain only one or a small number of glycoproteins in the membrane envelopes that they acquire when they bud from the cell surface. Furthermore, synthesis of the viral-envelope polypeptides represents a major fraction of the total protein synthesis in infected cells.

Although little direct evidence has accumulated, it is presumed that viral glycoproteins reach their destination via the same biogenetic pathways that are used by cellular membrane proteins. It has been shown that the viral-envelope polypeptides are synthesized on membrane-bound ribosomes and are co-translationally inserted into the endoplasmic reticulum (ER) membranes (Hay 1974, Morrison & Lodish 1975, Atkinson 1978, Lingappa et al 1978, Katz & Lodish 1979). During their synthesis and subsequent transfer from the ER through the Golgi apparatus to the plasma membrane, the polypeptides undergo several co- and post-translational modifications, such as proteolytic cleavage (Schulman & Palese 1977), glycosylation (Moyer et al 1976, Nakamura et al 1979, Hunt et al 1978) and addition of fatty acids (Schmidt & Schlesinger 1979), which are also characteristic of many cellular membrane proteins. It has been suggested that clathrin-coated vesicles are involved in the intracellular transport of the envelope glycoprotein G of vesicular stomatitis virus from the ER to the Golgi apparatus and from this organelle to the cell surface (Rothman & Fine 1980, Rothman et al 1980).

It has recently been observed that when polarized epithelial cells are infected with enveloped viruses (Rodriguez-Boulan & Sabatini 1978) virions assemble only on specific domains of the plasma membrane. Thus, influenza, sendai or simian virus 5 (SV5) bud exclusively from the apical surface of MDCK or MDBK cells, whereas vesicular stomatitis virions assemble only on the basolateral domains. The site of budding appears to be a characteristic property of each particular virus, independent of the type of polarized epithelial cell infected. On the other hand, the same enveloped viruses have been shown to bud indiscriminately from all aspects of the plasma membrane of non-polarized cells such as fibroblasts.

Several observations indicate that the asymmetric budding of enveloped viruses that occurs in cultured monolayers also takes place in natural epithelia. Thus, influenza and Newcastle disease viral particles have been observed budding from the free surface of cells in the chorioallantoic membrane of chick embryos (Murphy & Bang 1952, Bang 1953), and budding of mouse mammary tumour virus has been seen from the apical membrane of mammary cells (Pickett et al 1975). In addition, nuclear polyhedrosis baculoviruses appear to be produced from the basal aspects of infected insect intestinal cells (Hess & Falcon 1977). It is not difficult to envisage the implications that this asymmetric budding of enveloped viruses from epithelia must have for the spreading of viruses within the organism and for the transmission of viral disease.

A study of the distribution of viral envelope glycoproteins in infected epithelial monolayers, using immunofluorescence and ferritin immunoelectron microscopy (Rodriguez-Boulan & Pendergast 1980), indicated that polarized budding is preceded by the segregation of viral envelope proteins to the corresponding plasma membrane domains. Nucleocapsids, on the other hand, in cells infected with sendai virus appeared to be evenly distributed throughout the cytoplasm of MDCK cells. These observations led to the suggestion that the polarized distribution of envelope proteins is a main determinant of asymmetric budding.

The specificty of budding polarity does not depend on the oligosaccharide composition of the envelope glycoproteins. It has been observed that viruses assemble with the correct polarity in lectin-resistant cells that synthesize aberrant carbohydrate chains (Green et al 1981a,b). In fact, the polarity of budding was found to be unaffected by the absence of carbohydrate chains in viral-envelope proteins synthesized in cells treated with tunicamycin (Roth et al 1979, Green et al 1981a,b). The latter study also reported that the polarized distribution of viral glycoproteins, as assessed by immunolabelling, was unaffected by the deletion of oligosaccharide side-chains or by changes in specific sugar residues.

A study of the mechanisms that determine the asymmetric assembly of enveloped viruses would thus appear to be a valuable approach towards the elucidation of cellular processes which, in epithelial cells, must be responsible for the sorting-out and delivery of newly synthesized polypeptides to different plasma membrane domains.

Double infection of MDCK monolayers with viruses of opposite polarity

To compare directly the pathways followed by viral envelope glycoproteins of viruses with opposite polarities, from their site of synthesis to the site of viral budding, we developed conditions for the double infection of MDCK monolayers. Two combinations of viruses of opposite polarity were selected: SV5 and VSV, and influenza (WSN) and VSV. Successful double infection with SV5 and VSV was obtained when confluent monolayers of MDCK cells were first infected (1–5 plaque-forming units/cell) with SV5, the least cytopathic of these viruses, and 24–42 h later with VSV (10–15 p.f.u./cell). Because both WSN and VSV tend to shut off protein synthesis adherence to a more stringent protocol was required for successful double infection with these viruses. In this case, cells were first infected with WSN (10 p.f.u./cell) and 4.5 h later with VSV (10–15 p.f.u./cell). When superinfection was established at later times (6–7 h after infection with influenza) or when a higher multiplicity of influenza was used (25 p.f.u./cell) poor infection by VSV resulted. When these standard protocols were used, extensive double infection of MDCK cells was achieved with either pair of viruses. Five hours after superinfection with VSV, more than 95% of the cells could be labelled when specific antibodies against either of the two viruses were applied.

As expected, the synthesis of VSV proteins, as measured by pulse labelling with [^{35}S]methionine, increased substantially during the first 7.5 h after superinfection (Fig. 1). Over the same period, synthesis of SV5 or influenza proteins could still be detected in labelled doubly infected cultures. The rate of synthesis of SV5 proteins (Fig. 1a) appeared approximately constant during the first 7.5 h of superinfection with VSV, whereas synthesis of WSN proteins reached a peak at approximately 5.0 h of superinfection, decreasing thereafter (Fig. 1b).

Electron microscopic examination of doubly infected monolayers showed that, as long as tight junctions appeared intact, the polarity of viral budding characteristic of singly infected cells was maintained. The distinctive filamentous SV5 (Fig. 2) or the spherical influenza virions (Fig. 3a) were seen assembling at the apical surfaces of cells which, at the same time, showed VSV virions budding from the basolateral domains (Fig. 2, 3a). As VSV infection progressed, cells tended to become rounded and frequently their tight junctions were disrupted. In these cases, however, the polarity of viral budding was not completely lost. Although areas containing VSV virions could be identified on the free surfaces of the cells, these occurred in patches that were clearly distinct from other regions where SV5 or influenza virus was being assembled (Fig. 3b). In particular, in cells doubly infected with SV5 and VSV, SV5 virions tended to form clusters towards the apical pole of the cell, away from areas of the apical surface where budding of VSV was taking place (not shown).

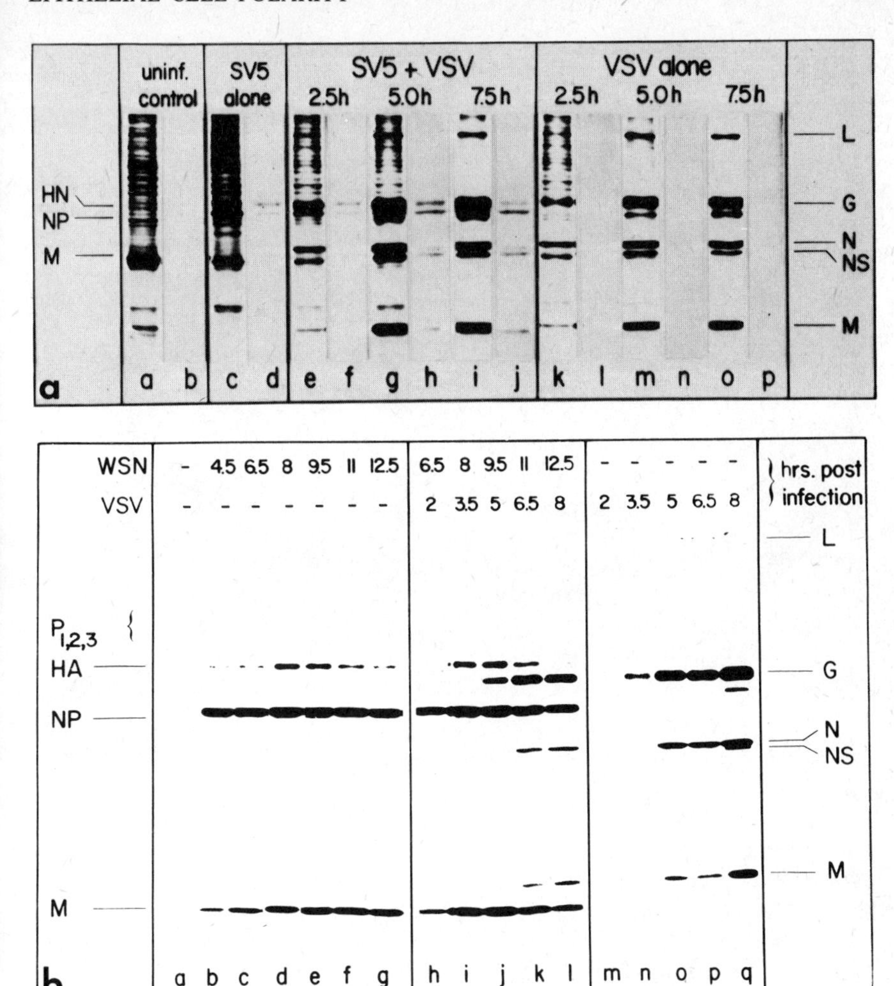

FIG. 1. ^{35}S-methionine labelling of virally infected MDCK cells. Cells were incubated with [^{35}S]methionine (50 μCi ml^{-1}) for 20 min and processed as previously described (Green et al 1981) for sodium dodecyl sulphate electrophoresis in a 10% polyacrylamide gel. (a) 24 h after infection with simian virus 5 (SV5) (5 plaque-forming units/cell) the cells were superinfected with vesicular stomatitis virus (VSV) (10 p.f.u./cell). Cultures were further incubated for the times indicated before radiolabelling. Tracks b,d,f,h,j,l,n,p represent immunoprecipitates made by using rabbit anti-SV5 antiserum. (b) After 4.5 h of infection with influenza virus (WSN) (10 p.f.u./cell), cultures (h–l) were superinfected with VSV (15 p.f.u./cell) and incubated for the times indicated. a–g: Cultures infected only with WSN. m–q: Cultures infected only with VSV.

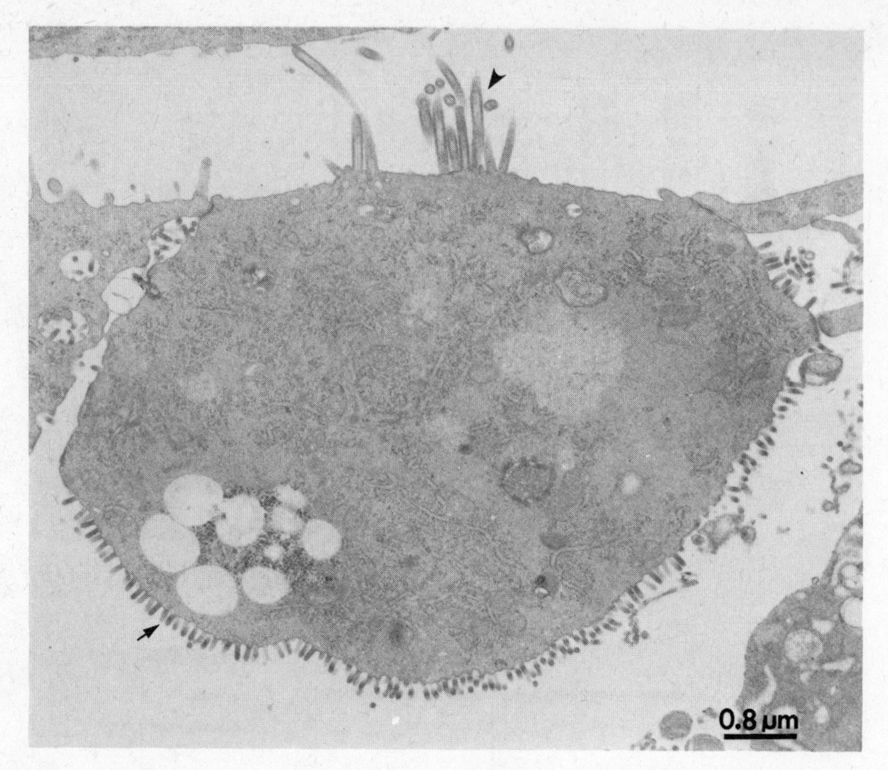

FIG. 2. MDCK cell doubly infected with SV5 and VSV. After 24 h of infection with SV5, cells were infected with VSV. 11 h later samples were processed for electron microscopy by standard techniques. The filamentous SV5 virions (arrow-head) appear only at the apical surface of the cell while VSV viruses assemble on the basolateral surface (arrow).

Intracellular pathways followed by viral glycoproteins

The distribution of the viral glycoproteins G of VSV and HA of influenza virus was examined in ultra-thin frozen sections of doubly infected cells, which were fixed in glutaraldehyde and immunolabelled according to a modification of the methods of Tokuyasu (Tokuyasu & Singer 1976, Tokuyasu 1978). The location of rabbit anti-G or monoclonal mouse anti-HA antibodies was determined by an indirect procedure using affinity-purified goat anti-mouse or anti-rabbit antibodies complexed to colloidal gold particles (Horisberger & Rosset 1977, Geuze et al 1981). With these techniques background labelling was very low when estimated from sections of uninfected or singly infected monolayers that were incubated with the non-corresponding antibody.

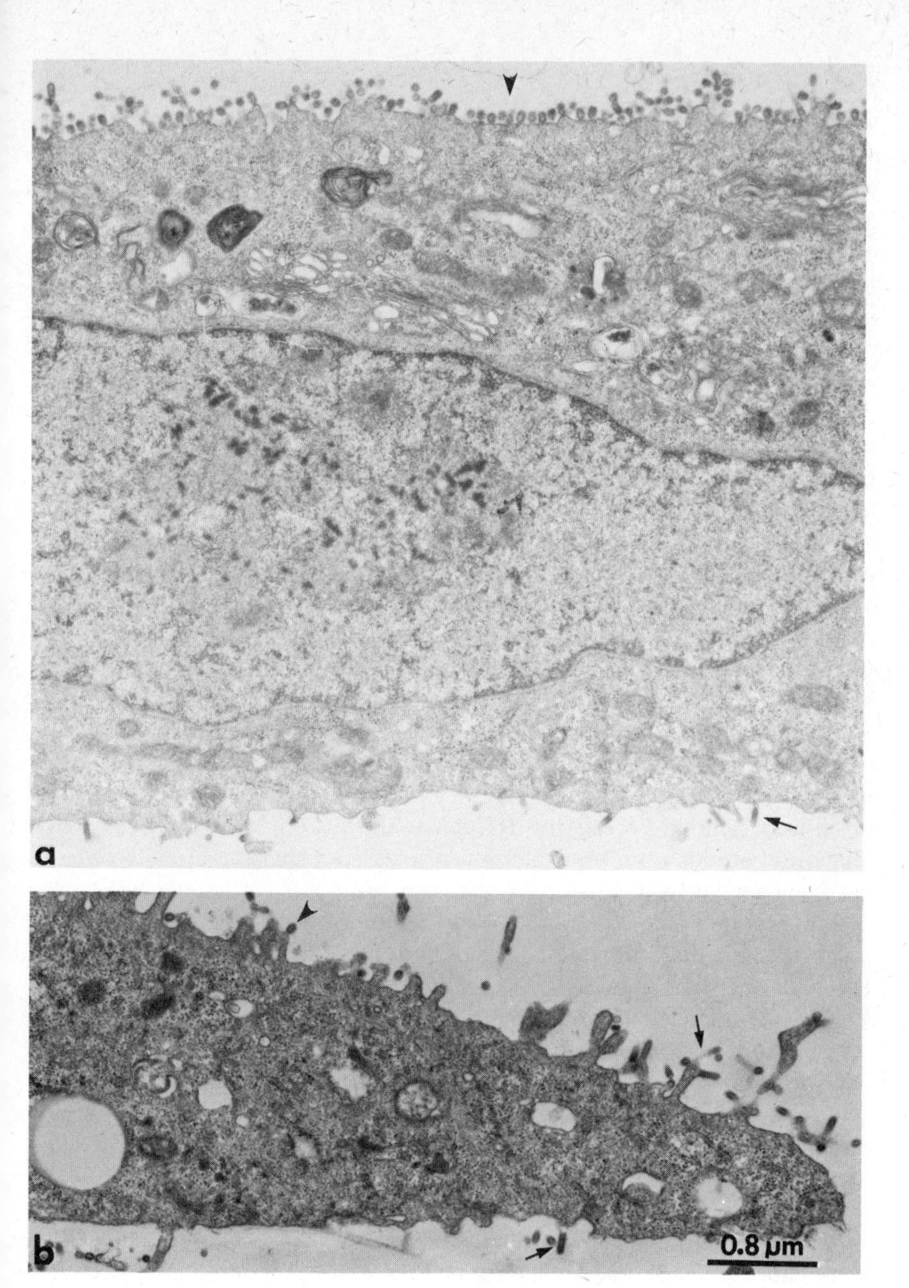

FIG. 3. MDCK cell doubly infected with influenza (WSN) and VSV. 4.5 h after infection with WSN, cells were infected with VSV and incubated for an additional 5 h (a) or 9 h (b). (a) Numerous influenza virions are present on the apical surface of the cell (arrow-head) and several VSV virions are seen budding from the basolateral plasma membrane (arrow). Although the cell in panel (b) does not have intact tight junctions, WSN (arrow-head) and VSV (arrows) bud from distinct regions of the plasma membrane.

Treatment of doubly infected cells with antibodies against either HA of influenza or G of VSV showed that, intracellularly, gold particles were preferentially distributed over the Golgi apparatus (Fig. 4a,b), throughout different Golgi cisternae, and in adjacent Golgi vesicles and vacuoles. There was no marked accumulation of gold particles over cisternae located towards one or other side of the Golgi stacks, or towards the middle region or the rims of the Golgi cisternae.

The concentration of anti-HA antibodies over the Golgi complex decreased as superinfection progressed. Five or more hours after VSV infection, when the cell surface was intensely labelled, very few anti-HA antibodies could be observed on the organelle, and these were located mainly towards the rims of the cisternae. In sections labelled with anti-G antibodies the protein was first localized in the Golgi apparatus 3 h after superinfection, when there were practically no gold particles on the cell surface. Labelling of the Golgi with anti-G antibodies remained prominent thereafter, when the G protein was also localized on the plasma membrane. The relative temporal changes, observed by immunolabelling, in the intracellular concentrations of HA and G proteins after VSV superinfection are consistent with the rates of synthesis of the respective viral proteins estimated from [^{35}S]methionine labelling experiments. The localization of the viral glycoproteins over the Golgi complex is in agreement with recent studies on the intracellular distribution of newly synthesized G protein of VSV (Bergmann et al 1981) and spike proteins of Semliki Forest virus (Green et al 1981a, b) in non-polarized cells. Our observations represent the first instance in which the HA glycoprotein has been shown to traverse the Golgi apparatus.

It should be emphasized that at no time in the course of double infection was a significant accumulation of viral glycoproteins detected over the rough ER, where the polypeptides are synthesized, or over any other organelle. This indicates that passage through the Golgi apparatus is the rate-limiting step in the transport of newly synthesized glycoproteins from the ER to the cell surface.

To investigate whether envelope glycoproteins of viruses of opposite budding polarity traverse different Golgi apparatuses that might serve exclusively one or another plasma membrane domain, we labelled thin sections of doubly infected cells with both types of antibody. To allow for the simultaneous visualization of both viral antigens within the same cells, the secondary goat antibodies against the mouse anti-HA were complexed to gold particles of 18 nm diameter, and those against the rabbit anti-G were complexed to 5 nm gold particles. In cells where both HA and G were detected intracellularly, Golgi apparatuses, irrespective of their location within the cell or proximity to either of the plasma membrane domains, were always labelled with both types of gold particle, which were present singly or

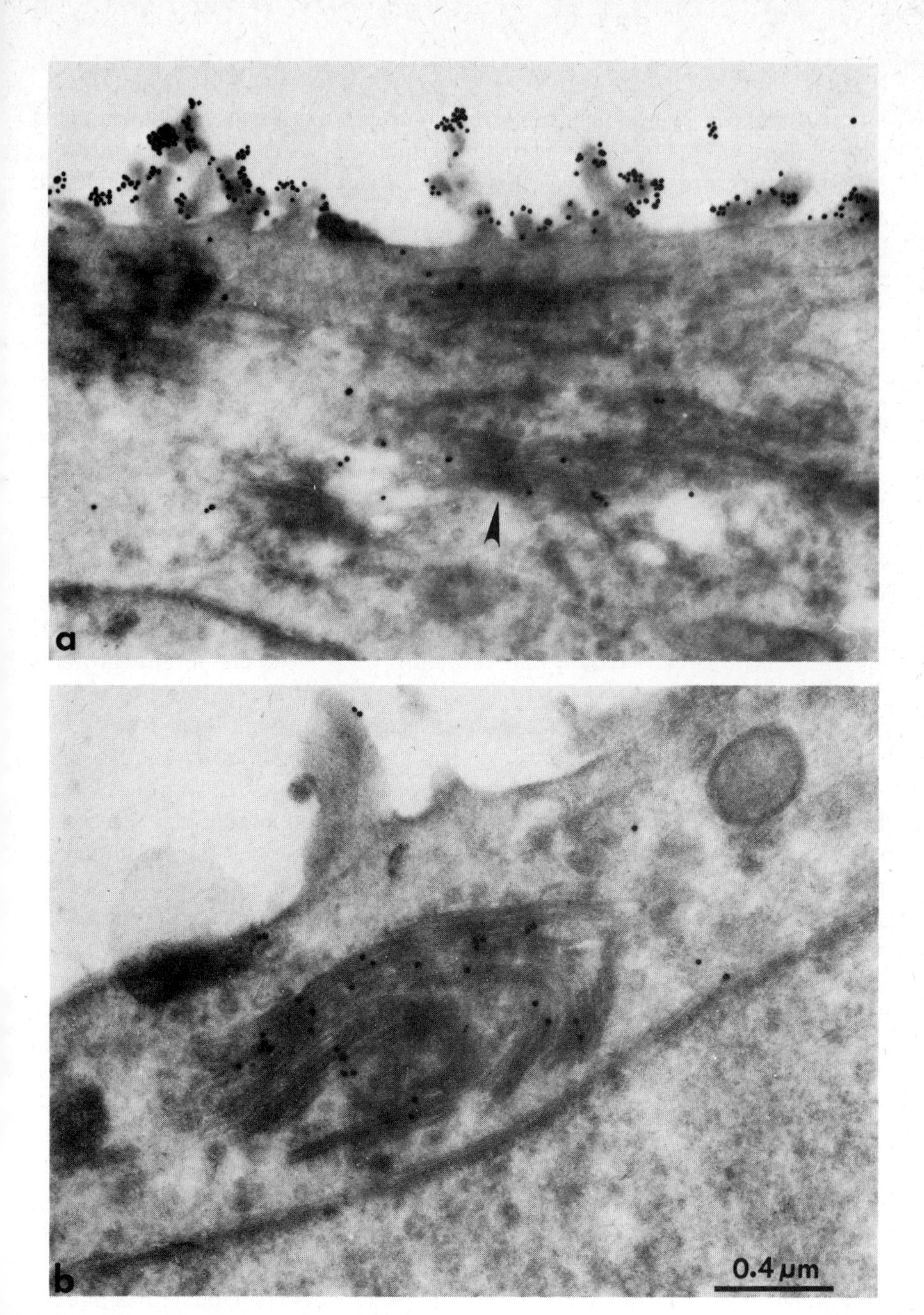

FIG. 4. Frozen thin sections of MDCK cells doubly infected with WSN and VSV, immunolabelled 3 h after VSV infection. (a) Monoclonal antibodies against the HA of influenza label the apical plasma membrane and the Golgi apparatus (arrow-head) of the cell. (b) Specific antibodies against the G protein are localized over the Golgi apparatus, although there is as yet little detectable labelling of the plasma membrane.

in small clusters (Fig. 5). Particles of different sizes were not intermingled within aggregates, but single particles or small aggregates of each kind were often seen in close proximity within the same cisternae (Fig. 6a, b, c). These observations clearly indicate that envelope glycoproteins of viruses with different polarity traverse the same organelle. Antibodies to the G protein showed no preferential distribution over the Golgi cisternae but those marking the location of HA were frequently found, as in the singly infected cells, over the peripheral regions of the cisternae.

Anti-viral glycoprotein antibodies were also localized on vesicles presumed to be involved in transport of the envelope glycoproteins from the Golgi apparatus to the plasma membrane. In most instances these vesicles contained gold particles of only one size (Fig. 6a). Occasionally, and in particular in larger vesicles, both types of gold particle were present within the same structure. It is worth emphasizing that due to the low level of labelling and the small size of most vesicles it is difficult to discard completely the possibility that those labelled with only one type of antibody carry the other antigen as well. It should also be noted that because only a fraction of the labelled vesicles may be involved in the recycling of plasma membrane proteins through the cytoplasm no inferences can be drawn about the direction of movement of individual cytoplasmic vesicles and about their role in the segregation of viral glycoproteins.

Because our results indicated that sorting-out of viral glycoproteins does not take place earlier than during their passage through the Golgi apparatus, the distribution of proteins in the cell surface was examined by immunolabelling. For this purpose thin frozen sections were used or intact cells were labelled, after scraping from dishes, and were then processed for conventional electron microscopy. With the progress of influenza infection HA accumulated predominantly on the apical domains of the plasma membrane, where 80–90% of the surface label was detected. This distribution of anti-HA antibodies on the cell surface was observed early in infection, when no budding particles were seen, as well as when viral production was intense (between 4.5–11.5 h after influenza infection) in singly and doubly infected cells.

In contrast to the rather strict correspondence observed between the sites of HA accumulation and influenza virus budding, significant amounts of the G protein were found not only on the basolateral surfaces but also on the apical domains of the plasma membrane of VSV-infected or superinfected cells (Fig. 7). In fact the apical labelling in these studies appeared as intense as labelling on the basolateral surfaces although, in the same cells, VSV budding was always restricted to the basolateral domains. These observations are difficult to reconcile with a mechanism by which the site of viral budding is determined exclusively by the preferential accumulation of envelope gly-

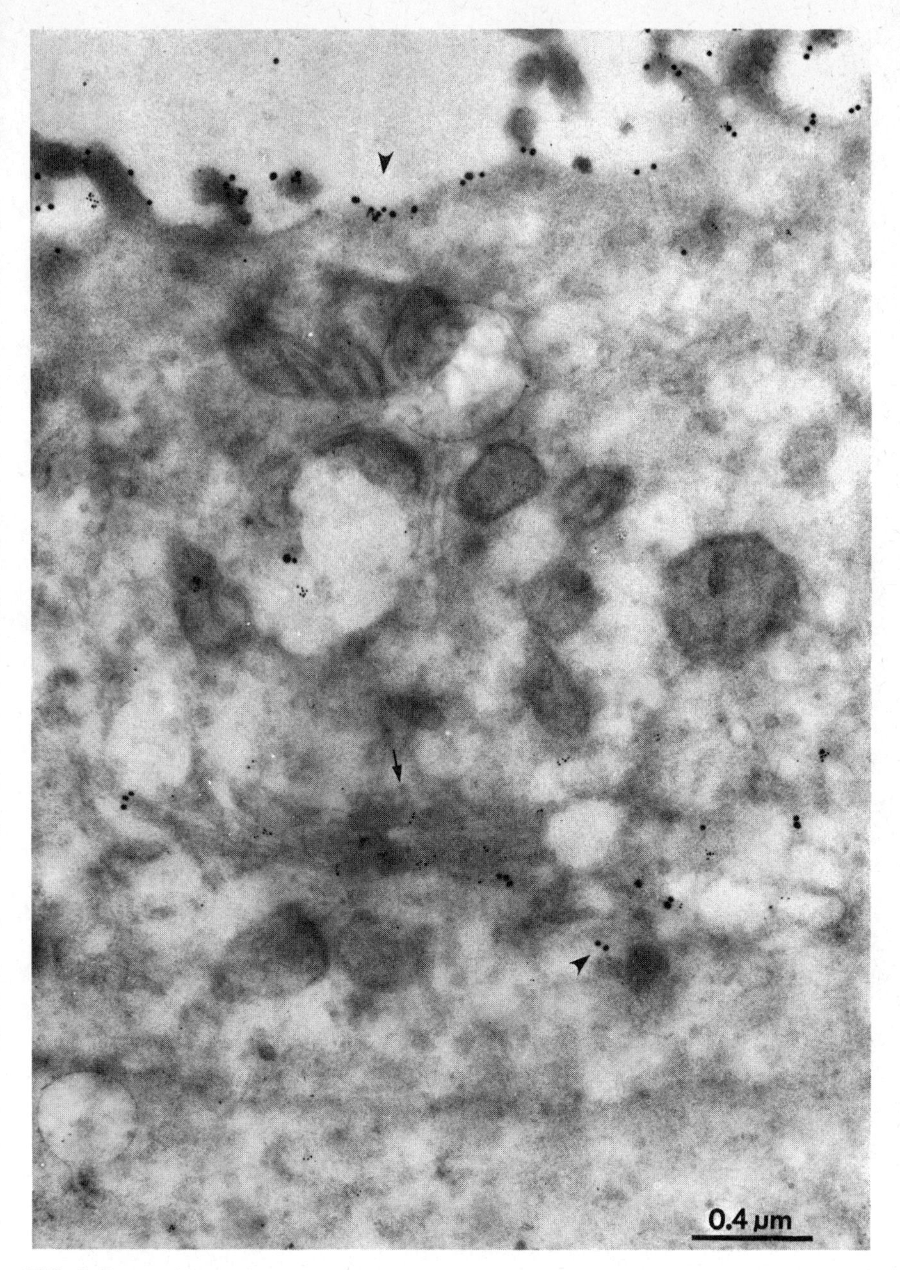

FIG. 5. Frozen thin section of a doubly infected MDCK cell labelled with antibodies against the HA of influenza (WSN) and the G protein of VSV (4 h post-VSV infection).

The apical plasma membrane shows substantial labelling with anti-HA antibodies (large gold particles, arrow-head) which are also seen in the peripheral regions of the Golgi cisternae. The G protein of VSV is simultaneously localized in the same Golgi apparatus (small gold particles, arrow).

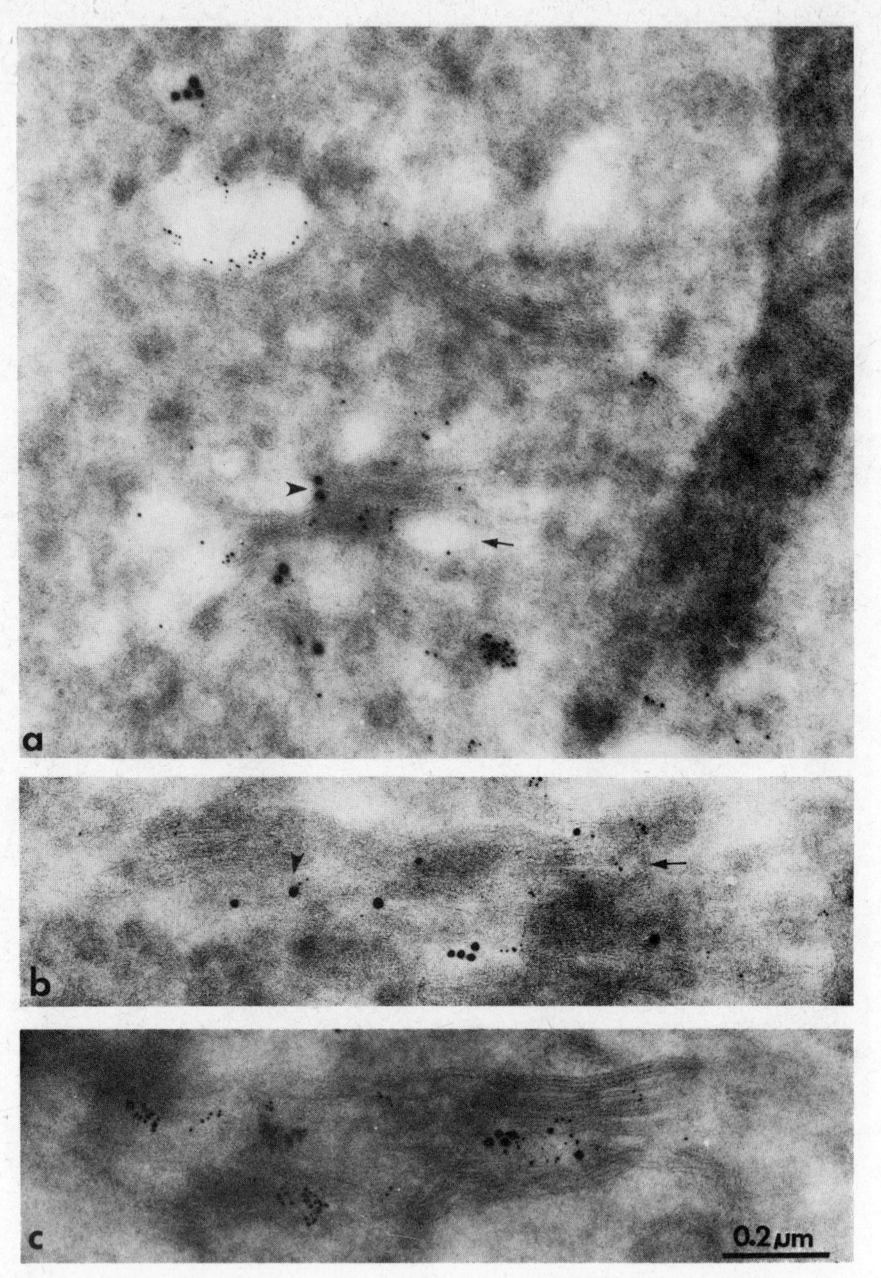

FIG. 6. Golgi apparatus in doubly infected MDCK cells immunolabelled with antibodies against envelope glycoproteins of the two different viruses. Frozen thin sections of doubly infected cells (4 h after VSV infection) have, within the same Golgi apparatus (a,b,c), cisternae labelled with particles of both sizes, corresponding to HA of influenza (large particles) and G of VSV (small particles). Cytoplasmic vesicles (a) contain mostly one type of gold particle.

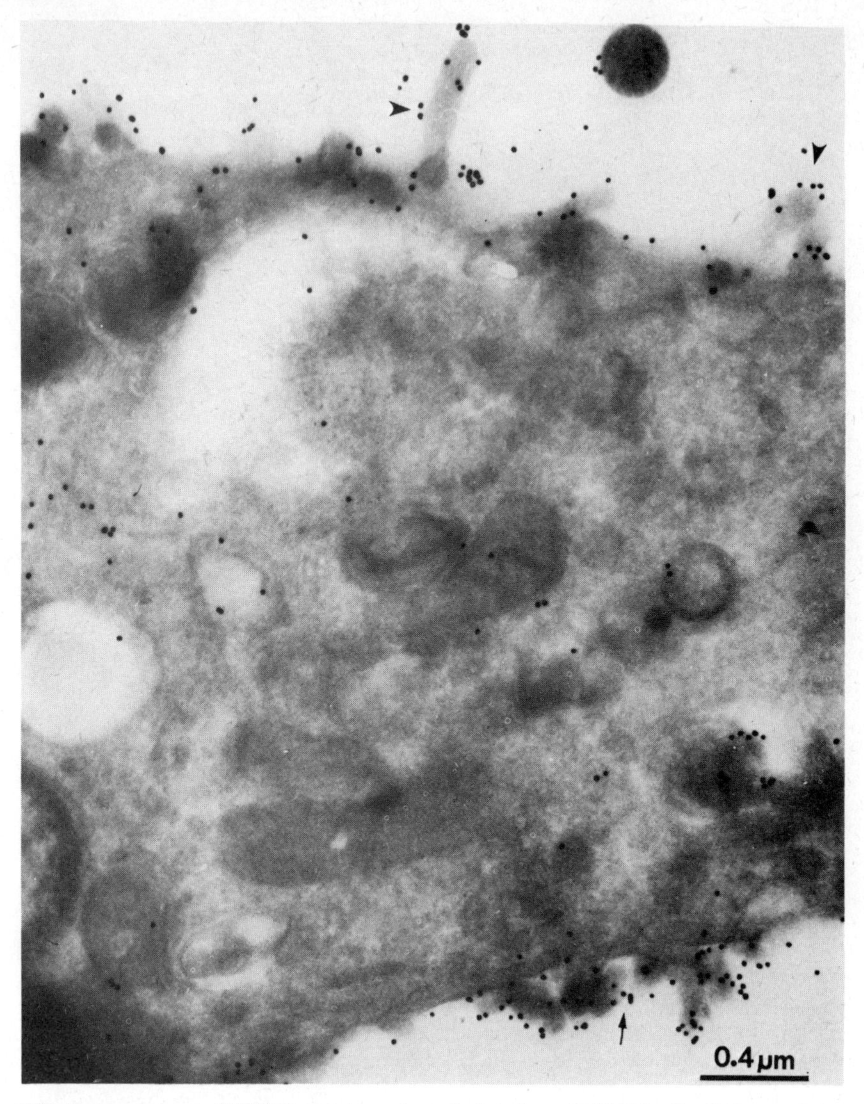

FIG. 7. Appearance of VSV G protein on the apical surface of MDCK cells. Cells were infected only with VSV (15 p.f.u./cell), and 5 h later frozen thin sections were prepared and labelled with anti-G protein antibodies. A group of labelled VSV virions is seen budding from the basolateral surface (arrow). Note that in this section, the G protein is also localized in the apical membrane (arrow-head) although no VSV budding occurs from this membrane domain.

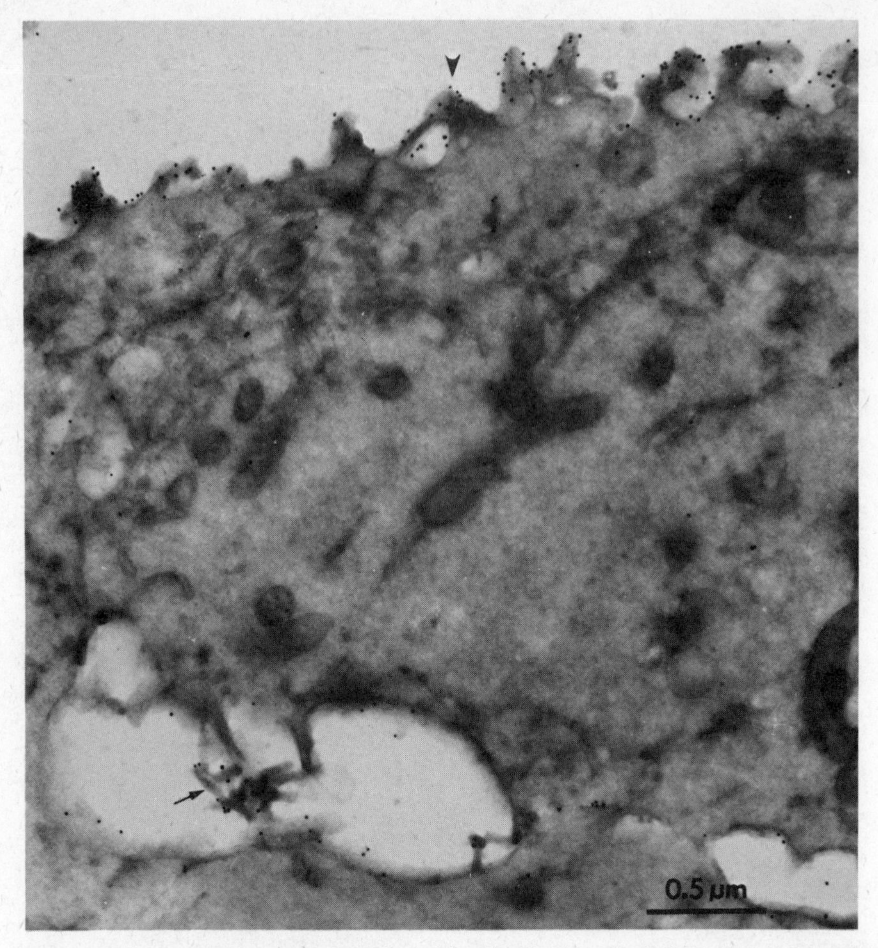

FIG. 8. Production of phenotypically mixed virions in MDCK cells doubly infected with WSN and VSV. 5 h after infection with VSV, sections were prepared and labelled with anti-HA antibodies. The label is localized predominantly on the apical membrane (arrow-head) but some labelling is also observed on the basolateral surface (arrow) where VSV nucleocapsids acquire an envelope that is frequently seen to contain HA molecules.

coproteins. Indeed, at least for the G protein, segregation to the site of budding may take place even after the envelope glycoprotein has reached the non-corresponding domain of the plasma membrane.

Cultured cells, when infected simultaneously with two different viruses, can produce aberrant particles known as phenotypically mixed virions and pseudotypes, in which one type of nucleocapsid is surrounded by an envelope composed partly or totally of glycoproteins of the other virus (McSharry et al

1971). Recently, using neutralizing antibodies against influenza, Roth & Compans (1981) monitored the formation of pseudotypes and mixed virions containing the VSV genome, in polarized MDCK monolayers infected with influenza and VSV. In this case the aberrant particles appeared in the medium only after a period of 1–2 h during which wild-type viruses were exclusively produced. As cytopathic effects became evident the production of aberrant virions increased dramatically, suggesting that opening of the tight junctions allowed the intermingling of viral glycoproteins in the cell surface.

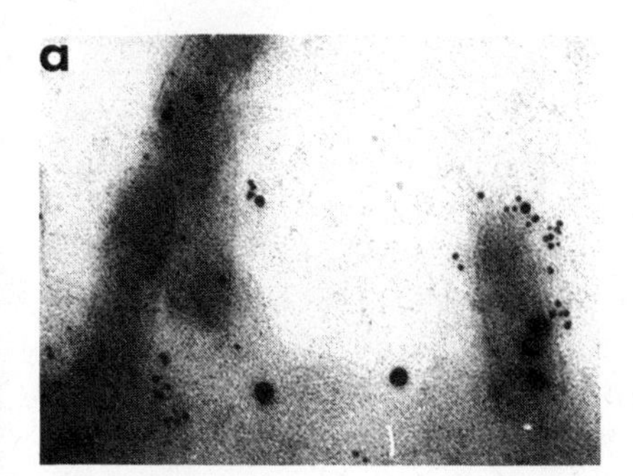

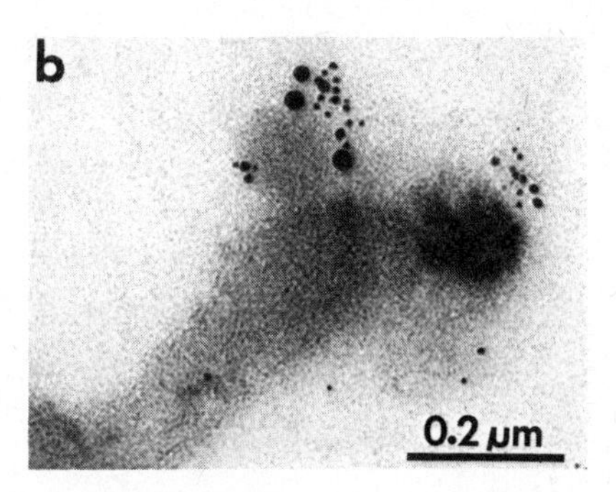

FIG. 9. Phenotypically mixed virions produced by doubly infected MDCK cells 4 h after VSV infection, labelled with both anti-G (small gold) and anti-HA (large gold) antibodies. (a) VSV virions bud from the basolateral membrane with a mixed envelope that contains HA molecules. (b) WSN virions , containing G protein, budding from a microvillus on the apical membrane.

By using the immunolabelling technique we were able to demonstrate the formation of mixed virions in doubly infected MDCK cells well before there was any evidence of loss of junctional integrity. Bullet-shaped virions (i.e. with a VSV nucleocapsid), labelled with varying amounts of antibody against the HA of influenza virus, were detected in the intercellular spaces and were seen budding from the basolateral domains of the plasma membrane (Figs. 8, 9a). The existence of such particles is consistent with the finding that a fraction of the HA may be located in the basolateral aspects of the plasma membrane of singly or doubly infected cells. Mixed-type virions with a rounded influenza nucleocapsid and with some anti-G antibodies labelling the envelope were detected budding from the apical surface (Fig. 9b), but such particles were rare.

Since the G protein was more abundant in the apical surface than HA was in the basolateral surface, these findings suggest that viral assembly on the basolateral surfaces was less discriminatory than assembly on the free surface. The relative paucity of observed VSV virions that were assembling at the apical surface may indicate that, though it is abundant, the G protein of this plasma membrane domain is not suitable for assembly. This could be due to the need for further maturation of the protein or to the absence from the apical surface of other proteins, such as the M protein, or even cellular proteins, which may play a role in assembly. We are now investigating the possibility that the apical location of G protein represents only a stage in its transfer to the correct site of assembly in the basolateral domain. We hope to define by immunolabelling techniques the extent to which a complex of G and M protein might manifest a polarized distribution.

Acknowledgements

We wish to acknowledge the gifts of monoclonal antiserum against HA from Dr R. G. Webster and of rabbit anti-G antiserum from Dr J. Lenard, and to thank Dr S. Papadopoulos for assistance during preparation of this paper, Ms H. Plesken for invaluable technical assistance and Ms Myrna Chung and Ms Sonia Martinez for typing the manuscript. This work was supported by National Institutes of Health grants GM 20277, AG 01461, AG 00378 and a National Institutes of Health postdoctoral fellowship to M.J.R.

REFERENCES

Atkinson PH 1978 Glycoprotein and protein precursors to plasma membranes in vesicular stomatitis virus infected HeLa cells. J Supramol Struct 8:89-109

ATTC 1981 American Type Culture Collection: Catalogue of stains II. ATCC, Rockville, Md, p 52

Bang FB 1953 The development of Newcastle disease virus in cells of the chorioallantoic membrane as studied by thin sections. Bull Johns Hopkins Hosp 92:309-322

Bergmann JE, Tokuyasu KT, Singer SJ 1981 Passage of an integral membrane protein, the vesicular stomatitis virus glycoprotein through the Golgi apparatus en route to the plasma membrane. Proc Natl Acad Sci USA 78:1746-1750

Cereijido M, Robbins ES, Dolan WJ, Rotunno CA, Sabatini DD 1978 Polarized monolayers formed by epithelial cells on a permeable and translucent support. J Cell Biol 77:853-880

Cereijido M, Ehrenfeld J, Meza I, Martinez-Palomo A 1980 Structural and functional membrane polarity in cultured monolayers and MDCK cells. J Membr Biol 52:147-159

Chesney R, Sacktor B, Kleinzeller A 1974 The binding of phloridzin to the isolated luminal membrane of the renal proximal tubule. Biochim Biophys Acta 332:263-277

Desnuelle P 1979 The tenth Sir Hans Krebs lecture. Intestinal and renal aminopeptidase: a model of a transmembrane protein. Eur J Biochem 101:1-11

Ernst SA, Mills JW 1977 Basolateral plasma membrane localization of ouabain-sensitive sodium transport sites in the secretory epithelium of the avian salt gland. J Cell Biol 75:74-94

George SG, Kenny AJ 1973 Studies on the enzymology of purified preparations of brush border from rabbit kidney. Biochem J 134:43-57

Geuze HJ, Slot JW, van der Ley PA, Scheffer RCT 1981 Use of colloidal gold particles in double labelling immunoelectron microscopy of ultrathin frozen sections. J Cell Biol 89:653-665

Green J, Griffiths G, Louvard D, Quinn P, Warren G 1981a Passage of viral membrane proteins through the Golgi complex. J Mol Biol 152:663-698

Green RF, Meiss HK, Rodriguez-Boulan E 1981b Glycosylation does not determine segregation of viral envelope proteins in the plasma membrane of epithelial cells. J Cell Biol 89:230-239

Handler JS, Steele RF, Wade JB, Peterson AS, Lawson NL, Johnson JP 1979 Toad urinary bladder epithelial cells in culture: maintenance of epithelial structure, sodium transport, and response to hormones. Proc Natl Acad Sci USA 76:4151-4155

Hay AJ 1974 Studies on the formation of the influenza virus envelope. Virology 60:398-418

Hess RT, Falcon LA 1977 Observations on the interaction of baculoviruses with the plasma membrane. J Gen Virol 36:525-530

Horisberger M, Rosset J 1977 Colloidal gold, a useful marker for transmission and scanning electronmicroscopy. J Histochem Cytochem 25:295-305

Hugon J, Borgers M 1966 Ultrastructural localization of alkaline phosphatase activity in the absorbing cells of the duodenum of the mouse. J Histochem Cytochem 14:629-640

Hunt LA, Etchison SR, Summers DF 1978 Oligosaccharide chains are trimmed during synthesis of the enveloped glycoprotein of vesicular stomatitis virus. Proc Natl Acad Sci USA 75:754-758

Katz FN, Lodish HF 1979 Transmembrane biogenesis of the vesicular stomatitis virus glycoprotein. J Cell Biol 80:416-426

Katz FN, Rothman JE, Lingappa VR, Blobel G, Lodish HF 1977 Membrane assembly *in vitro*: synthesis, glycosylation, and asymmetric insertion of a transmembrane protein. Proc Natl Acad Sci USA 74:3278-3282

Kinne R 1976 Membrane-molecular aspects of tubular transport. Int Rev Physiol 11:169-210

Kirby WN, Parr EL 1979 The occurrence and distribution of H-2 antigens on mouse intestinal epithelial cells. J Histochem Cytochem 27:746-750

Kyte J 1976 Immunoferritin determination of ($Na^+ + K^+$)ATPase over the plasma membranes of renal convoluted tubules. I: Distal segment. J Cell Biol 68:287-303

Lingappa VR, Katz FN, Lodish HF, Blobel G 1978 A signal sequence for the insertion of a transmembrane glycoprotein. J Biol Chem 253:8667-8670

Lojda Z 1974 Cytochemistry of enterocytes and of other cells in the mucous membrane of the small intestine. In: Smith DH (ed) Biomembranes 4A. Plenum Press, New York, p 43-122

Louvard D 1980 Apical membrane aminopeptidase appears at site of cell–cell contact in cultured kidney epithelial cells. Proc Natl Acad Sci USA 77:4132-4136

McSharry JJ, Compans RW, Choppin PW 1971 Proteins of vesicular stomatitis virus and of phenotypically mixed vesicular stomatitis virus–simian virus 5 virions. J Virol 8:722-729

Mills JW, MacKnight ADC, Dayer J-M, Ausiello DA 1979 Localization of ^{3}H-ouabain-sensitive Na^+-pump sites in cultured pig kidney cells. Am J Physiol 236:C157-C162

Misfeldt DS, Hamamoto ST, Pitelka DR 1976 Transepithelial transport in cell culture. Proc Natl Acad Sci USA 73:1212-1216

Morrison TG, Lodish HF 1975 The site of synthesis of membrane and non-membrane proteins of vesicular stomatitis virus. J Biol Chem 250:6955-6962

Moyer SA, Tsang JM, Atkinson PH, Summers DF 1976 Oligosaccharide moieties of the glycoprotein of vesicular stomatitis virus. J Virol 18:167-175

Murphy JS, Bang FB 1952 Observations with the electron microscope on cells of the chorioallantoic membrane infected with influenza virus. J Exp Med 95:259-268

Nakamura K, Nakane P, Brown WR 1979 Synthesis of the oligosaccharides of influenza virus glycoproteins. Virology 93:31-47

Perkins F, Handler JS 1981 Transport properties of toad kidney epithelia in culture. Am J Physiol 241:C154-C159

Pickett PB, Pitelka DR, Hamamoto ST, Misfeldt DS 1975 Occluding junctions and cell behavior in primary cultures of normal and neoplastic mammary gland cells. J Cell Biol 66:316-332

Richardson JCW, Simmons NL 1979 Demonstration of protein asymmetries in the plasma membrane of cultured renal (MDCK) epithelial cells by lactoperoxidase-mediated iodination. FEBS (Fed Eur Biochem Soc) Lett 105:201-204
Rindler MJ, Chuman LM, Shaffer L, Saier MH 1979 Retention of differentiated properties in an established dog kidney epithelial cell line (MDCK). J Cell Biol 81:635-648
Rodriguez-Boulan EJ, Pendergast M 1980 Polarized distribution of viral envelope proteins in the plasma membrane of infected epithelial cells. Cell 20:45-54
Rodriguez-Boulan EJ, Sabatini DD 1978 Asymmetric budding of viruses in epithelial monolayers: a model for the study of epithelial polarity. Proc Natl Acad Sci USA 75:5071-5075
Roth MG, Fitzpatrick JP, Compans RW 1979 Polarity of virus maturation in MDCK cells: lack of a requirement for glycosylation of viral glycoproteins. Proc Natl Acad Sci USA 76:6430-6434
Roth MG, Compans RW 1981 Delayed appearance of pseudotypes between vesicular stomatitis virus and influenza virus during mixed infection of MDCK cells. J Virol 40:848-860
Rothman JE, Fine RE 1980 Coated vesicles transport newly synthesized membrane glycoproteins from endoplasmic reticulum to plasma membrane in two successive stages. Proc Natl Acad Sci USA 77:780-784
Rothman JE, Bursztyn-Pettegrew H, Fine RE 1980 Transport of the membrane glycoprotein of vesicular stomatitis virus to the cell surface in two successive stages. J Cell Biol 86:162-171
Sabatini DD, Kreibich G, Morimoto T, Adesnik M 1982 Mechanisms for the incorporation of proteins in membranes and organelles. J Cell Biol 92:1-22
Schmidt MFG, Schlesinger MJ 1979 Fatty-acid binding to vesicular stomatitis virus glycoprotein: a new type of posttranslational modification of the viral glycoproteins. Cell 17:813-819
Schulman JL, Palese P 1977 Virulence factors of influenza A viruses: WSN virus neuraminidase required for plaque production in MDBK cells. J Virol 24:170-176
Schwartz IL, Shlatz LJ, Kinne-Saffran E, Kinne R 1974 Target cell polarity and membrane phosphorylation in relation to the mechanism of action of antidiuretic hormone. Proc Natl Acad Sci USA 71:2595-2598
Shlatz LJ, Schwartz IL, Kinne-Saffran E, Kinne R 1975 Distribution of parathyroid hormone-stimulated adenylate cyclase in plasma membrane of cells of the kidney cortex. J Membr Biol 24:131-144
Sigrist H, Ronner P, Semenza G 1975 A hydrophobic form of the small intestinal sucrase-isomaltase complex. Biochim Biophys Acta 406:433-436
Tokuyasu KT 1978 A study of positive staining of ultrathin frozen sections. J Ultrastruct Res 63:287-307
Tokuyasu KT, Singer SJ 1976 Improved procedures for immunoferritin labeling of ultrathin frozen sections. J Cell Biol 71:894-906

DISCUSSION

Cohn: Could you comment on the role of junctions in this system? If you were to explant these cells at low density, infect them and obtain viral replication, would junctions form?

Sabatini: The cells can become polarized either when they touch each other or when they touch the substrate. With Dr E.J. Rodriguez-Boulan (unpublished results) we have been able to maintain and infect the cells in suspension and plate them as single cells. One can sometimes see viruses of one type budding from the apical surface of the cell and those of another budding from the basal surface. Attachment of the cell to the substrate thus defines two regions of the plasma membrane. We have also studied cells within clumps that form from single cells in suspension, without attaching to the substrate. In these clumps the viruses bud to the outside or to the interior, depending on the virus. So, in establishing the limits of a membrane domain, the participation of a number of cellular features may be important, including the cytoskeleton, intercellular junctions and the relationship of the cell to the substrate. If we disrupt the junctions we get a more diffuse distribution of the viruses. But there is not a complete intermixing of the regions of viral budding.

Kornfeld: Has anyone tried putting vesicular stomatitis virus (VSV) G protein into liposomes and then fusing the liposomes with cells on a monolayer to see if the G protein migrates around?

Helenius: A related experiment has been done by K. Simons and his co-workers in Heidelberg (unpublished results). They have fused, to the apical surface of the cell, either influenza virus (which normally buds from that surface), or Semliki Forest virus (which normally buds from the basolateral surface). They then followed the fate of the antigens by using immunofluorescence and found that the influenza virus proteins stayed in the apical membrane, whereas the Semliki Forest virus antigens moved, within 20 min, to the basolateral side. This relocation seems to occur via vesicles, rather than over the tight junctions.

Meldolesi: Are you implying that VSV glycoprotein first goes to the apex and then is transferred to the basolateral plasmalemma, or that its plasmalemmal distribution is first random, irrespective of cell polarity, and then becomes polarized, by removal of the apically exposed molecules?

Sabatini: I'm implying that both mechanisms are possible. Either there is a random insertion of G protein on the cell surface or a polarized insertion in the apical membrane and then a distribution to the basolateral membrane. We cannot yet conclude, however, that the site of budding of the VSV is determined exclusively by the G protein. One would have to investigate the role of the M protein which is a peripheral membrane protein that binds to the G protein on the cytoplasmic side of the membrane. It is also possible that components of the cell membrane itself might help to determine the site of assembly.

Cuatrecasas: Do you believe that the VSV proteins are actually migrating along the membrane rather than through vesicles or cytoskeletal structures from the apical along to the basolateral borders?

Sabatini: At the moment I favour a possible role of vesicles in the redistribution of the G protein. We do see vesicles between the Golgi and plasma membrane, which contain only one type of protein and are near the apical surface. It would seem to be totally unproductive for the cell to make these segregated vesicles and then to deliver them to a membrane surface where there are both kinds of protein. I presume that at least some of those vesicles are coming in, having already selected the right kind of protein to transport to the other side. Tracer experiments like those reported by Dr Farquhar (see 157-183) might throw light on whether vesicles come from the surface rather than go to the surface.

Cuatrecasas: Do you think the tight junctions might form rather formidable barriers for the transfer of proteins?

Sabatini: I imagine that they do form formidable barriers but it is possible that they can disassemble and then reform.

Mellman: Are these cells similar to kidney epithelial cells *in vivo*, in which solutes are not normally transported across the epithelial layer but to lysosomes?

Sabatini: I don't know.

Cuatrecasas: Can these monolayers concentrate the sugars across a gradient? Can you accumulate glucose, for example, or amino acids?

Sabatini: Presumably this could be done but we haven't done it. The monolayers do transport water unidirectionally and develop an electrical potential (Cereijido et al 1978).

Cuatrecasas: Can you reverse the movement of water such that it goes to the luminal surface if you add, say, cholera toxin or prostaglandins, which are known to activate bicarbonate and water movement into the lumen?

Sabatini: We haven't done that.

Bretscher: For cells infected with influenza virus (Rodriguez-Boulan & Sabatini 1978) and labelled with the ferritin conjugates to antibodies against the spike protein, can you quantify the amount of ferritin on the apical side, as opposed to that on the basolateral side?

Sabatini: There is at least 10 times more labelling on the apical side than on the basolateral side. Rodriguez-Boulan & Pendergast (1980) previously concluded that G protein was exclusively localized to the basolateral domains. Our observations of the doubly infected cells led us to question the exclusive segregation of G protein, which does not seem to be stringent even in singly infected cells. Nevertheless, in the doubly infected cells I can safely say that 90% of the VSV virions bud from the basolateral domains. The reason that I cannot give a higher figure is that in doubly infected cells cytopathic effects may lead to junction disruption and some intermixing of budding viruses.

Raff: In tunicamycin-treated cells do either of these envelope proteins get to the cell surface and, if so, do they get to the appropriate location?

Sabatini: Green et al (1981a) have studied the role of glycosylation in the sorting-out process by utilizing tunicamycin-treated cells and 2-deoxyglucose-treated cells, as well as lectin-resistant MDCK mutants, which synthesize glycoproteins with aberrant oligosaccharides. In all these cases, including the complete absence of carbohydrate, the polarity of viral budding wasn't affected. Some of our recent work (Sabban et al 1981) dealt with the biosynthesis of band 3. We showed the arrival of the protein at the plasma membrane by trypsinization on the outside of the cell. We could recognize a number of tryptic fragments that were generated. In the cells treated with tunicamycin, the unglycosylated band 3 arrived at the surface normally, and trypsinization generated exactly the same pattern of peptides, except for one peptide, like the carbohydrate, which had a higher mobility. So here we have another example of a protein that completely lacks sugar and is still transported to the plasma membrane normally.

Cohn: Can you modulate the face of the cell by exogenous proteins, lectins or coating substrates?

Sabatini: We haven't done this.

Raff: In the doubly infected cells, were there vesicles containing only the influenza haemagglutinin as well as vesicles containing only the VSV G protein? If segregation occurred only after Golgi-derived vesicles fused with the apical membrane then one might not expect to see vesicles containing the haemagglutinin alone.

Sabatini: It is difficult to answer because we were on the tail of infection of the first virus. Just because we see both proteins in the same cell they need not have been synthesized simultaneously. It is even possible that there is a pool of one protein recirculating in the cell. I would take it that the proteins found throughout the Golgi are indeed newly synthesized proteins coming through the Golgi. But I admit that we must recognize the limitations of these experiments. Unlike the system used by Bergmann et al (1981) ours is not a synchronized system because we did not use a temperature-sensitive mutant to follow a unidirectional wave of protein coming through. Added to this is the difficulty of recognizing which of the vesicles represent recycling and which represent primary delivery to the surface.

Rothman: You described three kinds of vesicle from your electron microscopy. One contains one viral glycoprotein, a second contains the other viral glycoprotein, and a third contains both types of glycoprotein. If the probability of staining each viral glycoprotein antigen were low, even if all vesicles had both antigens, wouldn't you expect to see vesicles of all three types as a sort of statistical artifact?

Sabatini: It is certainly possible that in vesicles where we see only a few colloidal gold particles of one kind we have simply failed to label the other kind of antigen. But we have also seen some vesicles that are reasonably large and seem to contain only one of the antigens.

Rothman: I realize that the technique is very demanding, but it might be useful to have a statistical analysis of the results to relate them to the outcome expected of a Poisson distribution.

Sabatini: The frozen section technique is a very difficult technique in which the yield, related to the amount of effort put in, is small! We are not yet in a position to use it for a quantitative analysis.

Cohn: If you took a monolayer and inserted a non-specific label in the upper surface (e.g. lactoperoxidase iodination) would you expect randomization of the label with time?

Sabatini: Drs M.J. Rindler and I.E. Ivanov are currently doing that type of experiment in our laboratory. We are trying to iodinate the G protein that reaches the luminal surface and then to see whether the viruses (which always bud from the basolateral domain) will contain the protein that was labelled in the luminal face. This would definitely prove that the luminal protein is eventually transferred to the site of viral budding.

Widnell: If the vesicles that contain only G protein are recycling then, if the apical side of the cell is exposed to anti-G antibody, these vesicles contain antibody: G protein should presumably pick up antibody when it is located in the apical membrane and should then re-enter the cell. Have you tried anything like that?

Sabatini: We are considering experiments of that type. Daniel Louvard (1980) has done related experiments. He used antibodies against leucyl aminopeptidase, an enzyme that is located exclusively in the luminal surface, and he was able to clear the enzyme completely from the surface of the cell. He then followed the reappearance of the interiorized marker on the surface and saw that the enzyme was reinserted in the periphery of the cell surface, near the junctional region.

Palade: Some of us have hoped that the use of virus probes to explore pathways of traffic through the Golgi complex will reveal the sites of entry into this system as well as the sites of exit. As you pointed out, the cells should be synchronized and investigated under pulse–chase conditions. Could such a protocol tell us where in the Golgi complex the viral proteins first appear and which Golgi subcompartment is the last to contain viral antigens?

Sabatini: That was one of the primary aims of the work of Bergmann et al (1981), in which they followed the movement of G protein to the cell surface after it had accumulated in the Golgi at the non-permissive temperature.

Palade: But in those experiments of Bergmann et al (1981) the synchronization of antigen transport did not appear to be as sharp as needed for revealing the site of entry. The extent, intensity and generalization of labelling suggested that the entry was missed. The viral protein was not present everywhere, but it was detected in a very large fraction of Golgi cisternal membranes. Despite the inherent promise of the protocol, the synchronization achieved by temperature shift was not perfect. Is it possible to obtain better synchronization at the end of the transport process by inhibiting protein synthesis, for instance? This may identify the last compartment occupied by the viral antigen in the Golgi complex.

Sabatini: Experiments of that sort have been tried with cycloheximide to prevent further synthesis and to synchronize transport (Green et al 1981b). We are currently using cycloheximide to chase the G protein in MDCK cells.

Palade: Could you obtain mixed (double) labelling on one plasmalemmal domain and, at the same time, specific labelling for a single viral antigen on the rest of the plasmalemma?

Sabatini: There is rather a specific distribution of the influenza virus HA to the apical surface. That same surface contains G protein of VSV. But although we have seen G protein in the basolateral surface we have not observed much influenza virus glycoprotein there.

Palade: Can you double-label the apical domain of the plasma membrane, while labelling the basolateral domain only with the VSV antibody?

Sabatini: I believe that could be possible.

Palade: Since the bond of the antibody to the gold particle is electrostatic, is it possible that there is some antibody exchange between small and large gold particles?

Sabatini: I don't think so. We have done many controls.

Hopkins: We have looked at this point too and, at least in the test tube, there is no evidence of exchange.

Sabatini: If singly or doubly infected cells are treated, for example, with a mixture of antibodies, only one of which is conjugated to gold, we don't detect any transfer of gold between antibodies.

Palade: So the sorting can take place past the Golgi complex, and therefore we must add still another layer to an already complex system.

Rothman: Indeed, this is precisely what one would predict from the localization of the terminal glycosyltransferases to the trans cisterna; the glycosyltransferases must act in glycoproteins destined both for apical and for basolateral plasma membranes (as discussed in my paper p 120-137).

REFERENCES

Bergmann JE, Tokuyasu KT, Singer SJ 1981 Passage of an integral membrane protein, the vesicular stomatitis virus glycoprotein, through the Golgi apparatus en route to the plasma membrane. Proc Natl Acad Sci USA 78:1746-1750

Cereijido M, Robbins ES, Dolan WJ, Rotunno CA, Sabatini DD 1978 Polarized monolayers formed by epithelial cells on a permeable and translucent support. J Cell Biol 77:853-880

Green RF, Meiss HK, Rodriguez-Boulan E 1981a Glycosylation does not determine segregation of viral envelope proteins in the plasma membrane of epithelial cells. J Cell Biol 89:230-239

Green J, Griffiths G, Louvard D, Quinn P, Warren G 1981b Passage of viral membrane proteins through the Golgi complex. J Mol Biol 152:663-698

Louvard D 1980 Apical membrane aminopeptidase appears at site of cell–cell contact in cultured kidney epithelial cells. Proc Natl Acad Sci USA 77:4132-4136

Rodriguez-Boulan EJ, Pendergast M 1980 Polarized distribution of viral envelope proteins in the plasma membrane of infected epithelial cells. Cell 20:45-54

Rodriguez-Boulan EJ, Sabatini DD 1978 Asymmetric budding of viruses in epithelial monolayers: a model system for study of epithelial cell polarity. Proc Natl Acad Sci USA 75:5071-5075

Sabban E, Marchesi V, Adesnik M, Sabatini DD 1981 Erythrocyte membrane protein band 3: its biosynthesis and incorporation into membranes. J Cell Biol 91:637-646

Receptor-mediated transport of IgG across the intestinal epithelium of the neonatal rat

RICHARD RODEWALD and DALE R. ABRAHAMSON*

Department of Biology, Gilmer Hall, University of Virginia, Charlottesville, Virginia 22901, USA

Abstract The absorptive epithelium of the neonatal rat is developmentally specialized to transfer maternal immunoglobulin G (IgG) intact to the circulation while other milk proteins are digested. The epithelial cells of the duodenum and proximal jejunum which are responsible for IgG transfer represent a particular striking experimental model for study of receptor-mediated intracellular transport. Receptors located on the luminal plasma membrane selectively bind the Fc region of IgG. The IgG enters the cell by constitutive endocytosis within coated vesicles and is then released at the basolateral plasma membrane. Morphological evidence supports a model in which IgG crosses the cell as a ligand–receptor complex that dissociates only on exposure to a pH 7.4 environment found at the basolateral cell surface. Although uptake of IgG at the luminal plasma membrane is highly selective, small but significant amounts of other proteins enter the cell apparently non-selectively. Nevertheless, these latter proteins are not transferred across the cell. Double-tracer experiments indicate that IgG and these other proteins enter the cell simultaneously within the same endocytic vesicles, but that non-membrane-bound proteins are removed from the IgG transport pathway by an as yet poorly defined mechanism and sequestered within small apical vacuoles and lysosomes.

Epithelial tissues responsible for immunoglobulin transport in mammals represent striking examples of receptor-mediated endocytosis in which highly polarized cells transfer large amounts of specific macromolecules across tissue barriers. These tissues fall into two main functional categories. The first includes the extensive array of epithelia found in adults that transfer secretory immunoglobulin A from the serosal compartment to the outer mucosal surface to protect against the entry of foreign antigens. The second category consists of the specialized epithelia of the fetal yolk sac, placenta and

*Present address: Department of Medicine, Harvard Medical School, Boston, MA 02115, USA

1982 Membrane recycling. Pitman Books Ltd, London (Ciba Foundation symposium 92) p 209-232

neonatal small intestine that are responsible for transferring maternal immunoglobulin G (IgG) to the circulation of the developing animal. This transferred IgG provides important short-term humoral immunity for the immunologically naive newborn.

The purpose of this review is to summarize our studies on transport of IgG in the neonatal rat across the small intestine, the major site of maternal IgG transfer in this species. The neonatal intestine is most remarkable in its ability to transport selectively large amounts of intact IgG at the same time as other milk proteins are rapidly digested. The specificity of IgG transfer led us to choose this tissue as a model to study selective endocytosis with the assumption that our results would provide insights into immunoglobulin transport across other epithelia, and into other examples of selective endocytosis.

Transport pathway

Intestinal transfer of IgG in the neonatal rat is confined to the duodenum and proximal jejunum and ceases three weeks after birth (for reviews see Brambell 1970, Rodewald 1980a). During this time the physiology of the small intestine differs dramatically in many respects from that of the adult. Significant in the context of IgG transport is the fact that luminal proteolysis is sharply depressed, presumably to favour the transfer of intact IgG molecules. Milk proteins destined to serve the nutritional needs of the newborn are instead degraded intracellularly within highly specialized absorptive cells that line the ileum and distal jejunum. Both these distal cells and the proximal cells involved in IgG transfer are replaced abruptly at three weeks of age by absorptive cells of apparently adult form and function.

An important feature of the proximal absorptive cell not found in its adult counterpart is the presence of a system of small vesicles in the apical cytoplasm (Rodewald 1973). The vesicles are irregularly shaped tubules, 100–200 nm in diameter, which appear to form from coated pits at the bases of microvilli (Fig. 1). The cytoplasmic coat, which is similar to the clathrin coat of other coated vesicles, is often very difficult to discern owing to the densely packed filaments and other material within the terminal web. Most tubules, however, once they have formed and have passed through the terminal web, clearly lose this cytoplasmic coat. In addition to the endocytic tubules, a second class of smaller coated vesicles, 50–100 nm in diameter, is found in large numbers in the supranuclear region and at the lateral margins of the cell.

Evidence for the transcellular route of IgG transfer and for the participation of these vesicles comes from extensive transport studies we have done on

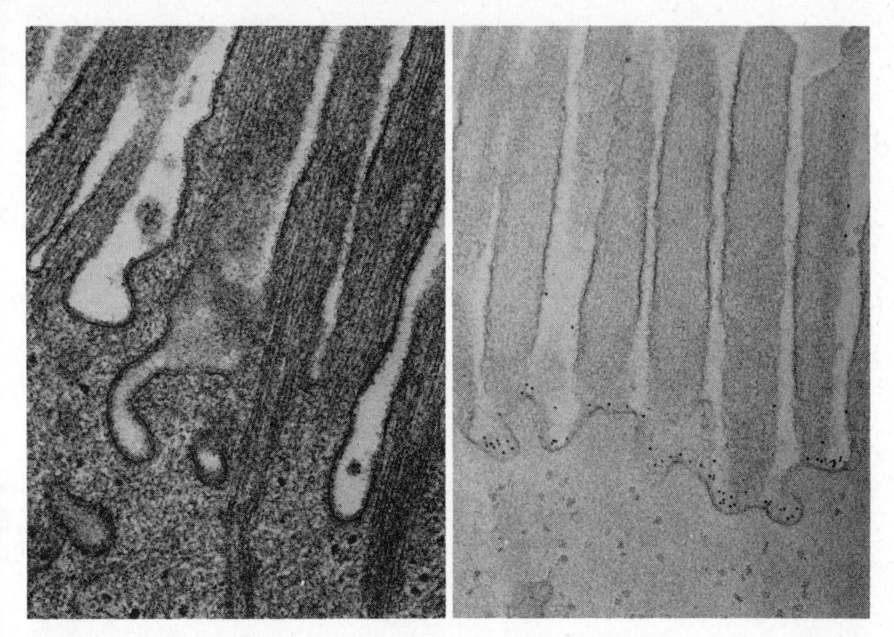

FIGS. 1 and 2. (1, left) Endocytic pits at the apical surface of an intestinal absorptive cell from the neonatal rat. (2, right) Apical surface of an absorptive cell exposed to immunoglobulin G–ferritin (IgG–Ft) (1.5 mg ml^{-1}; section stained with Pb only).

intestinal sacs, ligated *in situ* and injected with IgG that was conjugated to either horse-spleen ferritin (IgG–Ft) or horseradish peroxidase (IgG–HRP), as tracers for electron microscopy (Rodewald 1970, 1973, 1976a,b). More recently we have mapped an identical pathway, using immune complexes composed of HRP and anti-HRP IgG (Abrahamson et al 1979). Each of these IgG tracers binds to the luminal plasma membrane of the absorptive cells, often at high density within the endocytic pits (Fig. 2). Binding is highly selective and requires the Fc portion of the IgG molecule. By 10 min, many of the tubules within the terminal web contain tracer, and on continued incubation, extremely large numbers of vesicles throughout the apical cytoplasm become filled (Fig. 3). The tracers, however, do not enter the Golgi cisternae. After a period as short as 30 min, there is evidence for exocytosis of the IgG tracers from small coated vesicles at the lateral plasma membrane (Fig. 4).

Control experiments have indicated that very small but detectable amounts of unconjugated tracers may also enter these cells, even though the tracers do not bind to the luminal membrane. This uptake is particularly evident when cells are exposed to free HRP, a tracer easily visualized at very low

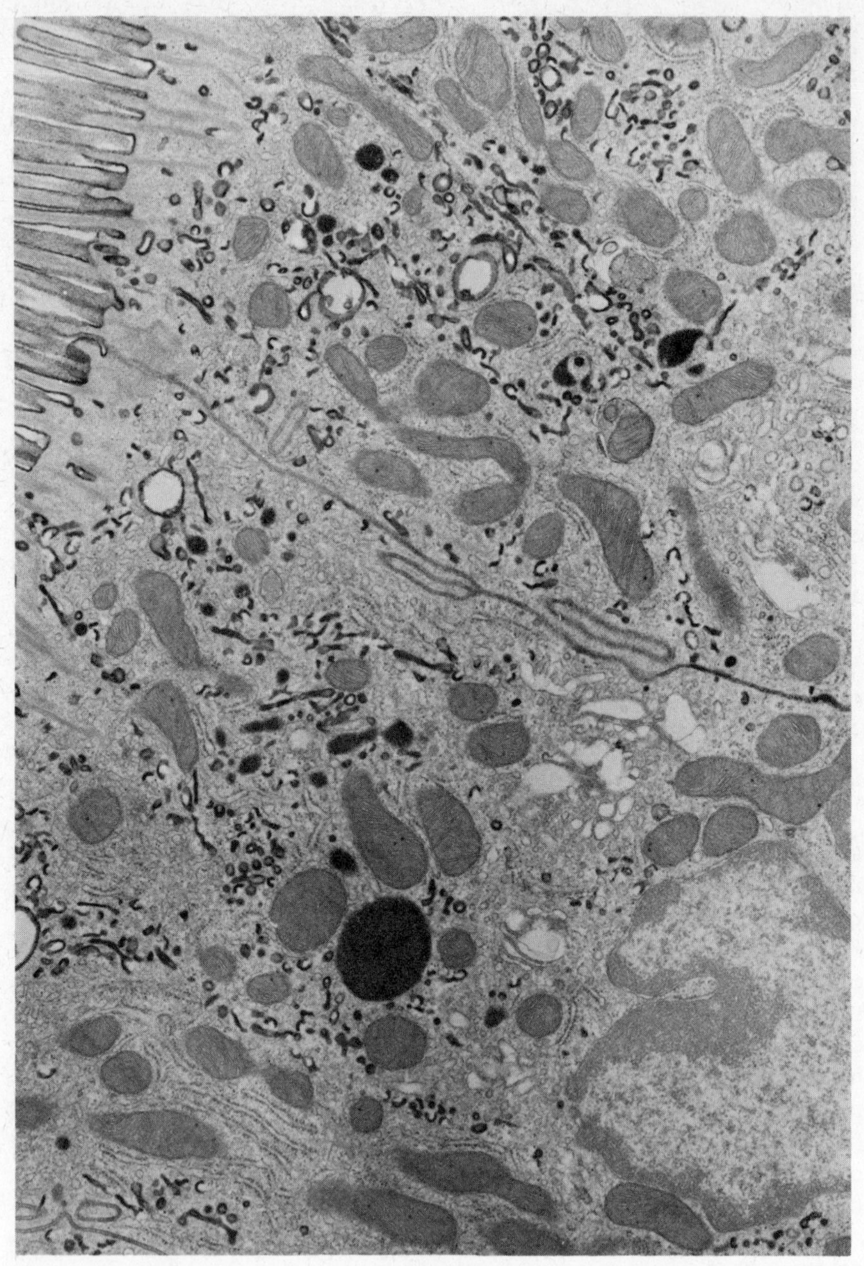

FIG. 3. Absorptive cells incubated with immune complexes composed of horseradish peroxidase (HRP) and anti-HRP IgG for 60 min. The tracer has bound to the microvillar membrane and is found in numerous vesicles throughout the cytoplasm but not within the Golgi complex.

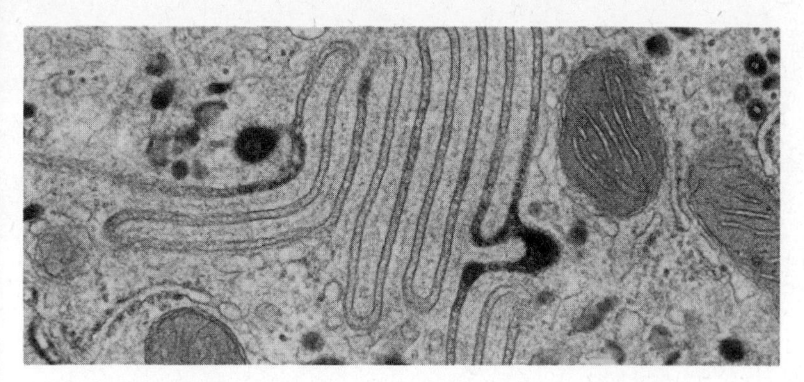

FIG. 4. Release of HRP–anti-HRP IgG complexes from coated pits at the lateral cell surface.

concentrations (Fig. 5). Even so, this tracer, which enters presumably by non-selective fluid uptake, is not successfully transferred across cells. Instead it is concentrated in the apical cytoplasm within small vacuoles, up to 1 μm in diameter, and in lysosome-like vesicles. Only the latter vesicles and not the small vacuoles can be shown, by histochemical means, to contain acid phosphatase and aryl sulphatase activities typical of lysosomes (Abrahamson & Rodewald 1981). We should note that the vacuoles and lysosomes, in addition, accumulate significant amounts of the IgG tracers, which suggests that a portion of even the IgG that enters the cells is degraded within lysosomes, in agreement with physiological studies (Morris & Morris 1974, 1976).

These results, summarized only briefly here, have led us to formulate a model for the pathway of transport in which IgG binds to specific receptors on the luminal plasma membrane, enters the cell by endocytosis within the apical tubules, and then is transferred to the small coated vesicles for discharge at the lateral surface (Rodewald 1973). The central involvement of membrane receptors agrees with the receptor model for transport proposed in detail by Brambell (1970) on the basis of extensive physiological studies by him and his collaborators. However, our model differs in important respects from Brambell's original hypothesis. First, we postulate that absorptive cells that line the duodenum and proximal jejunum are responsible for transport. Brambell (1970) and co-workers, in contrast, believed that transport occurred across more distal regions of the small intestine. Second, we propose that the principal selective event is the binding of IgG to receptors on the luminal cell membrane during endocytosis. This view differs from that of Brambell who proposed that IgG is selectively released from cells only after non-selective uptake of luminal proteins. However, as suggested by our results with free peroxidase, and as evident from more recent experiments, we now know that

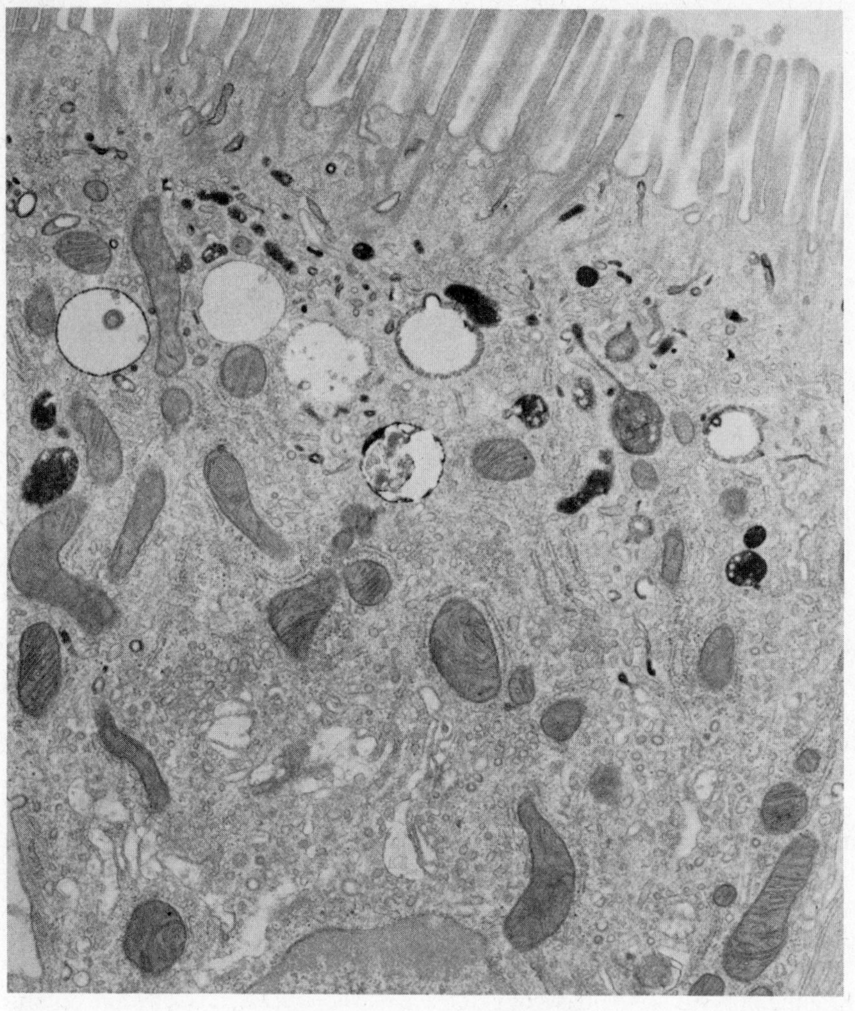

FIG. 5. Absorptive cell incubated for 2 h with HRP alone at the same concentration (1.5 mg ml^{-1}) as in Fig. 3. Tracer does not bind to the apical surface and is limited to the vesicles and small vacuoles in the region immediately below the terminal web.

the luminal membrane is not the only site of selection and that further selection probably occurs within the cell.

The receptor

Our current knowledge of the Fc receptor, the properties of which largely define the overall specificity of this transport system, is still rudimentary,

TABLE 1 Characteristics of the intestinal IgG receptor

1. Receptors recognize the Fc portion of IgG
2. $K_a = 1.3 \times 10^8\ M^{-1}$ and $5.5 \times 10^6\ M^{-1}$ at pH 6.6, 37 °C (Wallace & Rees 1980)
   $K_a = 2.6 \times 10^7\ M^{-1}$, $n = 3.2 \times 10^5$ sites per brush border at pH 6.0, 4 °C (R. Rodewald & J. P. Kraehenbuhl, unpublished)
3. Binding is pH-dependent and occurs maximally at pH 6.0–6.5. There is no detectable binding at pH 7.4
4. Receptors are found on both the luminal and the basolateral regions of the plasma membrane

although several very important features of receptor–ligand binding are already apparent. These features are summarized in Table 1.

The affinity of the receptor has recently been estimated by Wallace & Rees (1980) by measuring the binding of ^{125}I-labelled rat IgG to brush borders isolated from the absorptive cells. Their results suggest that the receptor exists in both high- and low-affinity forms whose binding is inhibited by both high pH and high ionic strength. Wallace & Rees (1980) postulated that the two forms resulted from different states of aggregation of the receptor on the membrane surface. Our results with a very similar assay system indicate, however, that there is probably only a single, high-affinity class of receptor expressed on the cell membrane and that IgG binding to this receptor is not influenced by ionic strength. We estimate from Scatchard analysis that there are approximately 3×10^5 receptors on the luminal surface of each cell (R. Rodewald & J. P. Kraehenbuhl, unpublished).

The most intriguing feature of IgG binding to the receptor is its extreme dependence on pH, a characteristic first reported by Jones & Waldmann (1972). Binding occurs with high affinity at pH 6.0–6.6 while virtually no binding can be detected at pH 7.4 or higher. The probable significance of this pH dependence was first noted at a Ciba Foundation symposium by Waldmann & Jones (1973). They suggested that IgG might be transferred across the cell as a receptor–ligand complex and that, if so, exposure of the complex to an environment at pH 7.4, as would be expected at the basolateral cell surface, would cause release of the ligand. We showed that, indeed, the normal pH of the luminal contents in the proximal intestine is between 6.0 and 6.5 (Rodewald 1976a). This would therefore permit high-affinity binding to luminal receptors. Furthermore, in morphological experiments we demonstrated that rat IgG–HRP bound selectively in a pH-dependent manner to both luminal and basolateral membranes of isolated absorptive cells at 0 °C (Rodewald 1980b). Thus, we could show that the same receptor, as defined functionally by its specificity for and pH-dependent binding of IgG, is expressed on both luminal and abluminal surfaces of cells. The presence of abluminal receptors has been shown independently by Wild & Richardson (1979).

A model in which IgG crosses the cell as a receptor–ligand complex and then dissociates from its receptor during a pH change could explain in a very simple way the net IgG transfer that occurs across the epithelium against a concentration gradient. However, the model raises several pertinent questions about the movements of the receptor within the cell which we are currently attempting to answer. Are the receptors present in each of the vesicular compartments involved in IgG transfer, as would be predicted by this model? Is there a net movement of receptors to the basolateral membrane, or are the receptors recycled as in other examples of receptor-mediated transport? If the latter is true, by what pathway do the receptors return to the luminal membrane? We have one hint at present that the receptors are probably recycled. We have shown that coated vesicles similar in appearance to those responsible for the release of IgG also form at the abluminal surface and move apically, as if they shuttle back and forth from this membrane (Rodewald 1980b). Unfortunately, we cannot easily do retrograde transport experiments with IgG conjugates to trace the possible inward movements of the receptors themselves, since the abluminal pH in intact tissue does not permit binding of IgG.

Intracellular selection of IgG

The specificity of receptor binding and the concentration of the receptors within the small endocytic tubules at the luminal surface are both key features that enhance the selective uptake of IgG by this system. The results with free HRP, however, certainly show that these features are still not sufficient to exclude all non-specific entry within the fluid volume of the endocytic vesicles. Perhaps the most surprising observation is that material which enters non-selectively does not seem to accompany the IgG across the cell. The inference is that there is a second, internal site of selection where additional sorting of IgG from other proteins takes place.

We have designed experiments to answer several specific questions about the nature of the non-specific uptake and internal sorting. First, we wished to determine whether the rate of bulk fluid endocytosis is stimulated by the presence of specific ligand. To answer this question, we exposed intestinal segments *in situ* to high concentrations of either free HRP or ferritin, which served as markers for the fluid contents of vesicles, in the presence and absence of IgG. After a short (10 min) incubation period, we counted the number of endocytic vesicles that contained tracer to estimate the initial rate of fluid uptake. In the experiments with HRP, we did parallel biochemical experiments in which we isolated and lysed absorptive cells that had been exposed to tracer and then we measured with an enzymic assay the amount of

HRP released. The results of these experiments (Abrahamson & Rodewald 1981) indicated clearly that the initial rate of HRP uptake is not influenced by whether IgG is present. Thus, endocytosis appears to be constitutive in these cells; i.e. it is not modulated by the interaction of receptor with ligand.

Although we assumed in our initial interpretation of these experiments that IgG and fluid-phase tracers would enter the cell in the same endocytic vesicles, a second model would also explain the results. This model, proposed by Moxon et al (1976) to describe IgG transport in the rabbit yolk sac,

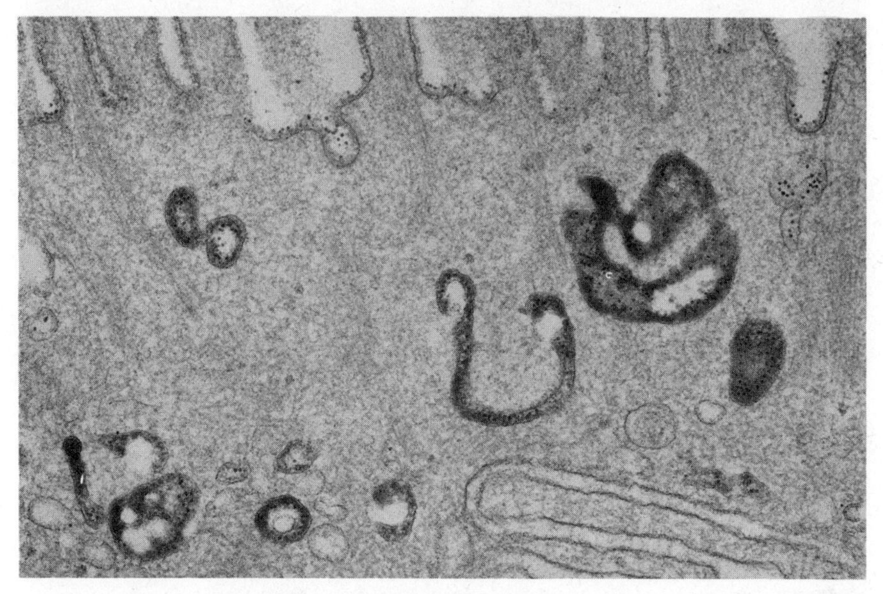

FIG. 6. Apical surface of a cell incubated with a mixture of IgG–Ft (1.5 mg ml^{-1}) and HRP (1.5 mg ml^{-1}) for 30 min. Only the IgG–Ft binds to the cell surface. Nevertheless, most endocytic vesicles within the terminal web contain both tracers (see Abrahamson & Rodewald 1981).

postulates that fluid-phase endocytosis and receptor-mediated endocytosis occur concomitantly but separately within two distinct classes of endocytic vesicle. To differentiate between the two models, we did transport experiments (Abrahamson & Rodewald 1981) in which cells were exposed to two tracers simultaneously: HRP was used again as a marker for fluid-phase uptake, and IgG–Ft was used as a tracer for selective endocytosis of IgG. A low concentration of IgG–Ft was used to ensure that virtually all the IgG–Ft that we might visualize within cells would result from receptor-mediated endocytosis.

Figs. 6 and 7 illustrate several important results of these double-tracer experiments. We found that only the IgG–Ft was bound to the luminal plasma

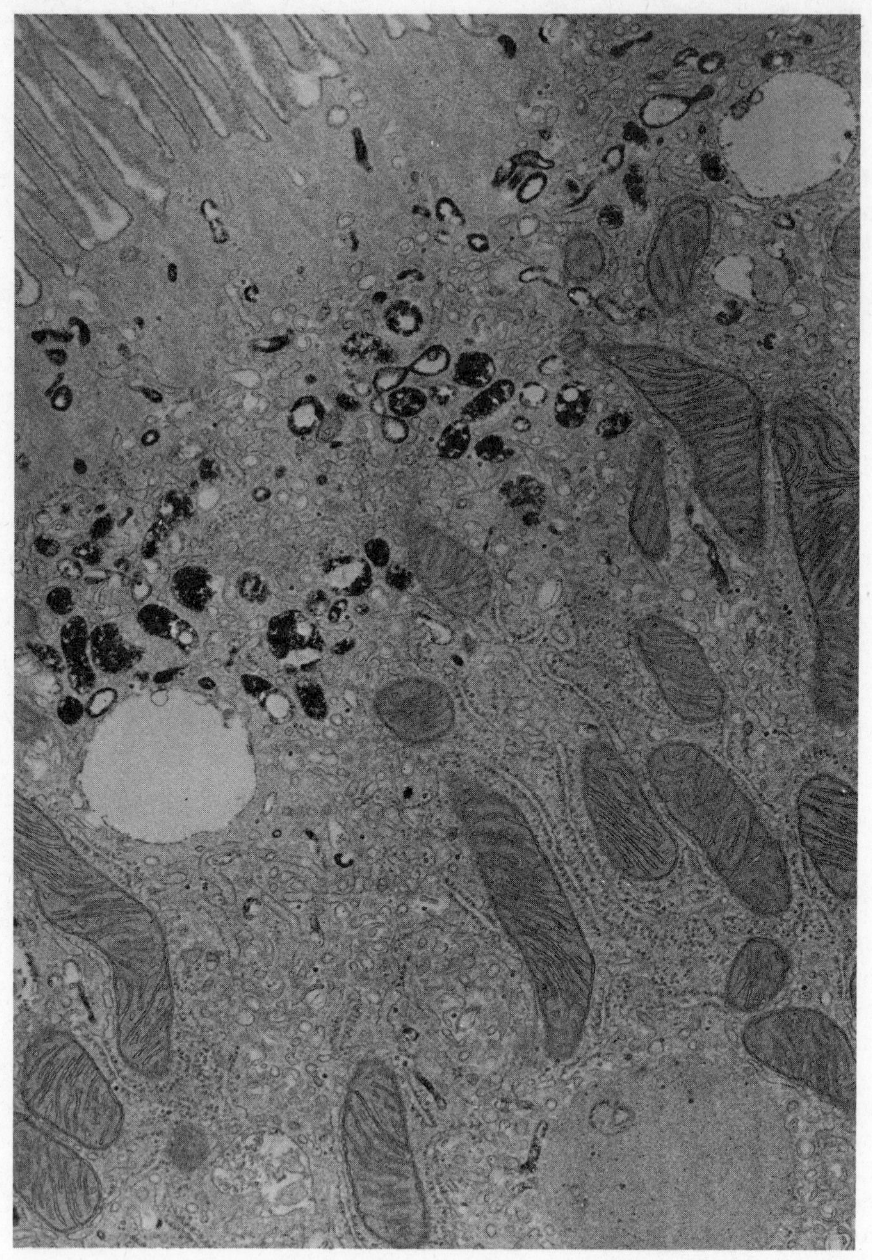

FIG. 7. Absorptive cell incubated with IgG–Ft and HRP for 60 min. The HRP that enters the cell is confined to the apical cytoplasm. Unlike IgG–Ft, the HRP is not released from the cell at the basolateral membrane (see Abrahamson & Rodewald 1981).

membrane, no doubt to the specific IgG receptors. However, within the terminal web, most endocytic vesicles contained both IgG–Ft and HRP. This indicates that the tracers did enter the cells together, within the same vesicles. Nevertheless, the two tracers were sorted within the cell. HRP remained confined to the apical cytoplasm and entered the small vacuoles and lysosomes—results identical to what was observed when cells were exposed to this protein alone. More basally within the same cells, however, most of the small vesicles that we found with tracer in them contained just the IgG–Ft conjugate, and only the conjugate was released into the intercellular spaces. The polarized distribution of the tracers within the cells suggests that the sorting of the HRP from the IgG–Ft took place in the apical cytoplasm, probably just below the terminal web in the vicinity of the small vacuoles and lysosomes. Although these and earlier experiments indicated that IgG also enters these compartments, the important point is that apparently only IgG passes beyond this level in the cell in any significant quantity.

Transfer of cationic ferritin

Further evidence for our model of IgG transport and, specifically, for this internal selection site has come from our recent experiments with cationic ferritin (CFt) (Abrahamson & Rodewald 1980, and unpublished). This highly positively charged tracer (isoelectric point = 8.5–9.0) adheres to anionic membrane sites, which are common on most cells. This ability has been exploited successfully in several experimental cell systems to trace the general flow of membrane components from the cell surface into internal compartments (Farquhar 1978, Herzog & Farquhar 1977, Herzog & Miller 1979, Ottosen et al 1980). When CFt is injected into the lumen of the proximal intestine, it binds avidly and at high density to the luminal membrane of the absorptive cells (Fig. 8). Considerably more CFt binds to the membrane, including that in the endocytic pits, than of the much more specific IgG–Ft tracer. We do not know if CFt binds to the IgG receptors, but there is no doubt that it binds to other membrane components.

In transport experiments, CFt enters the cell very rapidly, as did the IgG tracers, and it labels a large proportion of the endocytic tubules in the apical cytoplasm (Fig. 9). The amount of internalized CFt increases rapidly with time, and after 30 min of continuous exposure the tracer labels vesicles throughout the cell. Several features of the transport are noteworthy. The most interesting observation is that a portion of the CFt is transferred across the cell to the abluminal surface (Fig. 10). A similar finding has been reported by Jersild (1980). The appearance of the tracer within many of the small coated vesicles near the lateral cell surface suggests that a portion of the

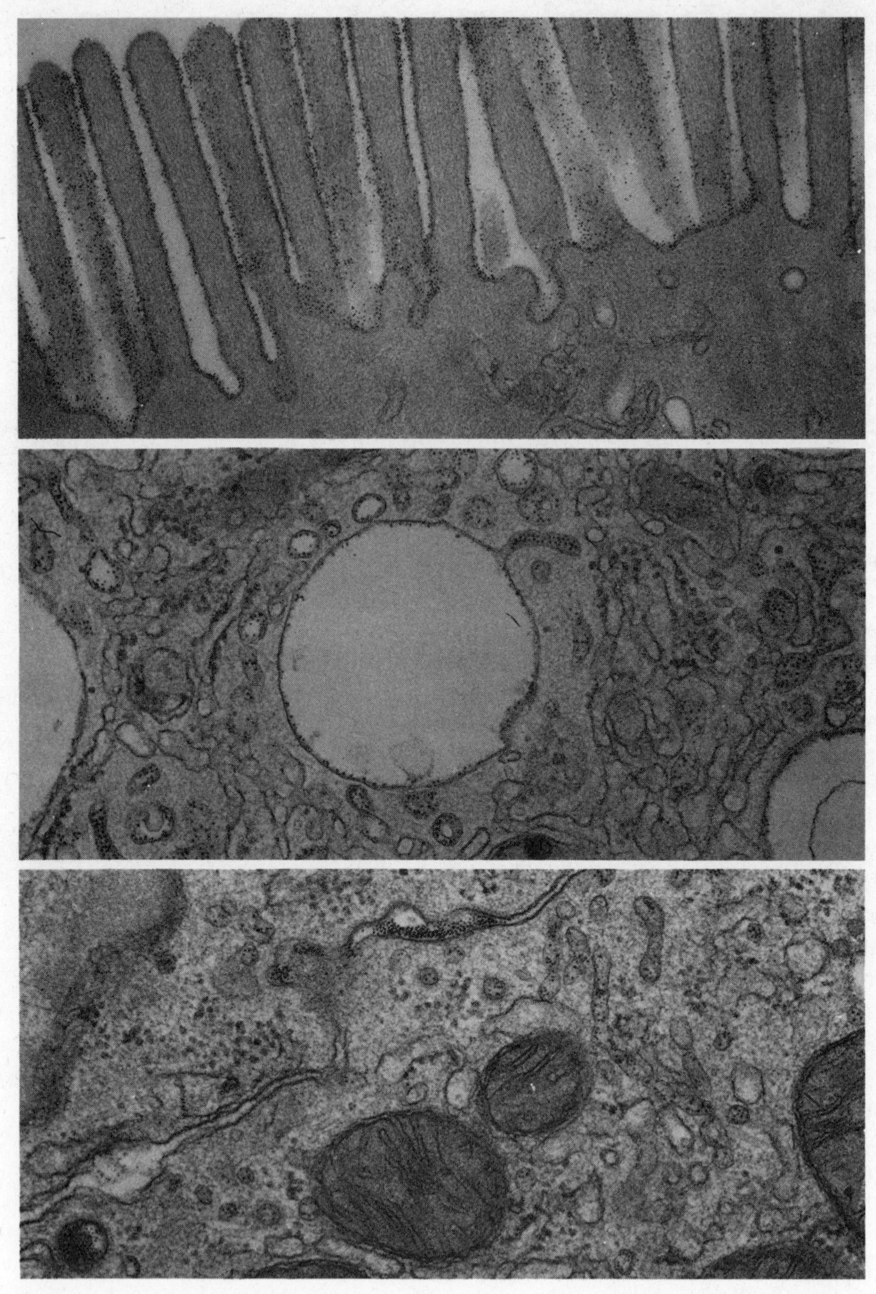

FIGS. 8–10. (8, upper) Apical surface of a cell exposed to cationized ferritin (CFt) (2 mg ml^{-1}). (9, middle) Apical cytoplasm of a cell incubated with CFt for 40 min. Tracer is found in abundance

internalized CFt traces the normal pathway for selective IgG transfer. Interestingly, the CFt within the intercellular spaces often forms clusters that remain bound to the cell membrane. This is readily apparent when the tracer is found near the bases of cells where the plasma membranes of adjacent cells are not closely apposed.

A second intriguing observation is that a portion of the tracer remaining inside the cell enters the small vacuoles and lysosomes of the apical cytoplasm. Because of the intensity of labelling of the vesicle membranes, profiles of CFt-laden endocytic tubules attached to apical vacuoles are easy to find (Fig. 9). We do not see similar connections between the tubules and apical lysosomes, however, even though these latter organelles also contain numerous tracer molecules. In addition, CFt found in the vacuoles is normally attached to the vacuole membrane, whereas within lysosomes the tracer is distributed in the internal matrix. These observations and the previous findings that most HRP is sequestered preferentially within these compartments suggest to us that the vacuoles may represent either precursors of secondary lysosomes or elements involved in transport of endocytic material to lysosomes.

A third, more cautionary observation is that in these cells small amounts of CFt gain access also to the Golgi compartment, an organelle conspicuous for its lack of involvement in the transport of all other tracers that we have studied. As early as 30 min after luminal exposure, we can identify small numbers of CFt molecules within Golgi saccules, although there is no apparent accumulation after longer times. We have not been able to determine the route of entry into this compartment or the final destination of the tracer found here. Its presence, however, certainly makes more ambiguous the simple interpretations about transport of CFt to lysosomes or even to the abluminal surface. These observations serve well to remind us that in the intestinal cells CFt may bind to a large number of membrane components that enter and follow diverse pathways within the cells.

Double-tracer studies with cationic ferritin

Notwithstanding these problems of interpretation, further studies on the transport of CFt in the presence of HRP have provided results consistent with

---

within endocytic tubules. One tubule is fused with the small vacuole in the centre of the micrograph. (10, lower) Lateral surfaces of cells exposed to CFt for 30 min. Tracer is present within the intercellular space and within tubules and coated vesicles of the neighbouring cytoplasm.

our hypothesis of an internal selection site. In these experiments, we exposed intestinal sacs to these two tracers for up to 2 h in a manner analogous to the double-tracer studies with IgG–Ft. As in the previous experiments, both CFt and HRP entered the cell within the same vesicles and, as before, there was no segregation of the two tracers within distinct compartments during endocytosis (Fig. 11). Nevertheless, the transport pathways diverged within

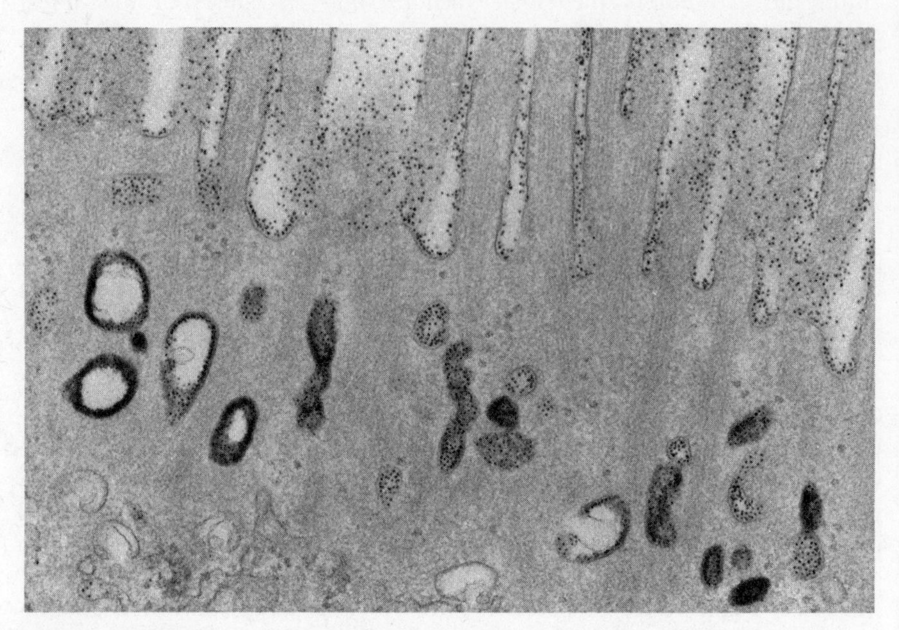

FIG. 11. Apical surface of cell incubated with CFt (2 mg ml^{-1}) and HRP (1.5 mg ml^{-1}) for 15 min. Only the CFt binds to the apical membrane. Most endocytic vesicles, however, contain both tracers.

the cell and in every respect appeared identical to the pathways followed by the tracers when administered separately (Fig. 12). HRP was restricted predominantly to apical regions and was concentrated within the vacuoles and lysosomes. Only CFt permeated more deeply within the cell and reached the intercellular spaces where it again accumulated on the abluminal cell membrane.

Cationic ferritin thus provides a further example of a molecule that can adhere to the surface of the endocytic vesicles and can be selectively sorted from other vesicle contents within the cells. We argue, therefore, that it is the continued binding to the membrane that is necessary for any ligand to be transported across the cell. For IgG conjugates and immune complexes, we have provided strong evidence that IgG binds specifically but reversibly to

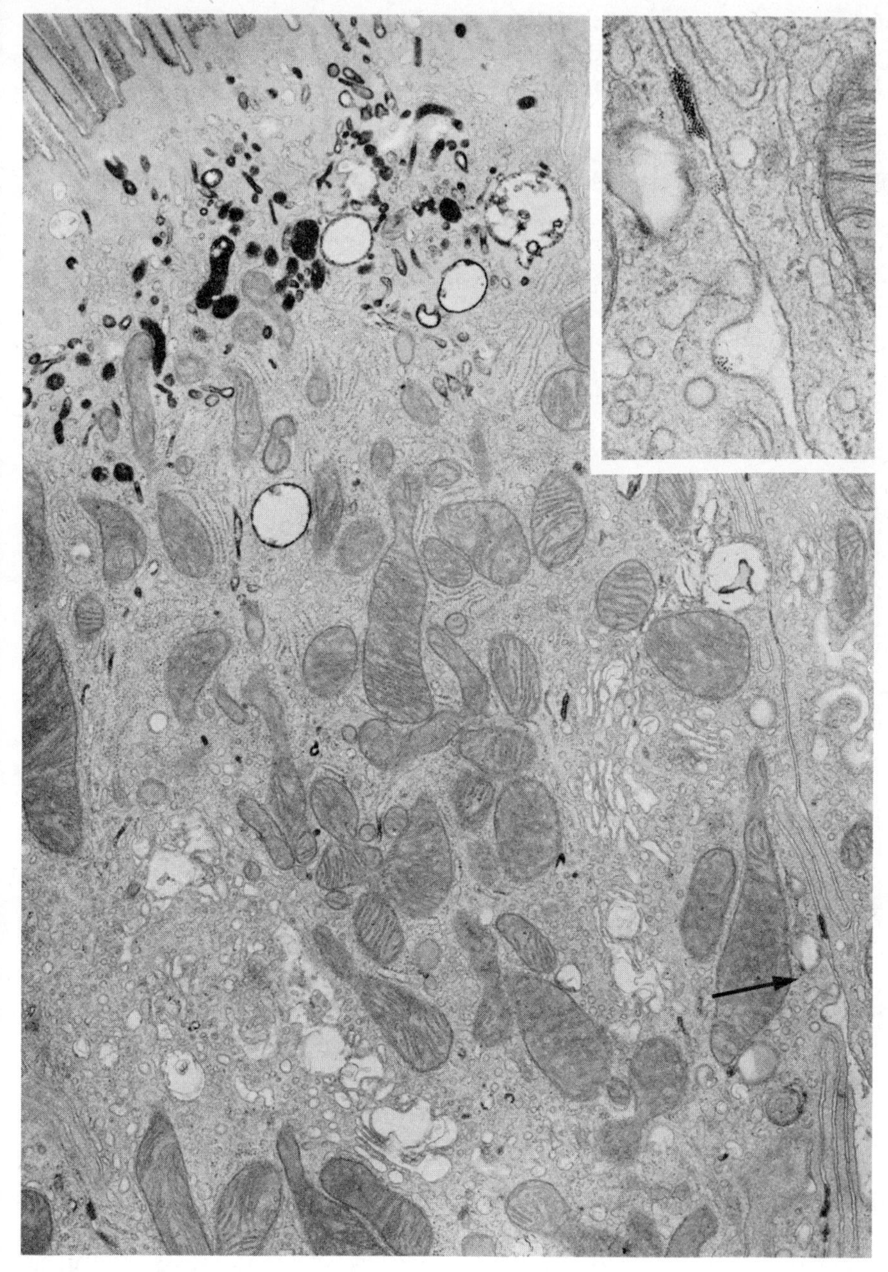

FIG. 12. Absorptive cell incubated with CFt and HRP for 60 min. HRP is limited predominantly to the most apical region of the cell. Only CFt is transferred across the cell to the intercellular space (see arrow in inset).

membrane receptors. We predict, therefore, that the IgG must be transported as receptor–ligand complexes at least beyond the site of internal sorting before any shift in pH or other influence might cause dissociation.

Site of internal selection

The most intriguing unresolved questions that arise from our studies are where and how does the internal sorting of vesicle contents occur. From the interconnections we observe between the endocytic tubules and apical vacuoles, and the accumulation of fluid-phase tracer within the vacuoles, we

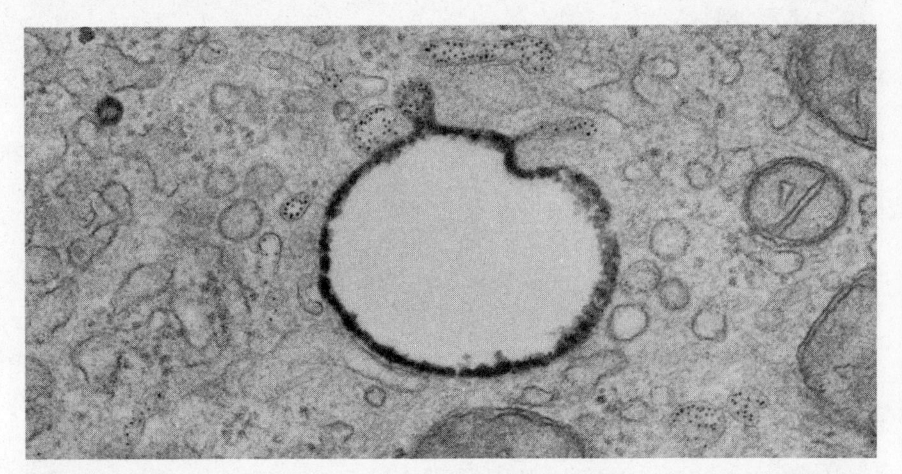

FIG. 13. Apical region of a cell exposed to CFt and HRP for 20 min. A small vacuole is shown with two attached tubules, one with both tracers and one with only CFt. The micrograph illustrates a possible model for sorting in which fluid contents are selectively discharged from the tubules into the vacuoles while membrane-bound contents remain within the tubules.

strongly favour the notion that these vacuoles are the sorting site. Indeed, in double-tracer experiments with CFt and HRP we have observed, at intermediate incubation periods of 15 and 20 min, several examples of apical vacuoles that contained HRP but few or no particles of CFt. This suggests to us that peroxidase may enter this compartment more rapidly than the membrane-bound tracer. However, since the amout of HRP cannot be easily quantified within the vacuoles, this evidence for differential rates of entry is not yet compelling. The working model for internal sorting that we nevertheless favour is one in which endocytic vesicles fuse momentarily with the apical vacuoles, long enough to transfer most of their fluid contents and possibly even a portion of their membrane-bound components to the vacuoles. The endocytic vesicles then detach and transfer their remaining bound contents to the exocytic coated vesicles. The small vacuoles, on the other hand, either transfer their contents to lysosomes or become lysosomes. In this model, the ligand transported across the cell would never be exposed directly to the lysosomal compartment, a feature which would ensure that most IgG as well as its receptor would not be degraded. A possible example of sorting is depicted in Fig. 13.

We believe that the internal sorting of IgG which we observe may be an expression of a widespread capability of endocytic cells. It is intriguing that vesicles morphologically similar to the apical vacuoles of the intestinal cells have been implicated in transport within several other cell types involved in selective endocytosis (Haigler et al 1979, Wall et al 1980, Willingham & Pastan 1980, Van Deurs et al 1981). The role of these small vacuoles in any ligand-sorting that might take place in these other systems, however, is not at all clear. The principal advantage to a cell of having internal sorting sites as we describe them is that it would allow the cell to utilize common transport pathways to carry different vesicle contents to different membrane compartments. Our studies emphasize that an important determinant of the final destination of a substance is its interaction with components of the vesicle membranes and, in particular, the reversibility of this interaction at the site of sorting.

Acknowledgements

This work was supported by grants from the United States Public Health Service (NIH AI-11937) and the Charles E. Culpeper Foundation.

REFERENCES

Abrahamson DR, Rodewald R 1980 Selective transfer of vesicle contents during endocytosis in the neonatal rat intestine. J Cell Biol 87:310a (abstr)

Abrahamson DR, Rodewald R 1981 Evidence for the sorting of endocytic vesicle contents during the receptor-mediated transport of IgG across the newborn rat intestine. J Cell Biol 91:270-280

Abrahamson DR, Powers A, Rodewald R 1979 Intestinal absorption of immune complexes by neonatal rats: a route of antigen transfer from mother to young. Science (Wash DC) 206:567-569

Brambell FWR 1970 The transmission of passive immunity from mother to young. Elsevier/North-Holland Biomedical Press, Amsterdam

Farquhar MG 1978 Recovery of surface membrane in anterior pituitary cells. Variation in traffic detected with anionic and cationic ferritin. J Cell Biol 77:R35-R42

Haigler HT, McKanna JA, Cohen S 1979 Rapid stimulation of endocytosis in human carcinoma cells A-431 by epidermal growth factor. J Cell Biol 83:82-90
Herzog V, Farquhar MG 1977 Luminal membrane retrieved after exocytosis reaches most Golgi cisternae in secretory cells. Proc Natl Acad Sci USA 74:5073-5077
Herzog V, Miller F 1979 Membrane retrieval in epithelial cells of isolated thyroid follicles. Eur J Cell Biol 19:203-215
Jersild RA Jr 1980 Restricted mobility and endocytosis of anionic sites of newborn rat jejunal brush border membranes. J Cell Biol 87:94a(abstr)
Jones EA, Waldmann TA 1972 The mechanism of intestinal uptake and transcellular transport of IgG in the neonatal rat. J Clin Invest 51:2916-2927
Morris B, Morris R 1974 The absorption of ^{125}I-labelled immunoglobulin G by different regions of the gut in young rats. J Physiol (Lond) 241:761-770
Morris B, Morris R 1976 Quantitative assessment of the transmission of labelled protein by the proximal and distal regions of the small intestine of young rats. J Physiol (Lond) 255:619-634
Moxon LA, Wild AE, Slade BS 1976 Localisation of proteins in coated micropinocytotic vesicles during transport across the rabbit yolk sac endoderm. Cell Tissue Res 171:175-193
Ottosen PD, Courtoy PJ, Farquhar MG 1980 Pathways followed by membrane recovered from the surface of plasma cells and myeloma cells. J Exp Med 152:1-19
Rodewald R 1970 Selective antibody transport in the proximal small intestine of the neonatal rat. J Cell Biol 45:635-640
Rodewald R 1973 Intestinal transport of antibodies in the newborn rat. J Cell Biol 58:198-211
Rodewald R 1976a pH-dependent binding of immunoglobulins to intestinal cells of the neonatal rat. J Cell Biol 71:666-670
Rodewald R 1976b Intestinal transport of peroxidase-conjugated IgG fragments in the neonatal rat. In: Hemmings WA (ed) Maternofoetal transmission of immunoglobulins. Cambridge University Press, London, p 137-149
Rodewald R 1980a Immunoglobulin transmission in mammalian young and the involvement of coated vesicles. In: Ockleford C, White A (eds) Coated vesicles. Cambridge University Press, London, p 69-101
Rodewald R 1980b Distribution of immunoglobulin G receptors in the small intestine of the young rat. J Cell Biol 85:18-32
Van Deurs B, von Bülow F, Møller M 1981 Vesicular transport of cationized ferritin by the epithelium of the rat choroid plexus. J Cell Biol 89:131-139
Waldmann TA, Jones EA 1973 The role of cell surface receptors in the transport and catabolism of immunoglobulins. In: Protein turnover. Elsevier/North-Holland, Amsterdam (Ciba Found Symp 9) p 5-18
Wall DA, Wilson G, Hubbard AL 1980 The galactose-specific recognition system of mammalian liver: the route of ligand internalization in rat hepatocytes. Cell 21:79-93
Wallace KH, Rees AR 1980 Studies on the immunoglobulin-G Fc-fragment receptor from neonatal rat small intestine. Biochem J 188:9-16
Wild AE, Richardson LJ 1979 Direct evidence for pH-dependent Fc receptors on proximal enterocytes of suckling rat gut. Experientia 35:838-840
Willingham MC, Pastan I 1980 The receptosome: an intermediate organelle of receptor-mediated endocytosis in cultured fibroblasts. Cell 21:67-77

DISCUSSION

Raff: Have you really provided evidence for intracellular sorting? Isn't it still possible that the sorting is going on at the plasma membrane? Perhaps only a small fraction of the newly formed endocytic vesicles contain the Fc receptors, and these may go all the way to the basolateral surface and discharge, while most of the vesicles formed from the plasma membrane may take anything that is in fluid phase.

Rodewald: The best evidence is not from the cationized ferritin experiments but from the double-tracer experiments with peroxidase and immunoglobulin G–ferritin (IgG–Ft) (Abrahamson & Rodewald 1981). A large majority of the vesicles that contain a fluid-phase marker also contain receptor and specific ligand.

Raff: Yes, but you said that 60% of IgG–Ft goes to lysosomes, and is therefore following the wrong pathway. Therefore the doubly labelled vesicles may be those that are carrying this 60% along with the fluid-phase markers.

Rodewald: That's possible, but 40% would be a rather low estimate of the amount that could cross the cell intact, and it would be extremely inefficient to destroy such a large percentage of the receptors. Transport efficiency would depend on the concentration of the IgG and on whether or not the receptors were saturated.

Sly: At what concentrations of IgG–Ft does the inefficient transfer occur? Is it at concentrations in excess of saturation?

Rodewald: Sixty per cent of the IgG can be degraded, as revealed by experiments using iodinated IgG (Morris & Morris 1977) at concentrations probably several orders of magnitude above the association constant of the receptor.

Sly: So the endocytosed ligand that is in excess of receptor is not bound, and is behaving like the peroxidase.

Rodewald: Dr Raff's point is still valid because in the morphological experiments we find the IgG tracers entering the lysosomal compartment even when we use concentrations of these tracers at which most must enter the cell by receptor-mediated endocytosis.

Palade: Perhaps the question can be formulated otherwise. Do you find in the apical part of the cell a mixed population of vesicles (i.e. mixed in terms of types of load present in the expected proportion)?

Rodewald: I can answer that partially. I don't see two morphologically distinct classes of vesicle.

Raff: Did you see some vesicles in the apical region that had only ferritin and no horseradish peroxidase (HRP)?

Rodewald: Yes, but very few. They are very similar, morphologically, to the vesicles that contained both tracers.

Palade: HRP cannot be fixed in the lumen, and therefore cannot be fixed in a vesicle in continuity with the lumen. But it *can* be 'fixed' (i.e. imprisoned) in closed compartments.

Schneider: Did you test biochemically whether the IgG which is transferred through the intestinal cells is still intact? If the sorting station is the lysosomes, there could be partial digestion of the IgG that is transferred. Did you check that?

Rodewald: No, but many experiments have shown that the IgG that is transferred across the cells is identical to intact IgG.

Goldstein: Is there any biochemical evidence that the uptake of IgG is continuous over many hours, thus implying a recycling of receptors? Or does the uptake reach a peak and then decline rapidly?

Rodewald: I don't think anyone has done experiments to look at that.

Meldolesi: You mentioned that the number of receptors is about 300 000 per cell. Does this figure account only for the receptors exposed at the apex, or for all receptors present, including those interiorized within the cell?

Rodewald: That figure is for the receptors on the apical surface. Presumably there would be many more, perhaps the majority, within the cell on the membranes of the endocytic vesicles.

Meldolesi: How many receptors are on the basolateral membrane?

Rodewald: I haven't measured the number on that surface.

Cuatrecasas: It might be interesting to look simultaneously at two different ligands that undergo receptor-mediated endocytosis, rather than to look at only one of these and another which is fluid-phase. In fibroblasts, and possibly other cells, many ligands are co-internalized. For example, viruses, EGF and α_2-macroglobulin can go into the same vesicle. It would be interesting to see whether a total segregation occurs. Would the viruses then go to the basolateral border, or would they be present initially in the same vesicle, as expected, and subsequently be disaggregated such that the IgG would go only to the basolateral membrane and the virus only to the lysosome?

Rodewald: I would like to think that the latter happens. There could be two destinations, even for two ligands that bind to separate receptors within the same vesicle. For instance, in the intestinal cells the IgG remains bound to its receptor at low pH. However, another receptor could dissociate its ligand at low pH, and this might occur at a sorting site if the pH there is low enough.

Cuatrecasas: I find it difficult to understand that 'stickiness' is the signal for sorting. In the intestinal lumen there are many polysaccharides and other things that will non-specifically stick or adhere strongly, and will be endocytosed. An awful lot of things would then go to the basolateral border and find themselves either in a lymphatic or in the circulatory system.

Rodewald: I do think that most of the selection is at the surface and that selection depends on whether a molecule binds to the surface of the endocytic pit. With free ferritin at concentrations as high as 25–50 mg ml^{-1}, very little ferritin is taken up into the cell.

Hopkins: How penetrable is the tubular system in the apical part of the cell? If you pre-fix the cells and apply peroxidase does any enzyme penetrate the tubules?

Rodewald: Very few profiles under those conditions are continuous with the apical surface.

Cohn: Have you waited a little longer and used older animals before doing the experiment? Does the Fc receptor on those cells disappear then?

Rodewald: Yes. By 22 days of age no receptors are demonstrable on the epithelial cells (Rodewald 1980). This corresponds to the age at which transport of maternal IgG ceases.

Cohn: And if you add immune complexes at that point what is their fate?

Rodewald: There is no uptake into the cells of either immune complexes or the other IgG tracers we have used (Rodewald 1973, Abrahamson et al 1979).

Cohn: The question is whether the receptor is the rate-limiting step of transport.

Rodewald: A lot of tracer experiments have been done with adult intestinal cells (Walker et al 1972, 1976), and there is evidence that some material crosses these cells. The amount that crosses is likely to be very small. Our morphological experiments suggest that very little material enters the adult cells from the lumen.

Cohn: So is the rate of uptake controlling the events?

Rodewald: I believe that a completely different cell type populates the adult intestine after three weeks of age. This cell type no longer expresses the receptor and no longer has this system of endocytic vesicles for transport of IgG (Rodewald 1973, 1980).

Mellman: Have you any evidence that the apical vacuoles have not fused with secondary lysosomes?

Rodewald: The only evidence we have that the apical vacuoles are not secondary lysosomes is that we never find aryl sulphatase or acid phosphatase activity in them.

Mellman: Are those enzymes good markers in these cells?

Rodewald: One *can* show those enzyme activities in the more densely staining vesicles.

Mellman: Secondly, in your morphological comparisons between the distribution of fluid HRP and HRP linked to IgG in the apical surface of the cells, does the distribution depend not so much on where the two types are localizing but on how much HRP is actually being internalized in each case? If more of the tracer were internalized when bound to IgG, it would provide a much more sensitive measure.

Rodewald: It is difficult to make any quantitative statement about the sorting when one uses these two tracers because the HRP is a qualitative marker. I would not put too much emphasis on the idea of differential entry of the HRP into the apical vacuoles in double-tracer experiments.

Schneider: Using both cultured hepatocytes and the liver *in vivo*, we have investigated the uptake, intracellular fate and transfer to bile canaliculi of several ligands taken up specifically by the liver: polymeric IgA (see p 114-115); galactosylated serum albumin (galBSA); and haemoglobin–haptoglobin complex (haem–hap). Five minutes after binding to hepatocytes the three ligands are associated with vesicles distinct from plasma membranes, Golgi and lysosomes and they equilibrate around 1.13 g ml^{-1} after isopycnic centrifugation on sucrose gradients. After that, two-thirds of cell-bound IgA but less than 5% of galBSA and haem–hap go to the bile, while one-third of polymeric IgA and more than 90% of the other two ligands gain access to lysosomes and are then digested (Limet et al 1982a,b). These results suggest that the sorting of these ligands takes place inside the cell.

Raff: Are they the same vesicles? Have you located them immunochemically?

Schneider: As I mentioned already there is no difference between the distribution of the three ligands on the basis of cell fractionation experiments. We are now doing morphological experiments to test this question (see Courtoy et al 1981, 1982) but we have observed that HRP–IgA conjugates are found in coated pits and profiles and then in vesicles comparable to those described by Wall et al (1980).

Geisow: Is it possible, in the isolated cell system, to use weak bases? On the basis of the pH dependence of the IgG–receptor interaction one would expect dissociation to occur.

Rodewald: We want to do inhibitor studies with lysosomotropic agents. Assuming that the sorting site we postulate is a pre-lysosomal compartment that could concentrate these agents, we might be able to increase the pH high enough to cause release of IgG from its receptor, thus blocking IgG transport at this sorting site. The problem is that we shall have to use isolated tissue, which is difficult to maintain *in vitro* for even short periods of time.

Geisow: If you *could* perturb the pH at the basal or at the luminal face, you might be able to use the receptor to follow recycling.

Rodewald: We have not done experiments on isolated cells to see if binding and uptake can take place from both luminal and basolateral surfaces. Incubation of the cells at pH 6.0 would not be physiological.

Helenius: Do you have any information about the pH in the apical vesicles?

Rodewald: No.

Helenius: A low pH in the apical vesicles might provide an easy explanation for the separation of *content* molecules and *bound* molecules. In some cases the binding of bound molecules could be sensitive to low pH and therefore the ligands could be released to become content molecules in the apical vesicles.

Rodewald: That could occur. An advantage of having a distinct compartment, such as these vacuoles or the intermediate vesicles that Ann Hubbard has mentioned (p 109-115), is that the ligands that are not going to enter the lysosomes need not be exposed to the lysosomal hydrolases. One would not need to postulate receptors that *a priori* are resistant to these enzymes.

Raff: Are the Fc receptors on the basolateral surface driven there by immunoglobulin in the lumen? If you keep the epithelium free from immunoglobulin do you still see receptors on the basolateral surface?

Rodewald: I don't know; we haven't done that experiment!

Rothman: You could use a virus like Semliki Forest virus, and coat it with a little anti-spike protein antibody. Then, the IgG receptor will pull in the virus, and you can see at what stage the virus fuses or not, as a way of measuring pH in endocytic compartments.

Rodewald: That would be a good experiment. Let me ask Dr Helenius if he ever sees fusion of viruses in vacuoles that don't appear to be lysosomes?

Helenius: Unfortunately we have not yet seen convincing images of viruses fusing inside intracellular vacuoles.

Palade: Do you ever see IgG in lysosomes?

Rodewald: Yes. That is, we see IgG–Ft conjugates in the lysosomes.

REFERENCES

Abrahamson DR, Rodewald R 1981 Evidence for sorting of endocytic vesicle contents during the receptor-mediated transport of IgG across the newborn rat intestine. J Cell Biol 91:270-280

Abrahamson DR, Powers A, Rodewald R 1979 Intestinal absorption of immune complexes by neonatal rats: a route of antigen transfer from mother to young. Science (Wash DC) 206:567-569

Courtoy PJ, Limet JN, Baudhuin P, Schneider Y-J, Vaerman JP 1981 Receptor-mediated endocytosis of polymeric IgA by cultured rat hepatocytes. Cell Biol Int Rep 5:57

Courtoy PJ, Limet JN, Baudhuin P, Schneider Y-J, Vaerman JP 1982 Ultrastructural aspects of transepithelial transfer of IgA through rat liver. Arch Int Physiol Biochim 90:B11-B12

Limet JN, Quintart J, Otte-Slachmuylder C, Schneider Y-J 1982a Receptor-mediated endocytosis of hemoglobin–haptoglobin, galactosylated serum albumin and polymeric IgA by the liver. Acta Biol Med Ger 41:113-124

Limet JN, Otte-Slachmuylder C, Schneider Y-J 1982b Uptake by hepatocytes, endocytosis and biliary transfer of polymeric IgA, anti-secretory component IgG, hemoglobin–haptoglobin and horseradish peroxidase. Arch Int Physiol Biochim 90:958-59

Morris B, Morris R 1977 The digestion and transmission of labelled immunoglobulin G by enterocytes of the proximal and distal regions of the small intestine of young rats. J Physiol (Lond) 273:427-442

Rodewald R 1973 Intestinal transport of antibodies in the newborn rat. J Cell Biol 58:198-211
Rodewald R 1980 Distribution of immunoglobulin G receptors in the small intestine of the neonatal rat. J Cell Biol 85:18-32
Walker WA, Cornell R, Davenport LM, Isselbacher KJ 1972 Macromolecular absorption. Mechanism of horseradish peroxidase uptake and transport in adult and neonatal rat intestine. J Cell Biol 54:195-205
Walker WA, Abel SN, Wu W, Bloch KJ 1976 Intestinal uptake of macromolecules. V: Comparison of the in vitro uptake of antigen–antibody complexes prepared in antibody or antigen excess. J Immunol 117:1028-1032
Wall DA, Wilson G, Hubbard AL 1980 The galactose-specific recognition system of mammalian liver: the route of ligand internalization in rat hepatocytes. Cell 21:79-93

General discussion II

Development of a cell line with a marked proliferation of crystalloid endoplasmic reticulum

Goldstein: With Michael Brown and Richard Anderson, I have recently obtained some unexpected results that may throw light on membrane biogenesis and organelle development. Several years ago, we decided to develop a cell line that would have a high level of hydroxymethylglutaryl-CoA reductase (HMG-CoA reductase, EC 1.1.1.34) so that we could clone the gene for this enzyme and study its regulation.

HMG-CoA reductase is a tightly regulated enzyme that is located on the endoplasmic reticulum (ER). It produces mevalonate which is used for the

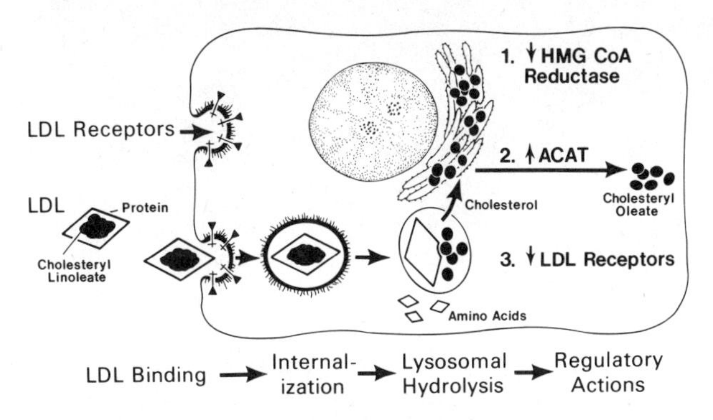

FIG. 1. (*Goldstein*) Sequential steps in the LDL pathway in cultured mammalian cells. HMG CoA reductase denotes 3-hydroxy-3-methylglutaryl CoA reductase; and ACAT denotes acyl-CoA:cholesterol acyltransferase.

synthesis of cholesterol and several non-sterol products in animal cells. Cells grown in the absence of plasma low density lipoprotein (LDL), their normal source of cholesterol, develop high levels of reductase; conversely, when LDL is present, the lipoprotein delivers cholesterol to cells through the LDL receptor pathway, and reductase activity is suppressed (Goldstein & Brown 1977) (Fig. 1). In the presence of compactin, a potent competitive inhibitor of reductase (Endo et al 1976), cells develop a compensatory increase in

1982 Membrane recycling. Pitman Books Ltd, London (Ciba Foundation symposium 92) p 233-245

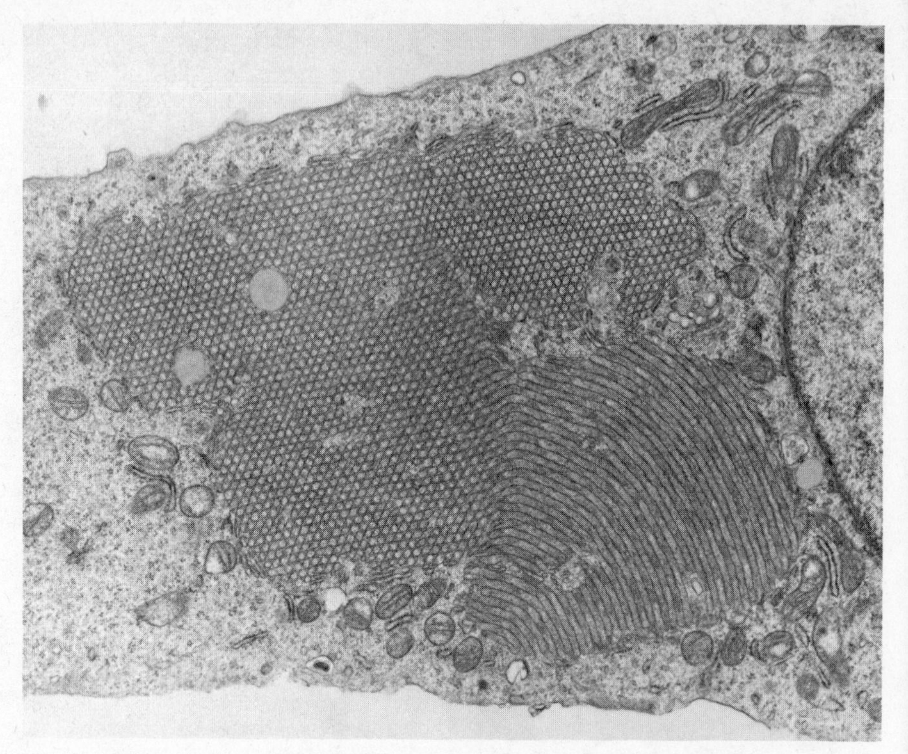

FIG. 2. (*Goldstein*) Crystalloid ER in UT-1 cells. This electron micrograph shows different orientations of membrane-lined tubules. Magnification, ×30 000.

reductase activity (Brown et al 1978). This regulatory response is triggered by a deficiency of cholesterol and other mevalonate-derived products. If the reductase activity cannot increase sufficiently to overcome the inhibition by compactin, the cells die (Brown & Goldstein 1980).

Studies with methotrexate, an inhibitor of dihydrofolate reductase, have established that a competitive inhibitor of an essential enzyme can be used to develop a cell line that has a marked increase in the amount of the enzyme (Schimke et al 1978). This is done by progressively raising the concentration of the inhibitor in the medium. After each increase, more than 90% of the cells die, but some cells develop increased enzyme levels and survive. In this way cultured cells with a 200-fold increase in dihydrofolate reductase levels were obtained. We have used a similar approach to obtain cells with high levels of HMG-CoA reductase. Chinese hamster ovary (CHO) cells were adapted to increasing concentrations of compactin in the absence of lipoproteins. We eventually obtained a clone of cells, designated UT-1, which has been grown for over one year in the presence of 40 μM-compactin (Chin et al

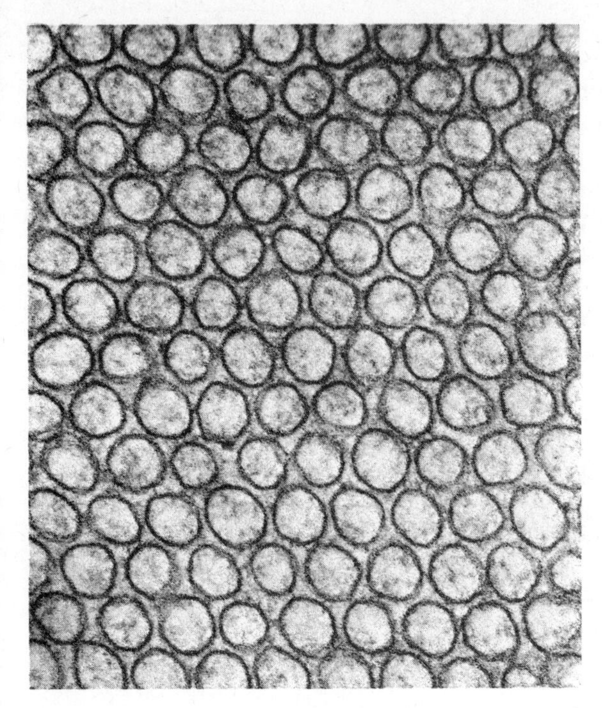

FIG. 3. (*Goldstein*) Cross-sectional view of the packed crystalloid arrangement of smooth tubules in UT-1 cells. Note the hexagonal arrangement of tubules. Magnification, ×81 000.

1982). Unadapted CHO cells growing in the absence of LDL are killed by compactin at 0.3 μM.

The HMG-CoA reductase activity in the UT-1 cells is 500-fold higher than that in the parental CHO cells. The reductase accounts for 2% of the total cell protein in the UT-1 cells, as calculated from enzyme specific activity and by immunoprecipitation with anti-reductase antibody after growth of cells in [^{35}S]methionine (Chin et al 1982). These immunochemical studies were done by Daniel Chin, Ken Luskey and Jerry Faust.

The surprising finding came when we examined the morphology of the UT-1 cells by electron microscopy. To accommodate the increased amounts of HMG-CoA reductase, the UT-1 cells developed a marked proliferation of tubular smooth ER membranes that were packed tightly in hexagonal arrays (Figs. 2 & 3). These crystalloid structures had sharp borders that were often surrounded by typical smooth and rough ER. Longitudinal profiles of these structures were sometimes continuous with cross-sectional profiles (Fig. 2). At high magnification, the hexagonal packing was evident (Fig. 3). The diameter of the tubules was approximately 86 nm, and the distance between tubule centres was 110 nm. The width of the membranes was 5–6 nm. No such crystalloid structures were seen in the parental CHO cells.

Interestingly we found that high levels of HMG-CoA reductase could be suppressed by 90% within 8–12 h of addition of LDL to the culture medium. Similarly, growth of the UT-1 cells in the presence of LDL for 24 h led to a disappearance of the crystalloid ER.

The UT-1 cells should prove useful, not only for cloning the gene for HMG-CoA reductase, but also for elucidating the mechanism by which induction of an ER enzyme stimulates the proliferation and packaging of smooth ER. Moreover, the rapid disappearance of the crystalloid ER in response to LDL-derived cholesterol suggests that UT-1 cells can be used to study the cellular mechanism for destroying an organelle and its enzyme when neither is needed any longer.

Sabatini: The images strikingly resemble, in their paracrystalline arrangement, areas of smooth ER seen in steroid-secreting cells, particularly in guinea-pig adrenal (Black 1972).

Goldstein: That's an astute comment! The fetal adrenal gland of the guinea-pig (Black 1972) and the ultimobranchial gland of the lemur (Sisson & Fahrenback 1967) have a crystalloid ER that closely resembles the structure we have produced in the UT-1 cells.

Meldolesi: You said that the enzyme disappears in a few hours but what happens to the membranes?

Goldstein: The crystalloid ER disappears in 24 h but we don't yet know how this occurs.

Blest: Another thing that the cytoplasmic tubules resemble is the transductive membranes of arthropod or many other invertebrate photoreceptors. In the arthropod photoreceptor, the transductive region consists of tubular microvilli of great regularity which are part of the plasma membrane. They can be assembled very rapidly indeed (Blest 1978, Stowe 1980), and their membranes, in addition to rhodopsin, have a very high sterol content (Zinkler 1974). Do you know the sterol content of the membranes of the cytoplasmic tubules?

Goldstein: No. We do know that the UT-1 cells have about half the total cholesterol content of the parent CHO cell. We are currently trying to purify the crystalloid ER and then we should be able to characterize its protein and lipid composition.

Sabatini: I have seen similar structures in some skeletal muscles after their denervation (unpublished observations).

Branton: Do you know the approximate coverage of the membrane with the reductase?

Goldstein: No. If the reductase accounts for 2% of the cell protein and if approximately 15–20% of the cell cytoplasm is occupied by crystalloid ER, then there are probably not many reductase molecules per tubule.

Branton: So it isn't the reductase itself which accounts for the total protein content of the membrane?

Goldstein: In cross-section, each ER tubule is surrounded by six other ER tubules, forming hexagonal arrays. The hexagonal array forms a very stable structure. It is conceivable that the reductase is the predominant protein present in the crystalloid ER, but we need to purify the organelle before we can answer that question.

Dean: Is the turnover of protein components in that array normal, rather as it would be in a phenobarbitone-induced ER before the regression?

Goldstein: The turnover of reductase is delayed about two-fold due to stabilization of the enzyme by the compactin. The marked increase in enzyme activity in the UT-1 cells is due primarily to an effect of enzyme synthesis.

Dean: What about the turnover of other protein components?

Goldstein: Those studies have not been done.

Cuatrecasas: Do you know whether cells with the dihydrofolate reductase have similar patterns?

Goldstein: Dihydrofolate reductase is a cytosolic enzyme, so I would not expect to see an expanded ER. Phenobarbitone induces microsomal enzymes in the liver, but the ER proliferation is not nearly so striking as that seen in the UT-1 cells.

Sabatini: Do you know if there is genomic amplification?

Goldstein: We are trying to clone the gene for HMG-CoA reductase to answer that question.

Sabatini: Is there a larger amount of mRNA in the cytoplasm, or is there a decrease in enzyme turnover?

Goldstein: On the basis of short one-hour pulse labelling of intact cells with [^{35}S]methionine, we know that the UT-1 cells have a synthetic rate for HMG-CoA reductase that can account for the marked accumulation of reductase protein in these cells. This finding suggests that the mRNA content is increased. However, we have not yet been able to obtain immunochemical evidence for synthesis of the reductase in a cell-free system programmed with poly A-rich RNA isolated from the UT-1 cells. Our antibody, which recognizes the mature and processed reductase synthesized by the intact cell, may not recognize reductase that is synthesized *in vitro*.

Sabatini: So you presumably do not immunoprecipitate a product of the expected size. Do you see some other products?

Goldstein: Many proteins are synthesized in our cell-free system, but none of them is specifically immunoprecipitated by our anti-reductase antibody. These cell-free studies have just begun and are preliminary at this point.

Blest: Have you made any freeze-fracture observations on these arrays?

Goldstein: No.

Bretscher: I believe that when *Escherichia coli* is infected with a mutant of filamentous bacteriophage that is defective in DNA maturation, an enormous

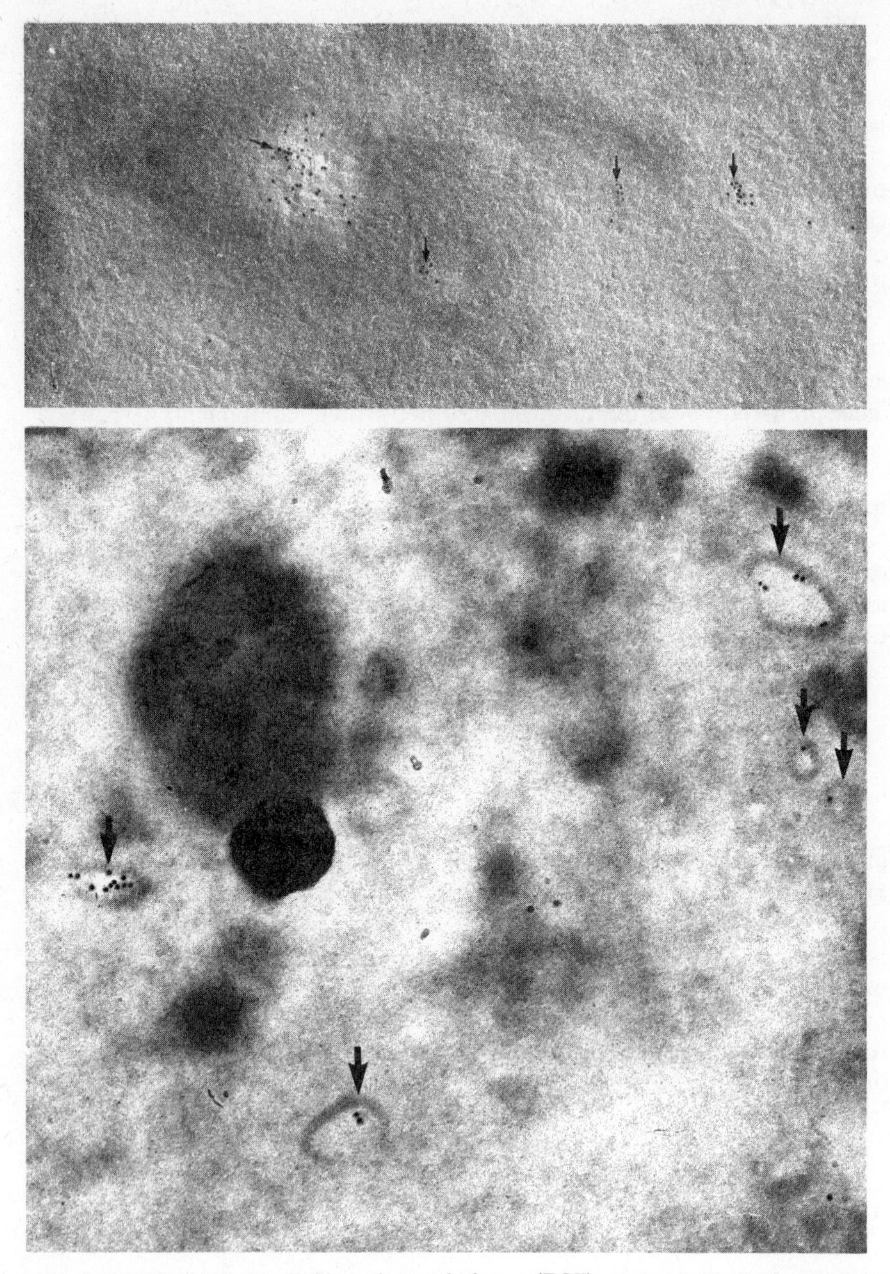

FIG. 4. (*Hopkins*) Upper: Epidermal growth factor (EGF) receptors localized with EGF biotin–avidin gold procedure on cells incubated for 15 min with EGF biotin. The 5 nm gold

amount of coat protein is made and inserted into the membrane. The cytoplasmic membrane also becomes highly extended but not in such an organized fashion (Schwartz & Zinder 1967). Presumably this extension occurs because too much protein is inserted into it.

Palade: There must be something special about the packed masses of smooth ER elements just described by Dr Goldstein; they do not fill the whole cytoplasm and therefore they are not tightly packed because of a lack of space. They are actually packaged within special elements of the rough ER, which seem to be half-rough and half-smooth.

Early events in the receptor-modulated endocytosis of epidermal growth factor and α_2-macroglobulin–protease complexes

Hopkins: As Dr Cuatrecasas described earlier (p 96-108), extensive studies have been done with fluorescein-labelled ligands. In our recent work we have been using gold colloid particles, complexed to a variety of ligands, to study receptor distribution on the cell surface. Because of the intense electron capacity of these particles it is possible to identify them on the surface of whole cell mounts examined by conventional electron microscopy. In whole-mount preparations an *en face* view of the cell surface is available, similar to that obtained in phase or fluorescence microscopy.

In our initial studies (Hopkins et al 1981) we used a two-step protocol involving biotinylated epidermal growth factor (EGF) and avidin-coated gold to localize EGF receptors on porcine granulosa cells.

In prefixed cells the receptors have a random, monodisperse distribution whereas in unfixed cells the receptors become aggregated. The aggregates are in two distinct size ranges of 2–6 and 6–60 particles (Fig. 4, upper). Within the larger aggregates tight groups of 2–6 particles are sometimes evident but it has not been possible to demonstrate a progressive movement of receptors from monodisperse through small to large aggregates.

More recently we have used similar methods to follow the binding of α_2-macroglobulin–protease (α_2MP) complexes to granulosa cells. In conditions in which the redistribution of HLA antigens is prevented (prefixation with 0.5% formaldehyde) we have found that α_2MP receptors have a non-random distribution. However, these receptors are not found in closely

---

particles occur in aggregates of two size ranges and within the aggregate layer a grouping similar to that of the smaller aggregates is apparent (arrow). Magnification: × 120 000.

Lower: α_2-macroglobulin–protease–gold complexes bind to cells prefixed in 0.5% formaldehyde. The particles form loosely associated groups within circumscribed areas (arrows) which we believe to be coated pits. Magnification: × 90 000.

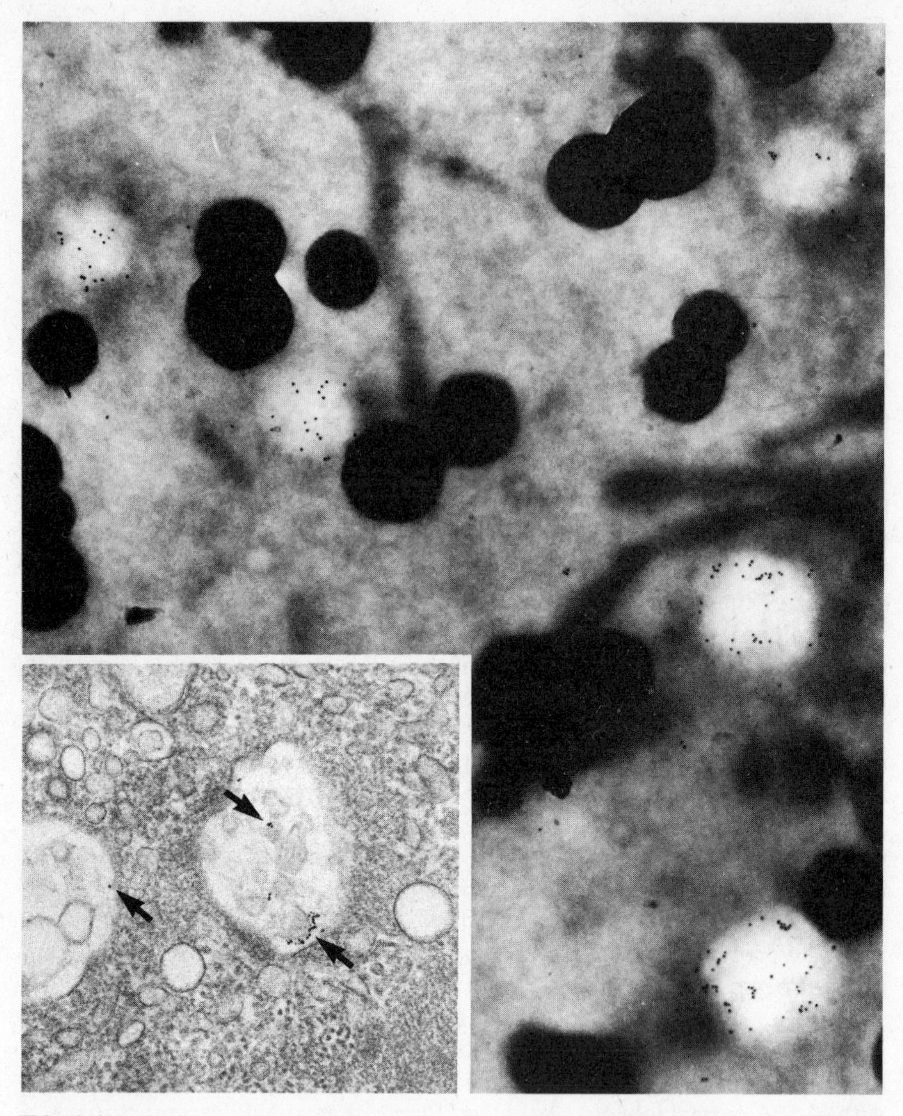

FIG. 5. (*Hopkins*) α_2-macroglobulin–protease–gold complexes in cells incubated for 15 min at 22 °C. 12 nm particles are associated with electron-lucent, often irregularly shaped areas. Dense spherical bodies are lipid droplets; filamentous densities are mitochondria. Inset: thin section of preparation treated similarly to the whole mount shown in the main figure. 5 nm gold particles are present within internal membrane-bound vesicles. Magnification: main section, × 60 000; inset, × 40 000.

grouped aggregates similar to those formed by EGF receptors but they exist instead as rather loosely grouped clusters within circumscribed areas on the surface (Fig. 4, lower). In cells incubated for 10 min with α_2MP–gold complexes without prefixation, the complexes accumulate within larger electron-lucent structures (Fig. 5). In thin sections taken from these preparations the circumscribed areas in which α_2MP complexes are found in prefixed cells are seen to correspond to coated pits whereas the larger electron-lucent structures (to which, in living cells, the complexes are rapidly transferred) are intracellular vesicles (Fig. 5).

A comparison of our observations with those in which fluorescent probes for EGF and α_2MP were followed (Maxfield et al 1978, 1979, Pastan & Willingham 1981, Schlessinger et al 1978) suggests that the microaggregates identified by gold complexes probably exist where there is an evenly distributed fluorescence. The punctate distributions interpreted in some fluorescence studies as surface aggregations are, in fact, likely to be vesicles that contain internalized ligand. While our observations suggest that EGF receptors become aggregated only when occupied by ligand, it seems that (like LDL) a high proportion of α_2-macroglobulin–protease receptors can aggregate within coated pits without binding ligand.

Sly: In your α_2-macroglobulin experiment is this where you see clustering? Does that involve warm-up or is it done in the cold?

Hopkins: That's done in pre-fixed preparations, i.e. preparations treated with aldehyde before we added the ligand. This prefixation protocol prevents redistribution of β_2-microglobulin antibody by second antibody.

Sly: Surely you are looking at different ligands in the EGF experiment? In one case you are using a fluorescent molecule bound to one EGF molecule, which is a monovalent interaction. But with gold there is a multivalent interaction, with the gold being surrounded by the EGF molecules.

Hopkins: The ligand we present to the cell is biotinylated EGF, and we follow that with avidin–gold. So we are presenting a monovalent ligand to the cell surface at the first stage.

Goldstein: One reason for the difference between your results and previous ones may be that you are using the α_2-macroglobulin–trypsin complex, whereas other studies, published by Willingham et al (1979), used α_2-macroglobulin that was not complexed with trypsin. Biochemical studies by Kaplan (1980) and by Van Leuven et al (1978) support your idea that the α_2-macroglobulin receptor recognizes the complex with trypsin better than the naked α_2-macroglobulin.

Palade: How large are the gold clusters when the labelling appears diffuse by immunofluorescence, and what is their size (measured by electron microscopy) before discrete spots are seen by immunofluorescence?

Hopkins: The smaller EGF receptor aggregates are around 80–100 nm diameter, and they would certainly occupy an individual coated pit. The very large clusters of up to 60 receptors are up to 1000 nm in diameter.

Palade: If they were comparable to individual coated pits they would be in the range 500–1000 Å (50–100 nm).

Coated vesicles—variations in morphology and function

Sabatini: Dr Rothman (p 120-137) suggested that the G protein is delivered to the cell surface by coated vesicles. What other examples are there of materials delivered to the cell surface by coated vesicles that fuse with the cell membrane? Franke et al (1976) have reported that large secretory granules in the mammary gland are partially coated.

Palade: Those vacuoles are large and irregular, and the coat covers only part of their surface (Franke et al 1976).

Helenius: Franke et al (1976) showed that these large secretory coated vacuoles contain casein micelles. Dr Rodewald's results (p 209-232) also indicated that immunoglobulin G (IgG) delivery to the basolateral surfaces might occur via coated vesicles.

Branton: In general, a large amount of plant cell-wall protein is deposited by coated vesicles. A new cell plate, for example, is formed in higher plants almost exclusively by coalescence of coated vesicles. This was not appreciated in the initial electron microscope examinations of cell-plate formation. This process may involve other materials, but protein is certainly involved in the nascent cell plates.

Mellman: Does that include the hydroxyproline-rich cell-wall protein, extensin?

Branton: Yes.

Bretscher: I believe it has been suggested that if a coated vesicle is to fuse with anything the coat must first be taken off so that the two bilayers can meet. In Dr Rodewald's work (p 209-232) a coated vesicle appeared to fuse with the basolateral membrane and to deposit something. But something else might have been happening: perhaps it is a coated pit retrieving membrane from the basolateral membrane.

Rothman: As a first approximation, one can reasonably adopt the view that a coated bud or pit is always *leaving* and not *arriving* at a membrane.

Sabatini: We have, however, seen repeated examples of images that were interpreted as depicting partially coated vesicles fusing with and delivering material into endosomes.

Pearse: Such partially coated structures could be incompletely uncoated vesicles fusing with a new membrane or with new coated pits that are forming on the endosomes.

Sabatini: The coat therefore is not only necessary to invaginate and make a vesicle, but the coat may also be partially retained while the vesicle is fusing through the uncoated portion.

Rothman: Another interpretation is that the coated portion of some endosomes represents a budding coated vesicle, leaving the endosome on the recycling arm of the pathway, retrieving receptors for delivery to the plasma membrane.

Sabatini: Although the coated vesicles are seen within the cell, these may not necessarily be involved in the delivery of products to the plasma membrane. At least some of the intracellular coated vesicles may be involved in the retrieval of membrane components from the cell surface. It would be inefficient for a vesicle to withdraw from the plasma membrane before allowing its contents to diffuse away. The work of De Camilli et al (1976) showed a rapid and specific withdrawal of vesicles from the membrane, once secretory granules had fused with the cell surface.

Palade: Morphological evidence indicates that coated vesicles located in the immediate vicinity of the plasmalemma (diam = 100–130 nm, coat included) have incomplete coats: the part facing the cell membrane is missing (Palade & Bruns 1968). Such appearances suggest that vesicles approaching the plasmalemma undergo a local depolymerization of their coats.

Dean: Have *in vitro* fusion experiments involving coated vesicles been done?

Branton: We've done some *in vitro* fusion experiments with coated vesicles loaded with the non-fluorescent dye carboxydiacetylfluorescein. We measured their fusion with lysosomes. The esterase in the lysosome hydrolyses the carboxyl groups and makes the dye fluorescent. One can thus quantify fusion by the increase in fluorescence. The fusion rate is some 20 times faster in the absence of coats than in the presence of coats. Controls show that we are not measuring leakage and general breakdown of the system. The coat seems to prevent fusion. The description that Dr Palade just gave, however, is frequently seen, and I see nothing wrong with the notion that the vesicle in the coated form carries a protein close to another membrane surface and then fuses by a *local* decoating.

Dean: How does the low but questionable rate of fusion between the coated vesicles and the lysosomes compare with the low but questionable rate of fusion that people achieve with lysosomes and liposomes? Could they be comparable? Could it be that to remove the coat is to make it into a very efficient fuser rather than that the presence of the coat makes it a poor fuser?

Branton: Yes, one can interpret it in both ways.

REFERENCES

Black VH 1972 The development of smooth-surfaced endoplasmic reticulum in adrenal cortical cells of fetal guinea pigs. Am J Anat 135:381-418

Blest AD 1978 The rapid synthesis and destruction of photoreceptor membrane by a dinopid spider: a daily cycle. Proc R Soc Lond B Biol Sci 200:463-483

Brown MS, Goldstein JL 1980 Multivalent feedback regulation of HMG CoA reductase, a control mechanism coordinating isoprenoid synthesis and cell growth. J Lipid Res 21:505-517

Brown MS, Faust JR, Goldstein JL, Kaneko I, Endo A 1978 Induction of 3-hydroxy-3-methylglutaryl coenzyme A reductase activity in human fibroblasts incubated with compactin (ML-236B), a competitive inhibitor of the reductase. J Biol Chem 253:1121-1128

Chin DJ, Luskey KL, Anderson RGW, Faust JR, Goldstein JL, Brown MS 1982 Appearance of crystalloid endoplasmic reticulum in compactin-resistant Chinese hamster cells with a 500-fold elevation in 3-hydroxy-3-methylglutaryl coenzyme A reductase. Proc Natl Acad Sci USA 79:1185-1189

De Camilli P, Peluchetti D, Meldolesi J 1976 Dynamic changes of the luminal plasmalemma in stimulated parotid acinar cells. J Cell Biol 70:59-74

Endo A, Kuroda M, Tanzawa K 1976 Competitive inhibition of 3-hydroxy-3-methylglutaryl coenzyme A reductase by ML-236A and ML-236B fungal metabolites, having hypocholesterolemic activity. FEBS (Fed Eur Biochem Soc) Lett 72:323-326

Franke WW, Lüder MR, Kartenbeck J, Zerban H, Keenan TW 1976 Involvement of vesicle coat material in casein secretion and surface regeneration. J Cell Biol 69:173-195

Goldstein JL, Brown MS 1977 The low-density lipoprotein pathway and its relation to atherosclerosis. Annu Rev Biochem 46:897-930

Hopkins CR, Boothroyd B, Gregory H 1981 Early events following the binding of epidermal growth factor to surface receptors on ovarian granulosa cells. Eur J Cell Biol 24:259-265

Kaplan J 1980 Evidence for reutilization of surface receptors for α macroglobulin–protease complexes in rabbit alveolar macrophages. Cell 19:197-205

Maxfield FR, Schlessinger J, Schechter Y, Pastan I, Willingham MC 1978 Collection of insulin, EGF and α_2-macroglobulin in the same patches on the surface of cultured fibroblasts and common internalization. Cell 14:805-810

Maxfield FR, Willingham MC, Davies PJA, Pastan I 1979 Amines inhibit the clustering of α_2-macroglobulin and EGF on the fibroblast cell surface. Nature (Lond) 277:661-663

Palade GE, Bruns RR 1968 Structural modulations of plasmalemmal vesicles. J Cell Biol 37:633-649

Pastan IH, Willingham MC 1981 Receptor mediated endocytosis of hormones in cultured cells. Annu Rev Physiol 43:239-250

Schimke RT, Kaufman RJ, Alt FW, Kellems RF 1978 Gene amplification and drug resistance in cultured murine cells. Science (Wash DC) 202:1051-1056

Schlessinger J, Shechter Y, Willingham MC, Pastan I 1978 Direct visualization of binding, aggregation, and internalization of insulin and epidermal growth factor on living fibroblastic cells. Proc Natl Acad Sci USA 75:2659-2663

Schwartz FM, Zinder ND 1967 Morphological changes in *Escherichia coli* infected with DNA bacteriophage fl. Virology 34:352-355

Sisson JK, Fahrenbach WH 1967 Fine structure of steroidogenic cells of a primate cutaneous organ. Am J Anat 121:337-368

Stowe S 1980 Rapid synthesis of photoreceptor membrane and assembly of new microvilli in a crab at dusk. Cell Tissue Res 211:419-440

Van Leuven F, Cassiman JJ, Van Den Berghe H 1978 Uptake and degradation of α_2-macroglobulin-protease complexes in human cells in culture. Exp Cell Res 117:273-282

Willingham MC, Maxfield FR, Pastan IH 1979 α_2Macroglobulin binding to the plasma membrane of cultured fibroblasts: diffuse binding followed by clustering in coated regions. J Cell Biol 82:614-625

Zinkler D 1974 Zum lipidmuster der photoreceptoren von insekten. Verh Dtsch Zool Ges 67:28-32

Structure of coated pits and vesicles

BARBARA M. F. PEARSE

MRC Laboratory of Molecular Biology, University Medical School, Hills Road, Cambridge CB2 2QH, UK

Abstract Purified coated vesicles typically contain clathrin and auxiliary structural proteins of about 100 000 and 50 000 relative molecular mass (M_r). A model is described for the packing of clathrin trimers into the characteristic pentagons and hexagons of the surface lattices of coats. In a coated vesicle, clathrin appears to bind to an inner 'core' of particles containing the 100 000 and 50 000 M_r proteins. These core proteins may, in turn, bind to receptors in the membrane of the vesicle. The vesicle itself surrounds specific content molecules bound to their respective receptors. Budding coated vesicles are believed to act as molecular filters in cells. The known structural features of coated membranes are discussed in terms of this apparent role.

Roth & Porter (1964) were first to describe the series of events in a cycle of coated vesicle function during the absorptive pinocytosis of yolk proteins by mosquito oocytes. Coated pits occur all over the plasma membrane, bind yolk proteins via specific receptors in the membrane and bud into the cytoplasm to form coated vesicles enclosing the yolk proteins. The coats then dissociate, thus releasing the vesicles to fuse with nascent yolk vacuoles. Meanwhile the coat components recycle to form further coated pits. The cycle time for the formation of a coated pit through to the uncoating of the coated vesicle is of the order of a minute or so, based on the rate of uptake of low density lipoprotein (LDL) (Anderson et al 1977) and Semliki Forest virus (Marsh & Helenius 1980). Thus, enormous amounts of material can be transported by this system. Perhaps the most striking example of this in the developing chicken oocyte, where a 20 mm diameter oocyte takes up about 1.5 g of very low density lipoprotein (VLDL) per day (Perry & Gilbert 1979).

Although absorptive pinocytosis at the cell surface is the role that is best characterized for the coated membrane system of eukaryotic cells (Goldstein et al 1979), coated vesicles also bud from internal organelles, such as the

1982 Membrane recycling. Pitman Books Ltd, London (Ciba Foundation symposium 92) p 246-265

Golgi apparatus (e.g. Friend & Farquhar 1967). The biochemical functions of this intracellular traffic are, at present, less clearly defined, but its existence indicates that coated vesicles are involved in extensive pathways where selected membrane components and ligands are transferred from one cell compartment to another. It is believed that in each cycle the coat is involved in the formation of a new vesicle.

If the sizes of various membrane-bound compartments of the cell are to be maintained, the surface area of membrane returned to a particular compartment must be equivalent to that removed by a budding coated vesicle. In addition, the set of proteins that gives a particular compartment its specific functions cannot be dispersed to other compartments. One can therefore think of cellular compartments as being connected in certain sequences by cyclical conveyor belts: the matrix of the conveyor belt is the lipid matrix of the membranes. The budding of coated vesicles generates a flow from particular membranes and concentrates certain molecules (e.g. specific receptors) for transport within that flow towards their destination. Other molecules characteristic of the parent membrane are excluded by the coated pit, which thus acts as a molecular filter (Bretscher et al 1980, Pearse & Bretscher 1981). This suggests a mechanism that would allow transport of material between cell compartments while preventing the intermixing of the essential components of those compartments.

My present purpose is to describe, as far as possible, the structural basis for this process of molecular filtration.

Purification of coated vesicles

Coated vesicles are characterized by the lattices of pentagons and hexagons on their cytoplasmic surfaces. They have diameters in the range 500 to 2500 Å and are composed of about 75% protein, 20–25% lipid and 1–2% carbohydrate. The bulk of the protein forms the coats which are approximately 150 Å thick. These enclose the typical lipid bilayers of the vesicles which contain a normal complement of phospholipids and a rather low content of cholesterol.

Because of their mean size and density, coated vesicles, like viruses, can be purified from cell extracts by differential centrifugation in density gradients. Originally, I used sucrose solutions as the separation media (Pearse 1975). However, the osmotic pressure effects exerted on the vesicles whenever they experience a change of sucrose concentration appear to cause partial disruption of the coated vesicles, resulting in a loss of some membranes together with the vesicle contents. This is particularly true of the larger coated vesicles (>1500 Å) which are completely destroyed after centrifugation on a sucrose density gradient. Therefore I recently developed a new procedure (Pearse

1982), replacing the sucrose gradients with isotonic D_2O and D_2O/Ficoll gradients. This enabled me to purify the larger coated vesicles from chicken oocytes, examples of which are shown in Fig. 1 (Pearse & Bretscher 1981). The surface lattice of the coats on these vesicles is particularly striking.

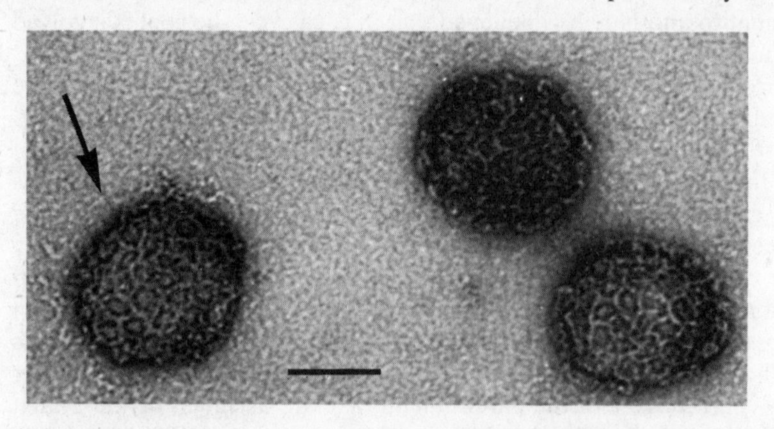

FIG. 1. Electron micrograph of coated vesicles from chicken oocytes. The sample was negatively stained in 1% uranyl acetate. The bar line represents 1000 Å. The arrow points towards the surface network on one vesicle where a pentagon/heptagon pair is clearly visible near the centre.

Coated vesicles from different cells and tissues are composed of a fairly standard set of structural proteins (Pearse 1978). Fig. 2 shows a typical analysis of a sample of coated vesicle proteins by electrophoresis on sodium dodecyl sulphate polyacrylamide gel (SDS gel). The most abundant protein is clathrin (Pearse 1975), which exists in solution as a trimer of polypeptides of 180 000 relative molecular mass (M_r) associated with three heterogeneous 'light' chains in the region of 35 000 M_r (Pearse 1978). These clathrin trimers (Ungewickell & Branton 1981, Kirchhausen & Harrison 1981) can be extracted from the coats by a number of mild procedures not expected to disrupt a lipid bilayer (e.g. extraction with 2M-urea; Schook et al 1979). Clathrin from a number of different sources appears to be rather highly conserved in terms of amino acid sequence (Pearse 1976). Other auxiliary structural proteins characteristic of coated vesicles are seen in these SDS gel analyses (Fig. 2), in particular those of 100 000 and 50 000 M_r. Also visible is a fine background of many other proteins, which may represent specific receptors and their ligands in the vesicles and may depend on the tissue of origin.

The coated vesicle structures, at least in terms of their protein–protein interactions, appear to be resistant to the detergent, Triton X-100. The contaminating membranes left after separation of coated vesicles on an isotonic D_2O gradient are often easily dissolved in 1% Triton X-100, and this

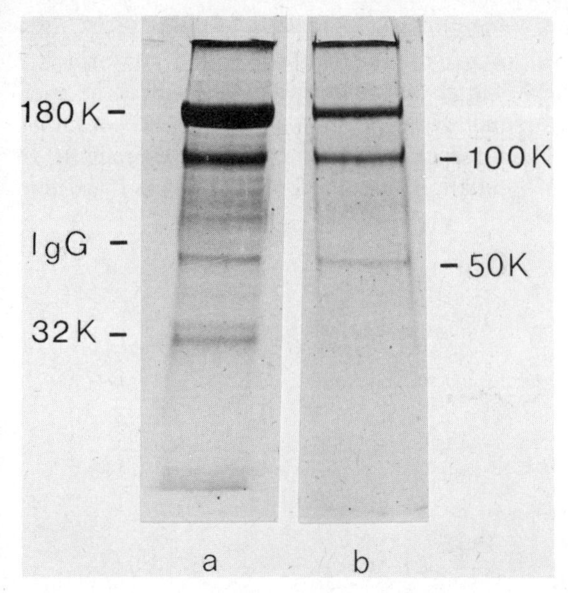

FIG. 2. Electrophoresis, on a sodium dodecyl sulphate polyacrylamide gel, of: (a) a sample of human placental coated vesicles; and (b) a sample of core particles from disrupted placental coated vesicles separated by gel filtration (see Fig. 5 for elution profile).

therefore provides a rapid and efficient means of preparing coated vesicle proteins. It eliminates the necessity for a lengthy density-gradient centrifugation step in their purification. Although the lipid bilayers of the original coated vesicles are affected by the Triton treatment, the coats remain intact and the molecules contained in the vesicles appear to stay trapped inside them (Pearse 1982).

Outer clathrin cage

The clathrin molecule participates in a highly polymorphic range of coat structures on both flattish membranes and vesicles of various sizes. On shallow coated pits the cytoplasmic surface network is essentially a lattice of hexagonal units. Occasionally a discontinuity occurs where a pentagon is coupled with a heptagon (Heuser 1980, and see Fig. 1). On more concave pits, pentagons are more frequent; when 12 pentagons are present, a closed polyhedron is formed (Euler's theorem). Stages in vesicle budding from the cytoplasmic surface of fibroblast plasma membranes have been visualized by rapid freezing and deep etching (Heuser 1980). The smallest polyhedral coats (500–700 Å diameter) are made up of 12 pentagons plus four or eight hexagons, while the larger ones contain larger numbers of hexagons (Crowther et al 1976). It is logical to expect the clathrin trimer to form the polyhedral vertices of the coat lattice because three edges meet at each vertex. This implies that the small coats, with four and eight hexagons, contain 28 and 36 clathrin trimers respectively. An estimated average M_r of 22×10^6 for a heterogeneous population of small empty coats is consistent

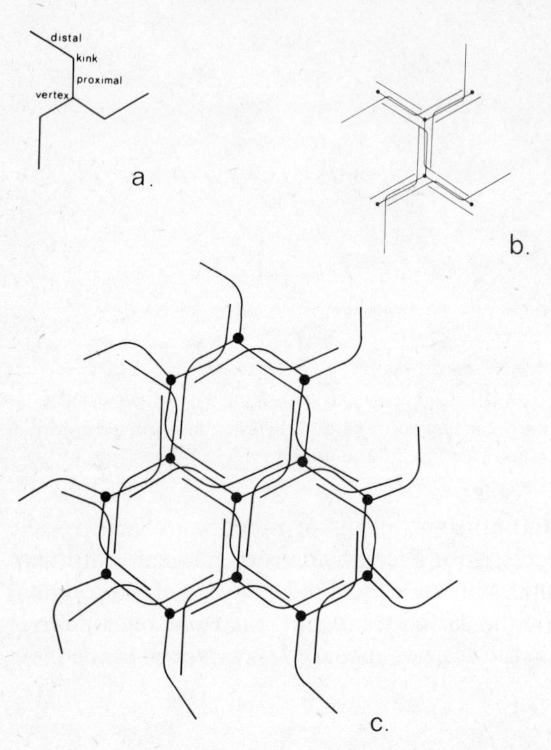

FIG. 3. The drawings show: (a) an individual triskelion, labelled with the terms used in the text; (b) how the proposed cross-over packing of the triskelions gives rise to the appearance of multiple lines along an edge and an equilateral triangle at each vertex from which the polygonal edges emerge in a skewed manner; and (c) three neighbouring hexagons, in which the skewing has been slightly exaggerated, to show that an edge is not perpendicular to the line joining the centres of the two polygons on either side of that edge.

with these particles containing about 30 trimers per particle (Crowther et al 1976). Larger coats of about 2000 Å diameter would be built of about 300 clathrin trimers.

The clathrin trimer has been termed 'triskelion' because of its remarkable conformation (Fig. 3), originally shown by rotary shadowing of fields of purified molecules (Ungewickell & Branton 1981). Each trimer consists of three thin legs, of about 430 Å length, radiating symmetrically from a central point. The length of the polygonal edge in the coat is about 186 Å. Thus each leg can run along two neighbouring polygonal edges. In individual triskelions there is a kink in each leg about 160 Å from the vertex, so that the distal part of the leg must veer off before reaching the neighbouring polyhedral vertex.

Features in low-dose electron micrographs of small fragments of clathrin cages lend support to the following cross-over packing model (Crowther & Pearse 1981; see Fig. 3). With this packing, each edge of a completed polyhedron is formed by four half-legs from four different triskelions: the proximal halves of two legs from neighbouring vertices run antiparallel and cross over each other; and the distal halves of two legs run from vertices one step further away on the polyhedral lattice. This can give rise to a particularly striking pattern around the vertex, consisting of an equilateral triangle oriented such that one vertex of the triangle points roughly at right angles to the polyhedral edge. There is a dot in the middle of the triangle corresponding to the material at the vertex of an individual triskelion. In the cross-over packing, individual triskelions are rotated clockwise from a position in which each leg points directly at a neighbouring vertex. Thus the edges appear to radiate from the triangle in a maximally skewed manner, and the polygonal holes (pentagons or hexagons) appear to be rotated relative to the underlying lattice. This model accounts for about 350 Å of the leg of the triskelion, leaving an additional 80 Å (approximately) at the distal tip which is not yet accommodated.

The molecular packing in the cages is governed by the general principles of quasi-equivalence first suggested for viral capsids (Caspar & Klug 1962). The whole arrangement of triskelions has local two-fold axes of symmetry, which are midway between the three-fold vertices of the polyhedra and normal to their edges. As there is a triskelion at each vertex, a pentagon or heptagon can be formed from a hexagon simply by removing or adding a triskelion. Only small distortions of the molecules are required to form the smaller or larger polygons. How pentagons might actually be introduced into the hexagonal nets of coated pits, as the coats deform, is a more complex problem which is not yet understood. Pentagons might form at an edge and diffuse by a series of dislocations into the hexagonal array, or dislocations could arise inside the hexagonal net, each giving rise to a pentagon and heptagon from two hexagons. In this latter case the heptagons would have to diffuse, by a series of dislocations, out of the hexagonal lattice to the edge, leaving behind pentagons and thereby inducing curvature.

The clathrin triskelion has a most unusual shape. It is a highly extended fibrous molecule which, when packed to form cages, interacts extensively not only with its nearest neighbours but also with quite distant molecules. The clathrin legs must be rather rigid to make the specific contacts required in the isometric shells. However, the triskelion must contain at least two major regions of flexibility to participate in the polymorphic range of surface lattices that it forms. The first region is a variable joint at the vertex of the triskelion. Thus the conical angle at the vertex can vary, as required, in aggregates ranging from almost planar (Heuser 1980) to quite sharply curved (Crowther et al 1976). The second region is a hinge in each leg of the triskelion at about 160 Å from the vertex. This is needed to accommodate the extended leg to the difference in angle between neighbouring edges in pentagons and hexagons. Thus the molecular design of its triskelion, coupled with the extended nature of its packing contacts, allows clathrin to form a wide range of shells combining mechanical strength with the flexibility needed to bud a vesicle from a membrane.

In suitable ionic conditions, clathrin triskelions polymerize within seconds. The assembly behaves like a condensation process (Oosawa & Kasai 1962) where, above a critical protein concentration, clathrin trimers self-associate to form cages, in equilibrium with a constant low concentration of monomer triskelions (Crowther & Pearse 1981). Rapid assembly and disassembly of the clathrin lattice is required in the continuous cycles of coated pit–coated vesicle function in cells.

In vitro, clathrin assembles to form approximately equal numbers of hexagonal and pentagonal units in the form of cages. However, clathrin in the cell presumably does not normally assemble on its own but on the surface of a membrane, by joining already existing lattices. The assembly is in association with the auxiliary coat proteins to which clathrin binds (Unanue et al 1981, Pearse 1982). The mechanism of budding is likely to involve these proteins and to require a source of energy to promote the process: it stops when the cells are poisoned by metabolic inhibitors or are chilled to 0 °C.

Inner structure of the coated vesicle

The outer clathrin cage of coated vesicles is disrupted in the presence of 2M-urea (Schook et al 1979). If samples of coated vesicles, extracted in 1% Triton X-100, are applied to electron microscope grids, washed with several drops of a urea-containing buffer and then stained with 1% uranyl acetate, an inner 'core' of material is revealed (Fig. 4; Pearse 1982). The 'cores' look more like protein aggregates than vesicles, suggesting that the lipid bilayers of the original coated vesicles have been affected by the Triton treatment.

Whether or not the integrity of the vesicle membranes is preserved, the coat structures trap molecules contained in the vesicles in their interior. Thus when placental coated vesicles are purified by sedimentation on isotonic density gradients or by extraction with 1% Triton X-100, they retain their

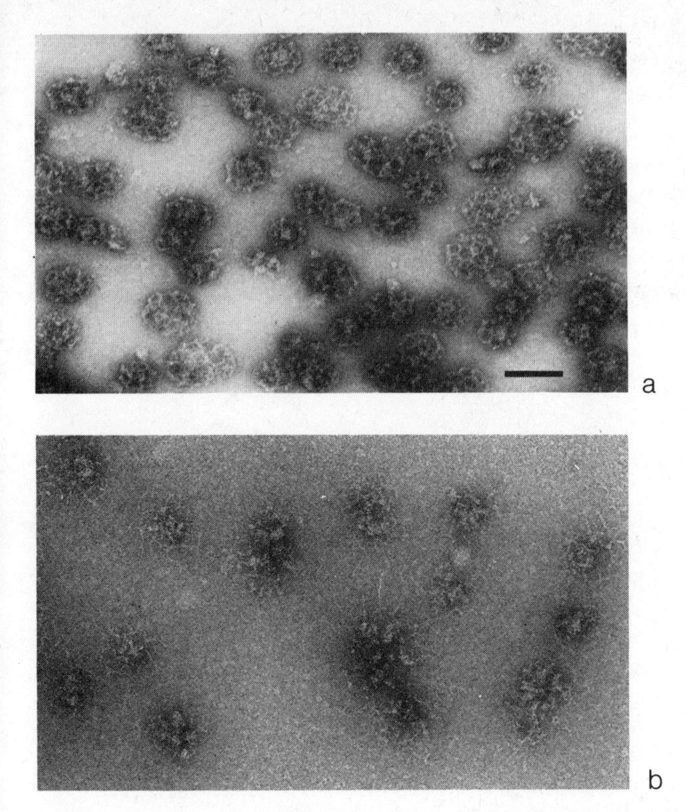

FIG. 4. Electron micrographs of coated vesicle preparations negatively stained in 1% uranyl acetate. Scale bar represents 1000 Å. (a) Human placental coated vesicles extracted in 1% Triton X-100. (b) Human placental coated vesicles, extracted in 1% Triton X-100, where the clathrin coats have been disrupted with urea before staining, to show the 'cores'.

contents intact (Pearse 1982). The most striking of these contents is ferritin, which colours the preparations brown. Electron microscopy shows that much of the ferritin is associated with coats and is not a contaminant. γ-Immunoglobulin and transferrin are also present in placental coated-vesicle preparations. Both proteins were detected by using appropriate antisera and ^{125}I-labelled *Staphylococcus aureus* protein A to stain nitrocellulose blots of samples of coated vesicle polypeptides previously separated by SDS gel electrophoresis (method of Towbin et al 1979).

If, after extraction in 1% Triton X-100 (which affects the lipid bilayers of the vesicles), the coats are disrupted in urea, the contents of the vesicles represented by ferritin, transferrin and most of the γ-immunoglobulin are released into solution. Coated vesicle preparations disrupted in this way can be fractionated by gel filtration on a column of Bio Gel A–15 m. Fig. 5 shows a typical profile of the absorbance at 280 nm of the eluate from such a column (Pearse 1982). The first peak (I) to elute contains particles corresponding to constituents of the 'cores' seen by electron microscopy (Fig. 4b). The bulk of the clathrin triskelions elutes in peak II, and ferritin elutes in peak III. Later fractions (e.g. IV) presumably contain other polypeptides including content molecules and receptor molecules.

Analysis on SDS gel electrophoresis shows that the 'core' particles (peak I) contain some clathrin and also contain the polypeptides of 100 000 and 50 000

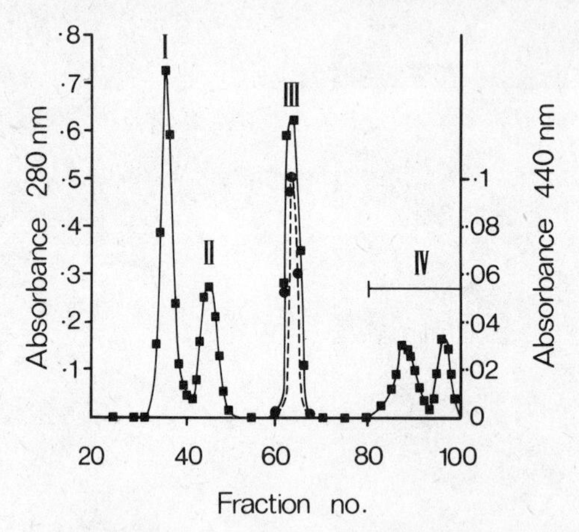

FIG. 5. A typical profile of disrupted human placental coated vesicle proteins after gel filtration on BioGel A-15 m. ■ absorbance at 280 nm. ● absorbance at 440 nm.
Peak I contains 'core' particles, peak II represents clathrin triskelions and peak III contains ferritin. Fractions labelled IV include other proteins amongst which are content and receptor molecules.

M_r (in ratio 2:1 as estimated by gel densitometry) which are the auxiliary structural proteins common to coated vesicle preparations (Fig. 2). The clathrin that remains bound to these complexes seems to be in equilibrium with clathrin trimers, most of which elute in peak II. If the triskelions from peak II are recombined with the core particles of peak I and returned to clathrin polymerization conditions, coated structures, which have the core particles inside clathrin cages, reassemble. The core particles from peak I alone also reaggregate in these conditions to form a relatively homogenous population of structures, reminiscent of small coated vesicles in size and shape but lacking the fine definition of the complete clathrin lattice.

The exact relationship of these core proteins to the membranes of the coated vesicles remains unclear although apparently most exist largely on the cytoplasmic surfaces of the vesicles (Unanue et al 1981). Probably the structural 'cores' interact in some way with receptors that span the lipid bilayer. I have recently shown (unpublished results) that the transferrin receptor, which is present in placental coated vesicles, is indeed tightly bound on these core complexes.

Not all the functions of coated vesicles derived from human placenta are known. The placenta contains several distinct cell types in which coated vesicles are likely to be involved in a variety of processes including secretion and lysosome formation. However, the predominant role of coated vesicles in the placenta near the end of pregnancy is probably the uptake of proteins from the maternal serum. Thus ferritin and transferrin act as vehicles for the iron (up to 4 mg per day) that is required by the fetus for haemoglobin synthesis. Passive immunity is conferred on the baby by transfer of certain classes of γ-immunoglobulin from the mother across the placenta (Brambell 1970), via coated vesicles. Thus any one coated vesicle is likely to contain a variety of different content molecules bound to their respective receptors. Ferritin appears to be essentially randomly distributed amongst the placental coated-vesicle population. As judged by electron microscopy, about 40% of the structures encompass one or more (up to six) molecules of ferritin, while the rest lack ferritin. Those associated with ferritin have an average of 1.4 ferritin molecules each. Transferrin and its receptor each account for about 0.25% of the total protein of placental coated vesicles. Thus there might be, on average, one or two molecules of transferrin per vesicle bound to transferrin receptor, which exists as a dimer. γ-Immunoglobulin accounts for about 2% of the total protein, suggesting an average of four of these molecules per vesicle. Many other ligands (e.g. LDL and insulin) and their receptors are also likely to be included in these coated vesicles.

The problem remains of how the 'core' of the coated vesicle is able, as it forms, to recognize and bind so many specific receptors in the membrane and yet to exclude other proteins belonging to the parent membrane. Perhaps the proteins of the 'core' form a sufficiently close-packed structure to allow no room for molecules characteristic of this original membrane.

Conclusion

To carry out its apparent function as a molecular filter, the coat of coated pits and vesicles must have the following features:

(1) mechanical strength coupled with the flexibility to bud a vesicle from a membrane;
(2) ease of rapid assembly and dissociation;
(3) the means of recognizing many different specific receptors to be transferred; and

(4) the ability to exclude other membrane proteins.

The intrinsic nature of the clathrin molecule and its packing into cages appears to confer the first of these properties. Both the outer clathrin cage and the inner structural proteins to which it is bound can be dissociated and reassembled *in vitro*. Perhaps these internal proteins form the 'core' of the molecular filter, binding specific receptors in the membrane but, by forming a close-packed structure, excluding others.

Acknowledgement

This work is supported by a fellowship from the Science Research Council.

REFERENCES

Anderson RGW, Brown MS, Goldstein JL 1977 Role of the coated endocytic vesicle in the uptake of receptor-bound low density lipoprotein in human fibroblasts. Cell 10:351-364

Brambell FW 1970 The transmission of passive immunity from mother to young. Front Biol 18

Bretscher MS, Thomson JN, Pearse BMF 1980 Coated pits act as molecular filters. Proc Natl Acad Sci USA 77:4156-4159

Caspar DLD, Klug A 1962 Physical principles in the construction of regular viruses. Cold Spring Harbor Symp Quant Biol 27:1-24

Crowther RA, Pearse BMF 1981 Assembly and packing of clathrin into coats. J Cell Biol 91:790-797

Crowther RA, Finch JT, Pearse BMF 1976 On the structure of coated vesicles. J Mol Biol 103:785-798

Friend DS, Farquhar MG 1967 Functions of coated vesicles during protein absorption in the rat *vas deferens*. J Cell Biol 35:357-376

Goldstein JL, Anderson RGW, Brown MS 1979 Coated pits, coated vesicles and receptor-mediated endocytosis. Nature (Lond) 279:679-685

Heuser J 1980 Three-dimensional visualization of coated vesicle formation in fibroblasts. J Cell Biol 84:560-583

Kirchhausen T, Harrison SC 1981 Protein organization in clathrin trimers. Cell 23:755-761

Marsh M, Helenius A 1980 Adsorptive endocytosis of Semliki Forest Virus. J Mol Biol 142:439-454

Oosawa F, Kasai M 1962 A theory of linear and helical aggregations of macromolecules. J Mol Biol 4:10-21

Pearse BMF 1975 Coated vesicles from pig brain: purification and biochemical characterization. J Mol Biol 97:93-98

Pearse BMF 1976 Clathrin: a unique protein associated with intracellular transfer of membrane by coated vesicles. Proc Natl Acad Sci USA 73:1255-1259

Pearse BMF 1978 On the structural and functional components of coated vesicles. J Mol Biol 126:803-812

Pearse BMF 1982 Human placental coated vesicles carry ferritin, transferrin and γ-immunoglobulin. Proc Natl Acad Sci USA 79:451-455

Pearse BMF, Bretscher MS 1981 Membrane recycling by coated vesicles. Annu Rev Biochem 50:85-101

Perry MM, Gilbert AB 1979 Yolk transport in the ovarian follicle of the hen. J Cell Sci 39:257-272

Roth TF, Porter KR 1964 Yolk protein uptake in the oocyte of the mosquito *Aedes aegypti*. J Cell Biol 20:313-332

Schook W, Puszkin S, Bloom W, Ores C, Kochwa S 1979 Mechanochemical properties of brain clathrin. Proc Natl Acad Sci USA 76:116-120

Towbin H, Staehelin T, Gordon J 1979 Electrophoretic transfer of proteins from polyacrylamide gels to nitrocellulose sheets: procedure and some applications. Proc Natl Acad Sci USA 76:4350-4354

Unanue ER, Ungewickell E, Branton D 1981 The binding of clathrin triskelions to membranes from coated vesicles. Cell 26:439-446

Ungewickell E, Branton D 1981 Assembly units of clathrin coats. Nature (Lond) 289:420-422

DISCUSSION

Rothman: You have suggested that the exclusion of bulk membrane protein from coated pits is related to the 'core' structure that you have proposed. In this scheme, you envisage a steric kind of exclusion due to a close-packed structure in the core. That would require a nearly complete coverage of the membrane surface by this protein. Is there enough of the 'core' proteins (50 000 and 100 000 M_r polypeptides) in coated vesicles to cover the vesicle surface so completely? Are these particles in fact *cores* inside the native coated vesicles or might they instead be *aggregates* resulting from the combined Triton and urea treatments used in their preparation?

Pearse: There probably is enough of the total proteins. In placental core preparations the 100 000 M_r protein is present in a 1:1 or 1:2 molar ratio with clathrin, but I have not done a proper stoichiometry on this yet.

Rothman: To follow up on the second point I made, can you count the core particles that are generated by extraction of coated vesicles in relation to the number of coats that you start out with?

Pearse: I hope to do that. If core particles are put in clathrin assembly buffer, they form structures like small coated vesicles that are depleted of clathrin.

Branton: In the reassembled units after dissociation with urea you found a stoichiometry of 1:1, but the stoichiometry of the 100 000:180 000 M_r is more like 1:3 *in the coated vesicles.*

Pearse: Yes; there is more clathrin in the coated vesicles, but even so there is still a substantial amount of those other proteins.

Branton: Much depends on the shape of the protein. If you place, for example, one of those proteins at each vertex of the coat polygons (which would be a position appropriate and consistent with the stoichiometry) then you can see a considerable amount of space from one of these protein molecules to the next. I doubt very much that a 100 000 M_r polypeptide would occupy over 200 Å.

Pearse: The possible close packing is not at the outer surface near the clathrin but it may be in the region of the membrane.

Bretscher: There would also, presumably, be a large number of receptors which would fill an unknown amount of space.

Pearse: Yes. These core particles are very complex and don't contain the 100 000 and 50 000 M_r proteins only. They probably also contain lipid, bound receptors and some bound ligands. They look like wedges cut through the coat structure.

Rothman: If the core structure is capable of causing exclusion it must do so largely by occupying an area on or in the membrane surface. If one looks not at the urea-treated vesicles but simply at the Tritonized coated vesicles (which retain most of their content and nearly all of the total protein that is present) does a negative stain penetrate that coat readily, implying space not occupied by a core surface between the lattice points of the overlying clathrin coat?

Pearse: The Tritonized coated vesicles do not look very different from untreated coated vesicles. The urea disruption of the clathrin cage reveals a rather solid-looking pile of material inside, but that could be aggregated or precipitated. Sections of coated vesicles often go through the coat and show an opaque central region (G. E. Palade, personal communication).

Rothman: If you were to draw a sphere at the level of the membrane, the fraction of that surface that would be occupied by the core protein would, presumably, be quite large.

Pearse: Yes; but at the moment I know very little about how these core particles relate to the membrane.

Hopkins: In sections, the coat often appears to consist of T-shaped structures rather than straight bristles. How does the arrangement of the triskelions relate to that?

Pearse: In projection in sections, the vertices of the structure can look either like T-shapes or like bristles. It probably depends on how one cuts the coat.

Branton: Depending on how one sections the coat, one can indeed see a variety of structures that resemble those in cross-sections of intact coated vesicles. In addition, if one removes the triskelions, leaving behind the 100 000 and the 50 000 M_r polypeptide and the membrane, one still sees projections on this membrane which resemble those seen between the 'coat' and the membrane. The 100 000 M_r polypeptide appears to be the membrane binding site at which the triskelions interact. When Dr Pearse isolated the urea extract there was also an interaction between the residue of triskelions and the 100 000 M_r polypeptide. We have found that selective removal of the 100 000 M_r polypeptide from the membrane will remove the binding site, which can be reconstituted by adding the 100 000 M_r protein back to the membrane.

Mellman: What conditions are necessary to strip the 100 000 M_r polypeptide from the membrane?

Branton: Slightly higher concentrations of Tris than are required to remove the triskelions.

Pearse: But doesn't one also extract the 50 000 M_r polypeptide as well?

Branton: Yes. To deal exclusively with a 100 000 M_r polypeptide, the membrane is treated with low concentrations of elastase, which is very selective in cleaving the 100 000 M_r polypeptide without touching the 50 000 M_r polypeptide.

Mellman: Are the elastase-clipped molecules still functional in the reconstitution system?

Branton: We have never tried to rescue what comes off with the elastase treatment.

Meldolesi: When vesicles lose their coats during their travel through the cell, do they also lose their set of core proteins?

Pearse: It is not yet clear what the exact relationship is between the 100 000 and the 50 000 M_r protein complex and the membrane. I would not want to state whether these polypeptides come off with the clathrin.

Meldolesi: So they might remain stuck to the membrane?

Pearse: I think at least some of the complex must come off.

Meldolesi: Secondly, you mentioned that the vesicles could be heterogeneous and that the impression of electron microscopists is that the vesicles are preserved to a different extent under suboptimal experimental conditions. Is there any hint, molecularly, to explain this observation, e.g. that coated vesicles located in the Golgi area are more fragile than those appearing close to the plasma membrane?

Pearse: The clathrin coat is similar on all the vesicles; the smaller structures seem much more stable on extraction than the larger ones (Pearse 1980). The larger endocytic coated vesicles may be lost because they never bud off from the membrane in the initial extract or because they are destroyed in the preparation.

Goldstein: Have you ever tested whether there are cholesterol esters inside these human placental coated vesicles after extraction with Triton.

Pearse: No, I haven't.

Goldstein: I ask this because physiological evidence suggests that the human fetus derives most of its cholesterol from the mother rather than from endogenous synthesis. Yet, we have never been able to demonstrate LDL receptors in unfractionated membranes of whole placenta. A clue to the presence of LDL receptors in these isolated coated vesicles would be the presence of a high cholesterol ester content inside the vesicles.

Jourdian: In relation to the numbers of these coated vesicles, how much protein (in g dry weight) would one have if one did subcellular fractionation and isolation of vesicles from placental tissue?

Pearse: We find 10 mg of coated vesicle protein in one placenta of 250 g wet wt (Pearse 1982). In most cells there appear to be several thousand vesicles. In the initial extracts clathrin is about 1% of the supernatant protein after removal of the mitochondria.

Sly: Do you see different receptors in the same coated vesicle?

Pearse: I can't answer that yet from these preparations. I think there must be different receptors in single coated vesicles because in Dr Goldstein's experiments most of the coated pits on the fibroblasts tend to contain LDL receptors (Anderson et al 1977), and yet we know that 101 other things are taken up by those same coated pits. In my experiments the ferritin is distributed over at least half the population of coated vesicles and yet several other things are also taken up. Any one coated vesicle could contain very many different receptors and different ligands. Which are present in the greatest numbers in any vesicle would depend on which ones got there first.

Hopkins: Your model suggests that clathrin does not interact directly with receptors on the external surface of the membrane. Do you think clathrin is likely to interact directly with any cytoplasmic components?

Pearse: Clathrin is on the outside but the coat dissociates almost immediately when the coated vesicle is formed. It is unlikely that clathrin will interact directly with cytoskeletal elements, and there is no evidence that it does.

Hopkins: Some of Dr Goldstein's work (p 77-95) showed that if one extracts cells in Triton and retains the sub-plasma-membrane cytoskeleton, the clathrin basket stays intact and remains in position on the cytoskeleton mesh. This may be due to protein–protein interactions.

Rodewald: You mentioned that clathrin inhibits the fusion of the coated vesicles with membranes. Does either the 100 000 M_r or the 50 000 M_r protein inhibit or enhance fusion of coated vesicles?

Pearse: In the region of fusion, all the cytoplasmic coat protein probably has to come away before the lipid bilayer of the coated vesicles can meet and fuse with the lipid bilayer of the next membrane compartment. A component may be present to aid the fusion, and when that component is uncovered the fusion can take place. It is analogous to a membrane-containing virus that fuses with a membrane; it has to uncoat to a certain extent before it fuses.

Raff: I am unclear about what you think happens when the coated vesicle loses its coat. In those cases where you can see that the coat has gone and yet the ligand is still associated with the vesicle membrane, do you think the core proteins are still associated with the vesicles?

Pearse: This has still to be investigated. Core proteins exclusively on the cytoplasmic surface of the vesicle are likely to come off the vesicles during uncoating.

Raff: But if the receptors are bound to the core proteins, then my explanation would be implied.

Pearse: I don't know *which* core protein is involved in binding to receptors. The core particles are very complex and largely uncharacterized. They still have clathrin on them and they apparently have receptors and some ligands too.

Palade: Could you elaborate on the nature of the two proteins (or group of proteins) of 100 000 and 50 000 M_r? Are they located on the cytoplasmic side of the vesicular membrane; are they peripheral membrane proteins; or are they integral membrane proteins that interact hydrophobically with part of the bilayer?

Pearse: A certain proportion of them are expressed on the cytoplasmic surface because that is where the clathrin binds, but I don't yet know whether some portion penetrates the membrane, or exactly to what they bind on the membrane.

Branton: Using the usual definitions of 'peripheral' one can say that they are peripheral membrane proteins: they certainly come off the membrane in 0.05 M-Tris, which is not a very harsh treatment.

Pearse: Were the original coated vesicles prepared by sucrose gradients?

Branton: Yes.

Pearse: Well perhaps the coats that easily come apart are the ones in which the membranes and the core structure have been disrupted by osmotic shock.

Branton: That is conceivable. I suppose I am applying standard definitions to 'peripheral' and 'integral'. Usually plasmolysis and treatments with Tris are not viewed as extracting intrinsic proteins. As you are, I am reluctant to say anything specific about the proteins, but one *can* say that they come off easily.

Bretscher: When they come off, are they single, discrete molecules or aggregates?

Branton: We have no evidence about that, but they do come off together with a 50 000 M_r polypeptide, and there is some indication from that, and from other evidence, of an interaction between the 50 000 and the 100 000 M_r polypeptides.

Sabatini: When clathrin molecules are reassembled into cages around membranes, out of an extract from coated vesicles, are the 100 000 M_r proteins absolutely essential?

Branton: Yes, in the sense in which you are asking the question. Some triskelions can reassemble into cages in the absence of any vesicles. Thus, in principle, a cage can be reassembled accidentally around any vesicle, e.g. a plain lipid vesicle. The triskelions that we use are semi-incapacitated, by treatment with Tris, so that they are reluctant to self-assemble into cages in the conditions we use. The triskelions thus reassemble around vesicles only if those vesicles contain the intact 100 000 M_r polypeptides.

Palade: Does this mean that the triskelions can assemble in a cage in the absence of either the 100 000 or the 50 000 M_r?

Branton: Dr Pearse showed several years ago that triskelions can self-assemble into cages. Formation of the complete cage requires light chains. In the absence of the light chains we get formation of half a cage, and cage closure doesn't occur unless the light chains are present.

Palade: In interpreting your results, Dr Pearse, you are assuming that coated vesicle proteins are required to interact with receptors and selected molecules which—as a result of molecular filtration (to use Dr Bretscher's expression)—find themselves clustered within a coated pit. The core proteins are the best candidates for carrying out the selection because of their location. In addition, clathrin triskelions and polypeptides of 30 000 M_r are required for completing the structure of the cages. So, what are the roles of these different molecular components? Is clathrin the stabilizer? Does clathrin define the morphology of the structure? Does clathrin define the size of the sample of membrane to be removed? Is the molecular *filter* function to be relegated to the core protein?

Pearse: This is a problem, but clathrin, which we know most about, is on the outside and forms the 'scaffold'. Yet we don't know what drives changes in that scaffold. That is almost certainly controlled by other proteins.

Bretscher: One might also ask why it is that the Golgi coated vesicles are so small and the endocytic ones so large. Could the vesicle size be determined by the light chains associated with the clathrin?

Pearse: It is possible that there are different light chains in the small coated vesicles from those in the larger ones.

Sabatini: Would this 100 000 M_r polypeptide that is required for the binding of the clathrin cage to the membrane bind only to certain membranes that have receptors, or can it bind to any vesicles?

Branton: We haven't investigated the interaction between the 100 000 M_r polypeptide and the membrane, and we have only barely begun to investigate the interaction between the clathrin and the 100 000 M_r polypeptide.

Pearse: It is a highly complex structure in which there are many interactions that mediate specific ligand binding right through the membrane to the outer clathrin cage. We have yet to find out which binds to what!

Meldolesi: Interest has shifted from clathrin to core proteins. Since we suspect that coated vesicles do several different things, would it be possible to isolate coated vesicles, as you have done from the brain and placenta, from tissues that have extremely specialized functions, and to compare the proteins of coated vesicles in different tissues? At this level we may find subtle differences that help to explain why certain receptors and ligands are specifically clustered.

Sabatini: I would expect the contrary, and that only the part that is exposed on the cytoplasmic side of the membrane is common to all receptors. Perhaps the specific binding is on the *other* side of the membrane.

Meldolesi: For Golgi coated vesicles, are we talking about receptors or about something else?

Sabatini: We are talking about two main functions in every cell.

Meldolesi: If they account for at least two functions, then coated vesicles must be heterogeneous, and it would be important to have direct information about this.

Goldstein: If the light chains *are* different, this could determine the functional difference between endocytic coated vesicles and the Golgi-derived coated vesicles. It would be interesting to compare the light chains of coated vesicles from the chicken oocyte with those from the brain. In the chicken oocyte the coated vesicles are clearly entering the cell.

Pearse: There are definitely different light chains in the brain from most other tissues, and I think that light chains from the chicken oocyte may be different again. However, the coated vesicle populations that I have isolated so far from chicken oocytes are rather heterogeneous and not exclusively endocytic.

Branton: How many light chains did you find when you isolated large vesicles from chicken oocytes?

Pearse: The snag is that I could not isolate a biochemical amount of those large (>2000 Å diameter) coated vesicles, which are presumably endocytic coated vesicles.

Sabatini: I would suggest another possible explanation for the differences in size of vesicles. The rigidity of the bilayer in the plasma membrane (which contains much cholesterol) may be different from the rigidity of the bilayer in the endoplasmic reticulum. Therefore the number of pentagons and the requirements for converting the planar lattice into an icosahedral one would be different. A smaller number of 50 000 M_r light chains may more easily induce curvature in one type of membrane than in another.

Palade: You mentioned that clathrin is a highly conserved protein; hence, its primary structure should be the same (or nearly the same) from species to species. How much do we know about the amino acid sequence in clathrin?

Pearse: My evidence about the sequence is simply from one-dimensional finger printing of mammalian clathrin derived from about three or four samples from different tissues from pigs, bullocks and mice. I cleaved the protein at cysteine residues and analysed the fragments on sodium dodecyl sulphate gels (Pearse 1976). This is not a highly sensitive method but the patterns of peptides are remarkably similar, whereas the patterns obtained from myosins from different tissues are completely different (Burridge & Bray 1975).

Branton: How extensively has immunological cross-reactivity been examined?

Goldstein: The polyclonal antibody that we made to the clathrin of coated vesicles from pig brain cross-reacts with clathrin from bovine adrenal cortex, human fibroblasts, Chinese hamster ovary cells and mouse macrophages (Anderson et al 1978, Mello et al 1980).

Sly: Do the antibodies facilitate the formation of the clathrin cage or inhibit it?

Palade: We (Merisko et al 1982) have used a system of *Staphylococcus aureus* cells coated with anti-clathrin and reacted with solubilized clathrin (urea or high pH extract of coated vesicles). Upon dialysis, reconstituted clathrin cages are found on the surface of the immunoadsorbent. This means that the antibody we have used recognizes a different sequence from the sequences involved in cage assembly. Apparently, the antibody facilitates cage formation, but we do not know what this means in terms of molecular interactions among cage components.

Sly: So are any antibodies known to inhibit cage formation?

Palade: I wouldn't be surprised!

Branton: It is interesting that some antibodies do not inhibit at all. Substantial regions of the leg of the triskelion are apparently not required for cage formation. For example, in our experiments, and those of Dr Rothman, one can clip off pieces with about 40 000 or 50 000 M_r from the distal region of each triskelion leg, but the central core of the triskelion will still reassemble into pentagons and hexagons.

Palade: There is a rather large difference between the complex electrophoretogram of coated vesicles obtained from placenta and the earlier, much simpler, electrophoretograms of coated vesicles prepared from brain. Coated vesicle fractions from rat liver also give a rather complex electrophoretogram, although they appear structurally quite homogeneous (Palade et al, unpublished observations). How can this difference be explained? Did earlier preparations consist largely of vesicle-free cages (Woodward & Roth 1979)?

Pearse: It is probably related to osmotic pressure effects during changes in the concentration of the sucrose solutions used in the early preparations. If the sucrose gets inside the structures, then when one removes the sucrose from the outer solutions one presumably obtains an 'explosion' that bursts open the cages and releases much of the content of the vesicles. The cage is then reformed, but the tight coats lack a good proportion of the internal structure, particularly the content molecules.

Sly: The early work of J. E. Heuser on nerve stimulation and retrieval of membrane from the synaptinemal complex showed that an enormous amount of homogeneous membrane is retrieved (see Heuser 1978). Perhaps the brain vesicles are principally involved in retrieval of that membrane, which would be quite homogeneous compared to the ones which are full of multiple heterogeneous receptors that one sees, for example, in liver or fibroblasts.

REFERENCES

Anderson RGW, Brown MS, Goldstein JL 1977 Role of the coated endocytic vesicle in the uptake of receptor-bound low density lipoprotein in human fibroblasts. Cell 10:351-364

Anderson RGW, Vasile E, Mello RJ, Brown MS, Goldstein JL 1978 Immunocytochemical visualization of coated pits and vesicles in human fibroblasts: relation to low density lipoprotein receptor distribution. Cell 15:919-933

Burridge K, Bray 1975 Purification and structural analysis of myosins from brain and other non-muscle tissues. J Mol Biol 99:1-14

Heuser JE 1978 Synaptic vesicle exocytosis and recycling during transmitter discharge from the neuromuscular junction. In: Silverstein SC (ed) Transport of macromolecules in cellular systems. Dahlem Konferenzen, Berlin (Life Sci Res Rep 11) p 445-464

Mello RJ, Brown MS, Goldstein JL, Anderson RGW 1980 LDL receptors in coated vesicles isolated from bovine adrenal cortex: binding sites unmasked by detergent treatment. Cell 20:829-837

Merisko EM, Farquhar MG, Palade GE 1982 Coated vesicles isolated by immunoadsorption on Staphylococcus aureus cells. J Cell Biol 92:846-857

Pearse BMF 1976 Clathrin: a unique protein associated with intracellular transfer of membrane by coated vesicles. Proc Natl Acad Sci USA 73:1255-1259

Pearse BMF 1980 Coated vesicles. Trends Biochem Sci 5:131-134

Pearse BMF 1982 Human placental coated vesicles carry ferritin, transferrin and γ-immunoglobulin. Proc Natl Acad Sci USA 79:451-455

Woodward MP, Roth TF 1979 Influence of buffer ions and divalent cations on coated vesicle disassembly and reassembly. J Supramol Struct 11:237-250

Endocytosis, the sorting problem and cell locomotion in fibroblasts

MARK S. BRETSCHER

MRC Laboratory of Molecular Biology, Hills Road, Cambridge CB2 2QH, UK

Abstract Fibroblasts endocytose lipid plus a subset of plasma membrane proteins over their entire surface and reinsert this into the plasma membrane at the cell's leading edge. This process is used to extend the fibroblast forwards. This circulation causes a flow of these endocytosed molecules over the cell's surface. Molecules, such as proteins, sitting in this flow can distribute themselves randomly by Brownian motion, but large objects (or small tethered ones) cannot. These large objects therefore cap. A mechanism is presented whereby this process could be used for locomotion using many weak interactions with the substrate. In addition it is suggested that the observed selectivity of coated pits may be sufficient to sort out proteins during transfer of membrane from one organelle to another so that the specific characters of the parent membranes are maintained.

Fibroblasts endocytose substantial amounts of their surface membrane, and presumably return it to the surface at the same rate. Here, I should like to construct an overall picture of this process and to develop two points, which relate to the problem of sorting membrane proteins and to cell locomotion.

The endocytic cycle

Much of our understanding of endocytosis in fibroblasts derives from the parallel studies of R. G. W. Anderson, M. S. Brown, J. L. Goldstein and their colleagues on the uptake of low density lipoprotein (LDL) and on the isolation and properties of coated vesicles by B. M. F. Pearse and others. LDL, bound to specific receptors which are largely localized in coated pits on the cell's surface, is endocytosed when the coated pit invaginates to form a

1982 Membrane recycling. Pitman Books Ltd, London (Ciba Foundation symposium 92) p 266-281

coated vesicle. The coated vesicle rapidly sheds its coat (probably in a few seconds), and the now internalized LDL is transported on to the lysosomes. The LDL receptors, as if by magic, are almost immediately returned to the cell's surface (Goldstein et al 1979). The magnitude of the whole process is impressive: fibroblasts take up the equivalent of their entire surface area every hour or so (Anderson et al 1978). This uptake by the coated pit–coated vesicle transition appears to occur uniformly over the surface of the cell.

The internalized membrane has to be returned to the cell surface, but where does this happen? And is it returned randomly over the cell's surface? There is one line of argument which suggests that this may not be so. It seems to me likely that all the membrane material destined for the plasma membrane of a fibroblast—that is, the endocytosed membrane on the one hand and newly synthesized membrane on its route from the endoplasmic reticulum (ER) on the other—is pooled at some (probably Golgi or post-Golgi) stage inside the cell and then taken together to the cell surface. If this is so, then the question becomes: where on the cell surface are newly synthesized proteins added? Here we may have a clue. Marcus (1962) studied the appearance of a newly synthesized viral protein in substrate-attached HeLa cells infected with Newcastle disease virus (admittedly, not normal fibroblasts). He found that the viral haemagglutinin first appears on the cell's surface at the cell's leading edge, and concluded that in uninfected cells this is where newly synthesized proteins are probably added to the plasma membrane. This, in turn, might mean that internalized membrane is also returned to the cell's surface at the cell's leading edge. For the present, I shall assume that this is so for fibroblasts.

This would mean that there is a flow of membrane across the dorsal (and ventral) surface of a polarized fibroblast. Since the source of membrane (the leading edge) and the sink where endocytosis occurs (all over the cell's surface) are not coincident, circulating membrane material will flow rearwards from the leading edge. The material which constitutes this flow will be that which is endocytosed and recirculated. This would include LDL receptors, other receptors found in coated pits (such as the receptors for transferrin, mannose phosphate, α_2-macroglobulin, insulin and so on) and lipid molecules. But what about other plasma membrane proteins? Do they circulate as well?

The evidence that other plasma membrane proteins do not circulate hangs on the finding (Bretscher et al 1980) that some plasma membrane proteins not only are not concentrated in coated pits, but are actively excluded from them. This was found when the distribution of two major fibroblast marker proteins—θ or Thy 1 antigen, and an 80 000 M_r protein recognized by a monoclonal antibody called H63 (Hughes & August 1981)—was monitored along the plasma membrane. This was done by a sandwich technique of immunoferritin labelling of fibroblasts chilled to 0 °C, followed by fixation and localization of ferritin particles along the plasma membrane in thin sections. The density of ferritin particles found along the plasma membrane adjacent to a coated pit was determined, as well as the length of that coated pit. From this could be calculated the number of ferritin particles expected in that coated pit if the particles were randomly distributed along the entire plasma membrane. This could then be compared with the number of ferritin particles actually found in that coated pit. Summed over many coated pits, the results for these two antigens are shown in Table 1. This shows that few

TABLE 1 Distribution of specific antibodies bound to 3T3 cells labelled at 0 °C

			No. of ferritin particles in coated pits	
Expt	*Antiserum*	*No. of coated pits examined*	*Expected*	*Observed*
1	Anti-Thy 1	77	439	16
	Control	77		12
2	Anti-H63	100	172	16
	Control	100		16

Cells were labelled at 0 °C with (Expt 1) rabbit anti-Thy 1 serum, followed by ferritin–goat anti-rabbit Ig; or (Expt 2) rat monoclonal anti-H63 antibody, followed by ferritin–rabbit anti-rat Ig. In controls the first-layer antibody was omitted. The mean contour length of the coated pits in both experiments was 0.28 μm. The average density of ferritin along the plasma membrane was: Expt 1, 20 particles/μm (control, 0.3 particles/μm); Expt 2, 7.2 particles/μm (control, 0.5 particles/μm).

ferritin particles are actually found in coated pits and that those found can roughly be accounted for by the background binding of ferritin in the absence of the first specific antibody. If we take the presence of ferritin to reflect the presence of either Thy 1 or H63 antigen, the results show that both antigens are excluded from coated pits. The precise degree of this exclusion is hard to tell, but the evidence suggests that they are excluded at a level of 99% or greater. That is, the density of these two surface proteins in coated pits is 1% or less of that along the rest of the plasma membrane.

This observation has, I believe, two profound implications. It demonstrates that coated pits can act as molecular filters for processes involving the sorting of proteins. In addition, it implies that the flow of membrane across a fibroblast's surface is not a generalized membrane flow, but rather is restricted to lipid and to those receptors that are collected into coated pits.

Relationship to cell locomotion

Abercrombie and his colleagues (1970a,b) proposed that a moving fibroblast continuously inserts recycled membrane at its leading edge. The suggested biological function of this is to extend the cell processes forwards, ahead of the cell. That something extraordinary is happening at the leading edge can be observed by all the ruffling and blebbing that occurs at just this region of the cell's periphery. This mobile behaviour is probably due in part, or wholly, to the rapid insertion of recycled membrane there. We now see that the recycled membrane is only lipid plus a subset of membrane proteins. What does this do to the average membrane protein which does not circulate, represented here by Thy 1 and H63?

At first sight it would seem that proteins sitting in this flow, but not part of it, would be swept away from the leading edge towards the rear of the cell. In fact, this probably does happen to a slight extent, but, because of the liquid nature of cell membranes, these proteins can diffuse around and therefore randomize themselves to some extent. To what extent depends on several factors: on how fast the flow is, on the length of the cell, and in particular on the diffusion coefficient of the protein. Elsewhere I have shown that, assuming a diffusion coefficient of 4×10^{-9} cm^2/s (the value determined by Poo & Cone [1974] for rhodopsin), diffusion over distances comparable to cellular dimensions is more important than flow (Bretscher 1976, 1982). In other words, both Thy 1 and H63 would be expected to be roughly randomly distributed over the cell's surface.

But if the diffusion coefficient of a protein were greatly reduced by some means or other, then flow would dominate: the protein could not diffuse against the flow and would be carried towards the cell's tail. Here, it is rather like a stick floating in, and being carried along by, a river. There are two contexts in which this may operate.

When protein antigens on the surface of a cell are cross-linked by an antibody, they form large, two-dimensional precipitates called *patches*. At 37 °C these patches rapidly accumulate at the rear of the cell to form a *cap* (Taylor et al 1971). This capping process has all the features that might be anticipated for a large aggregate being swept passively to the rear of the cell. (1) The antigens must be cross-linked to form a large patch. Only cross-linked molecules are thereby capped; those not extensively cross-linked, such as those cross-linked by an antibody which only dimerizes the antigen, do not cap. (2) Most molecules which are appropriately cross-linked cap, including glycolipids. This shows that direct contact of the cross-linked molecules with the cytoplasm is unnecessary. (3) The process requires metabolic energy as, doubtless, would a flow cycle involving endocytosis, transport through the cell and reinsertion in the cell membrane. (4) The process is polar, in the sense that caps do not form anywhere on a cell's surface, but always collect at the tail of the cell. (5) The capping process is fast: patches move rearwards on a fibroblast at about the same rate as that calculated for the flow (M. S. Bretscher, unpublished), which is about the same as the rate of rearward particle migration that is actually observed (Abercrombie et al 1970b, Harris & Dunn 1972).

In other words, the flowing lipid matrix of the membrane could naturally explain the process known as capping. Of course, capping is an *in vitro* phenomenon, induced by adding a protein cross-linking agent. It has no obvious counterpart of physiological relevance, but it it is possible that there is one. Consider a cell (not necessarily a fibroblast) sitting on a substrate, adhering to it by many weak interactions. So long as one of these weak interactions exists between a surface protein and the substrate, that protein cannot diffuse and will tend to cap. But as soon as the weak interaction comes apart, the protein can diffuse around; in particular, it can diffuse forwards, ready to be re-used and attach, weakly, to the substrate ahead of its original site. In other words, a cell could move forwards in this micropedal fashion; the difference between it and a regular millipede would be the number of 'feet' it has and the fact that the feet can diffuse forwards. Of course, I am not equating these 'feet' with the substrate attachment sites seen on the ventral surface of fibroblasts: these are large structures, attached to stress fibres, which are possibly required by a motile fibroblast to pull the mass of the cell forwards (Abercrombie et al 1970b).

The sorting problem

Since membrane proteins, eventually destined for a host of intracellular compartments (ER, Golgi stacks, lysosomes, apical or basolateral plasma membranes and possibly other compartments), are all initially inserted into the endoplasmic reticulum, it is clear that a sorting problem exists. The first stage of this is the separation of those proteins which are to remain in the ER from those that are to move on. How is this sorting done?

An ingenious proposal by Rothman (1981) is that the reason the Golgi exists in a stack of membranes is that the separation of these ER proteins from others is achieved by a form of counter-current distribution: several theoretical plates are needed because of the low degree of 'purification' achieved in each step. My own view is that this counter-current scheme is attractive, but that it is more likely that the stack is a device for concentrating cholesterol away from ER phospholipids than for fractionating proteins. Cholesterol is made in the ER, yet it is scarce there but rich in the post-Golgi compartments. It seems to me that it is more difficult to separate cholesterol from other lipids than it is to sort out proteins.

Whether several plates are needed to separate ER proteins from the others depends on the underlying mechanism. There is evidence, from Rothman & Fine (1980), that coated vesicles are involved in a transfer of newly synthesized coat protein of vesicular stomatitis virus (VSV) shortly after its synthesis. The question arises whether coated vesicles, budding from the ER, could have sufficient precision to select out those newly synthesized proteins which are to go on to the Golgi, leaving behind ER-resident proteins. Rothman has estimated that these newly synthesized proteins may account for only one part in 10^4 of the proteins in the ER, the remainder being ER residents.

From the studies on the LDL receptor, as much as 80% of the receptor may be found in coated pits, which account for about 2% of the cell's surface. The selection of LDL into coated pits is therefore about 200 (density of the LDL receptor in coated pits:density along the plasma membrane). As mentioned above, the equivalent selection of either Thy 1 or the H63 antigen is 10^{-2} or less. The difference in selectivity is thus over 2×10^4. LDL receptors are efficiently bound in coated pits, yet less than one part in 5000 of Thy 1 or H63 antigen is found in them. The specificity of selection by coated pits is thus large, and may be great enough to achieve the separation of post-ER proteins from ER-resident proteins in one step. Whatever else, it indicates that the mechanism of exclusion of Thy 1 or H63 antigens—presumably by packing the coated pit so tightly with selected receptors that there is no space for them—is a fascinating problem.

REFERENCES

Abercrombie M, Heaysman JEM, Pegrum SM 1970a The locomotion of fibroblasts in culture. I. Movements of the leading edge. Exp Cell Res 59:393-398

Abercrombie M, Heaysman JEM, Pegrum SM 1970b The locomotion of fibroblasts in culture. III. Movement of particles on the dorsal surface of the leading lamella. Exp Cell Res 62: 389-398

Anderson RGW, Vasile E, Mello RJ, Brown MS, Goldstein JL 1978 Immunocytochemical visualization of coated pits and vesicles in human fibroblasts: relation to low density lipoprotein receptor distribution. Cell 15:919-933

Bretscher MS 1976 Directed lipid flow in cell membranes. Nature (Lond) 260:21-23

Bretscher MS 1982 Surface uptake by fibroblasts and its consequences. Cold Spring Harbor Symp Quant Biol, in press

Bretscher MS, Thomson JN, Pearse BMF 1980 Coated pits act as molecular filters. Proc Natl Acad Sci USA 77:4156-4159

Goldstein JL, Anderson RGW, Brown MS 1979 Coated pits, coated vesicles, and receptor-mediated endocytosis. Nature (Lond) 279:679-685

Harris A, Dunn G 1972 Centripetal transport of attached particles on both surfaces of moving fibroblasts. Exp Cell Res 73:519-523

Hughes EN, August JT 1981 Characterization of plasma membrane proteins identified by monoclonal antibodies. J Biol Chem 256:664-671

Marcus PI 1962 Dynamics of surface modification in myxovirus-infected cells. Cold Spring Harbor Symp Quant Biol 27:351-365

Poo M, Cone RA 1974 Lateral diffusion of rhodopsin in the photoreceptor membrane. Nature (Lond) 247:438-441

Rothman JE 1981 The Golgi apparatus: two organelles in tandem. Science (Wash DC) 213:1212-1219
Rothman JE, Fine RE 1980 Coated vesicles transport newly synthesized membrane glycoproteins from endoplasmic reticulum to plasma membrane in two successive steps. Proc Natl Acad Sci USA 77:780-784
Taylor RB, Duffus WPH, Raff MC, De Petris S 1971 Redistribution and pinocytosis of surface immunoglobulin molecules induced by anti-immunoglobulin antibody. Nat New Biol 233:225-229

DISCUSSION

Sly: The assumption that endocytosis occurs randomly is not really tenable in view of the ordered structure of the coated pits that one sees in fibroblasts, as Dr Goldstein and his colleagues showed (Anderson et al 1978). In fibroblasts the endocytic vesicles form linear arrays, which does seem to suggest that coated vesicle formation and endocytosis are not random processes.

Bretscher: The point here is that, even if coated pits are not arranged totally randomly on the surface of motile cells, at a coarse level they are pretty evenly spread over the cell surface.

Goldstein: The linear arrays of clathrin-lined coated pits appear to be limited to spread cells, such as fibroblasts. When round cells, such as Chinese hamster ovary cells, are stained with anti-clathrin antibody, linear arrays of immunofluoresence are not detected, yet these cells show rapid internalization of receptor-bound low density lipoprotein (Goldstein et al 1979).

Bretscher: What really matters is the distribution as one goes backwards from the leading lamella. The density of pits per square micron of surface seems to be uniform.

Baker: If one were to irradiate the front end of the fibroblast with ultraviolet light, and if all the 'action' were going on there, it should stop the capping going on the back. Has that ever been done?

Bretscher: No, but you are quite right.

Meldolesi: Is your model implying that the membrane proteins that get interiorized by adsorptive endocytosis are more concentrated at the front of the cell and less concentrated at the back?

Bretscher: That may indeed be so. I would suspect that there might be a higher concentration of LDL receptors or transferrin receptors at the leading edge than at the back, and it might be a striking difference. Preliminary evidence on giant HeLa cells suggests this is so (M. S. Bretscher, unpublished results).

Mellman: If an entire surface complement of LDL receptors is internalized every few minutes, there should, in fact, be quite a large concentration at the leading edge. Given the overall diffusion rates that Dr Palade quoted earlier, can you calculate what the concentration gradient should be?

Bretscher: The life-time of an LDL receptor on the cell surface before it hits a coated pit is about 5 min or less. If one puts an LDL receptor at the leading edge, the extent of its free travel in 5 min will depend on its diffusion coefficient. I believe that the diffusion coefficient of all these molecules (despite what is measured) would be about 5×10^{-9} cm^2 s^{-1} at 37 °C. If that is true, then the LDL receptors would get to about the middle of a 50 μm-long cell in 5 min. By then it would have been endocytosed and reinserted at the leading edge. So there would be few LDL receptors near the tail of such a cell.

Geisow: By experimental intervention, it is possible to induce 'high and low tides' in the flow of cell surface membrane—for example, when one stimulates pinocytosis and thereby increases membrane flow. In this situation a monovalent antibody would allow one to see the distribution of a particular receptor; would one then see capping? How well is the diffusion coefficient of membrane protein balanced with the overall transit of plasma membrane on the surface?

Bretscher: It is quite well balanced.

Geisow: So a small perturbation of membrane flow rate might allow redistribution of membrane proteins?

Bretscher: Yes. One has to step-up the flow rate by a factor of 2 or 4 in order to see something. I have studied extensively the instantaneous distributions of concanavalin A receptors and freeze-fracture particles on motile chick heart myoblasts, to try to find a gradient of either of them along the long axis of the cell's membrane, and I could not find one (M. S. Bretscher, unpublished results).

Sly: Are you implying that recycling of receptors is in equilibrium with newly synthesized receptors on the cell surface?

Bretscher: No. But if one could label newly synthesized receptors and see where they appear on the cell surface, initially one would see them at the cell's leading edge. That would be analogous to the experiment by Marcus (1962). If one could look at recycled receptors when they appeared on the surface they, too, would be localized in that region. In fact they might come together, possibly in the same vesicle, to the leading edge.

Rothman: Is there any flow underneath the cell on the surface that is apposed to the plate?

Bretscher: Harris & Dunn (1972) showed several years ago that attached particles move on both the dorsal and ventral surfaces of motile fibroblasts. The difficulty is that there is not much space between the dish and the ventral surface of the cell. But when a particle did attach to the ventral surface, it moved backwards at the same rate as those on the top surface until it hit an adhesion site, at which point it stopped. Their conclusion was that the flow is not simply on the dorsal surface but also on the ventral surface.

Sabatini: The foot could be equivalent to a junction in between cells, which is perhaps an adhesion point to the substrate. Louvard (1980) has shown that the site of insertion of the recirculating material taken in from the surface is near the junction. Unpublished experiments with Dr E. J. Rodriguez-Boulan on the polarity of viral budding in individually plated epithelial cells suggest a boundary precisely in the neighbourhood of the adhesion site to the substrate. One could conceive that the neighbourhood of that site would be equivalent to the front of the fibroblast.

Bretscher: The feet of the cells appear, in the interference microscope, not as thin lines but as thick areas of about 1 μm diameter (Abercrombie & Dunn 1975). They are strictly two-dimensional structures, and I very much doubt that they are directly related to tight junctions.

Sabatini: Do you assume that the movement of proteins on the surface of the cell is not dictated by an association with cytoskeletal elements which may also relate to the cytoplasmic foot?

Bretscher: The average plasma membrane protein cannot be associated with the cytoskeleton. Even if all the actin inside the cell were used to hold the membrane proteins, there would not be enough actin to do the job.

Geisow: Would your model predict that an epithelial sheet of cells would curl? I mean by this that the movement of membrane from one pole to another in the outermost cells might cause the edges of the sheet to roll over. This might be experimentally testable.

Bretscher: I have no idea.

Cohn: Can you equate the movement of carbon particles with the movement of membrane? Has anybody repeated the experiments with a particle that might be more tightly attached?

Bretscher: Abercrombie et al (1970) did those experiments with carbon particles, molybdenum disulphide particles and cell debris, whilst Harris & Dunn (1972) used dowex beads. All the particles behaved in the same way on these cells. In addition, if patches are induced with concanavalin A on these cells at 0 °C, and if the rate of migration of these patches is then observed at 37 °C, they also go at the same rate (Abercrombie et al 1972, M. S. Bretscher, unpublished results).

Sabatini: Would you assume that the material which is interiorized over the whole surface and which then reappears at the front of the cell goes through a common station? Are you implying that this is the Golgi?

Bretscher: I don't know where it goes. The LDL and transferrin receptors clearly go by completely different routes through the cells because of their very different transit times. The LDL receptor probably doesn't have time in its brief transit through the cell to visit lysosomes, whereas it is likely that the transferrin receptor does so.

Sabatini: Would this mean that the concentrations of those receptors on the surface of the cell after recycling would be different and therefore that the diffusion would be at different rates?

Bretscher: It's possible that the LDL receptor and the transferrin receptor are returned to the cell surface in different places. Therefore if one could look at the instantaneous distributions of those receptors on the cell at any moment they might indeed be different.

Kornfeld: In the red cell there is evidence that the plasma membrane protein band 3 is connected to the cytoskeleton. One would have to postulate that receptors are different from other membrane proteins: one cannot make the general statement that no plasma membrane protein is directly connected to the cytoskeleton.

Bretscher: I believe that certain specific plasma membrane proteins of fibroblasts will act like the hole in a tent canvas into which the tent pole fits. There must be proteins that anchor the cytoskeleton to give the cell membrane its shape. The question is: do most plasma membrane proteins interact directly with the cytoskeleton? This idea has been suggested partly because fluorescence bleaching experiments show that the diffusion coefficient of any molecule on the surface of almost any type of cell is 10^{-10} cm^2 s^{-1} (see Cherry 1979). Yet this is far too low for a freely diffusing molecule (for which it should be about 100-fold higher), and therefore these workers have suggested that the protein must be attached to some immobile structure (i.e. cytoskeleton) for most of its time. I disagree with this interpretation, and think that the method is wrong.

Palade: The red blood cell is an extreme example of extensive stabilization (nearly immobilization) of intrinsic membrane proteins by interaction with an extensive infrastructure formed by spectrins and other peripheral proteins. Therefore, results obtained on red blood cells may not apply to other cell types.

Branton: Dr Bretscher, do you not believe the results of those measurements?

Bretscher: That is correct. Mike Edidin and Richard Cone (see Wey et al 1981) have measured the diffusion of rhodopsin by fluorescence bleaching, and this method gave the same answer, of about 3×10^{-9} cm^2 s^{-1}, as that obtained earlier by Poo & Cone (1974). They were supposedly validating the fluorescence bleaching method. All the other plasma membrane proteins that have ever been measured by that method give answers of about 3×10^{-10} cm^2 s^{-1}. However, Poo et al (1979) have looked at the diffusion coefficient of bulk concanavalin A receptors on myoblasts by electrophoresis and subsequent relaxation at 22 °C, and find that the average diffusion coefficient of those is

5×10^{-9} cm^2 s^{-1}. Therefore at 37 °C the values are probably about 10^{-8} cm^2 s^{-1}. There is a serious discrepancy in the actual methods of measurement. In Dr Poo's method the cells are unperturbed, and the results therefore probably more reliable.

Sabatini: You said there isn't enough actin to bind these proteins but at least in the red cell the connections between membrane and cytoskeletal proteins are indirect (Bennett & Branton 1977) and there might be sufficient actin to bind all proteins.

Bretscher: If one assumes that all the actin (the most abundant cytoskeletal protein) inside a fibroblast is bound only to the plasma membrane, the molar ratio of actin subunits to plasma membrane proteins is about 4:1. The diffusion coefficient measured by fluorescence bleaching suggests that all the plasma membrane proteins are anchored almost all the time. Given the ratio, that level of anchoring is not plausible.

Baker: So what is wrong with the fluorescence bleaching method?

Bretscher: Fluorescein bleaching is usually used in the method, but fluorescein is used in immunofluorescence studies because it can't be bleached easily. In order to bleach fluorescein, about 10^6 quanta are needed. One has to ask what happens to the other 999 999 quanta. I imagine that they excite the fluorescein, which then decays and produces free radicals that fix the cell.

Goldstein: What are the temperature correlations between endocytosis and capping?

Bretscher: They both stop at about 16 °C.

Blest: Did your experiments with the coated pits provide any evidence which might suggest the sequence of the events during binding and internalization of a ligand? Obviously, something has to stimulate the formation of a coated pit, and you have shown that it acts as a molecular sieve. Does a special domain start to act as a sieve, and then, as a secondary consequence, to induce the polymerization of clathrin? Alternatively, does a coated pit form and, by doing so, define the region that will function as a sieve?

Bretscher: I know of no evidence about the sequence of events. One imagines that a coated pit grows at its edges by accretion not only of clathrin subunits but also of receptors at the edges. When it gets to a certain size it somehow or other starts curving.

Blest: We have a preparation in which the tip of a microvillus belonging to a photoreceptor is pinched off in the course of membrane turnover by a coated pit in the plasma membrane of an adjacent glial cell (Blest & Maples 1979). The first event in the sequence is the transformation of the microvillus tip, which lengthens. We infer this to result from the local deletion of the microvillar cytoskeleton (Blest et al 1982a,b). Transformation precedes the induction of the coated pit, and this suggests that a topographically precise signal must be given to the glial membrane which causes the pit to form in the right place.

Hopkins: What about the positive control in your experiments, Dr Bretscher? Have you been able to use antibody and ferritin to demonstrate receptors in pits?

Bretscher: The positive control wasn't done at the same time in the same place. It was done by Dr Goldstein and his co-workers who detected low density lipoprotein (LDL) receptors with antibodies against the receptor in coated pits (Goldstein et al, this volume, p 77-95). We have subsequently shown that one can bind the anti-transferrin receptor antibody to the receptor in coated pits (Bleil & Bretscher 1982).

Hopkins: The question is surely not whether one can get ferritin in the pits, but whether the same amount of ferritin is found in the pit as one would expect.

Bretscher: We cannot say that we label *each* receptor that is present with ferritin, but when we labelled coated pits on HeLa cells, each section through a coated pit contained about 30–40 ferritins, so there is no doubt about the accessibility of that protein in a coated pit.

In relation to Dr Blest's previous comment, it is indisputable that coated pits act as molecular filters in some sense, but the question is: how accurately do they do this? Our (Bretscher et al 1980) experiments quantify that to some extent. Secondly, we need to know the mechanism by which that exclusion occurs. I would suggest that the only way of excluding molecules from a coated pit is by packing the coated pit with other things so that there is no space. The structure of the theta (θ) antigen, Thy 1, which is one of the markers we used, is relevant here. The Thy 1 has been sequenced by Alan Williams (see Campbell et al 1981), and it turns out to be a glycoprotein which can be folded up into a domain of an antibody molecule, but it has no membrane segment that is polypeptide: it has no extensive hydrophobic sequence in it. But it does have, on its C-terminal residue, a highly hydrophobic molecule which anchors the protein to the membrane. It is the only membrane protein that I know with this type of structure. As this molecule is excluded from a coated pit on fibroblasts, the exclusion is probably done by packing with receptors on the non-cytoplasmic surface of the coated pit and not within the bilayer or on the cytoplasmic side.

Goldstein: Have you calculated how many receptors would fit in a typical coated pit?

Bretscher: I have calculated the proportion of the surface area of coated pits that might be filled in with transferrin receptor on the HeLa cells. Given that this receptor has a subunit size of 90 000 M_r, and assuming that this subunit has a cross-sectional area of 4 000 Å^2, and that 2% of the cell surface is coated pits, the fraction of the surface area of a coated pit taken up by that one receptor, packed side by side, would be about 10% (i.e. a substantial portion).

Mellman: A less substantial portion would be occupied by the LDL receptor. If one estimates that the total number of LDL receptors per fibroblast is 10–20 000, and that the total number of coated pits is 1000 per cell (Marsh & Helenius 1980), there should be an average of 10–20 LDL receptors per coated pit. Thus, the density of receptors in coated pits (which occupy 2% of the cell surface area) would be similar to the density of any randomly distributed membrane protein present in 1 000 000 copies per cell (2% of the cell surface would include 20 000 molecules). A possible example of such a protein is the 90 000 M_r macrophage antigen 2D2C which is present at 1 300 000 per J774 cell (Mellman, this volume, p 35-58).

Branton: If there are a million LDL receptors per cell, wouldn't they occupy as much as 5% of the 2000 μm^2 surface area?

Bretscher: Yes; it could be as much as that.

Meldolesi: Would it be possible to test experimentally your idea that molecular exclusion accounts for the surface specificity of the cell? One could use different ligands and thus modulate the number of receptors that are expected to be clustered in coated pits. One could see whether mistakes are made at different rates by the cell when the load of ligands is small or when it is high.

Bretscher: I don't know how one would do that.

Rothman: You could add cytochrome b_5, a protein that is not normally there, to the outside of the cell. It would probably insert into the plasma membrane. I wonder whether it would later be found in the intracellular compartments?

Palade: Is it possible to estimate how much of the membrane area is taken off by proteins, and how much by lipids in an enveloped virion? In other words, can we find out whether extensive occupancy of the bilayer by viral glycoproteins accounts for the local exclusion of the cell's plasmalemmal proteins?

Helenius: In the Semliki Forest virus the protein content of the membrane is very high, about 60% by weight. However, less than 15% of the volume of the bilayer is occupied by protein (Utermann & Simons 1974). The bulk of the protein is outside the bilayer.

Palade: So the occupancy becomes a problem outside rather than within the bilayer. How crowded are the proteins on the virus surface, at some radial distance from the membrane?

Helenius: There is only 20–35 Å space between the spikes. In other enveloped viruses the distances may be even less than that.

Bretscher: The viruses are a very good analogy in that their spike proteins crowd out the surface extracellularly, and thereby prevent the inclusion of host proteins in the virus capsid.

Palade: If the geometry were reverted, the situation should be the same in coated vesicles.

Rothman: Using calf brain coated vesicles, Alan Matusomoto and John Rubenstein in my laboratory (unpublished) have calculated the lipid:protein ratio in vesicles freed of clathrin (after agarose gel purification). The coated vesicle preparations used were 97% coated vesicles and about 3% smooth vesicles. The vesicles were purified on a gradient of metrizamide. About 99% of the clathrin was removed, although most of the extrinsic 100 000 and 50 000 M_r polypeptides were still in their vesicle preparation. The lipid:protein ratio of these vesicles was about 0.5 mg phospholipid per mg protein. We also found that the cholesterol content was about 20 mole %, i.e. about 0.05 mg cholesterol per mg protein. Therefore, the total lipid:protein ratio was about 0.55 mg per mg protein. This represents a lower limit for the ratio in the membrane proper, because the protein value includes both content and extrinsic proteins (100 000, 50 000 M_r). One would have expected a much smaller lipid:protein ratio if a protein core were the basis for exclusion.

Palade: Is the preferred site of entry of endocytosed membrane along the advancing fringe of the cell?

Bretscher: I believe so!

Palade: Is this therefore connected with cell movement?

Bretscher: Absolutely, yes.

Palade: Then is vesicle entry connected with cell movement, in the sense that movement within the cell imparts directionality to vesicle displacement? In other words, does cell movement sweep the vesicles in that particular direction?

Bretscher: We don't know. I would imagine that parts of the cytoskeletal apparatus are involved in directing the endocytic vesicles from the Golgi or elsewhere back to the leading front. A fibroblast that has a polarized direction of movement will begin to ruffle all around its edges when it is put into colchicine (to disrupt the microtubules, supposedly). So microtubules may provide that particular axis of polarity. But I don't know how the vesicles get to the leading edge. Given that a cell endocytoses substantial amounts of its surface, the cell could use that property to project its leading edge forward ahead of any structures that it may have.

Palade: Is there enough membrane put into the advancing fringe by re-entering coated vesicles to allow cell movement?

Bretscher: Yes, within a factor of about two. The backwards flow rate on the dorsal surface agrees with the calculated flow rate, given the rate of endocytosis.

Goldstein: The complete amino acid sequence of one receptor that resides in coated pits has been reported by Drickamer (1981). He used the chicken asialoglycoprotein receptor of liver. This protein has an M_r of 26 000 and the C-terminus of the molecule is believed to face the outside of the cell. About

75% of the receptor protein extends outside the cell, and one carbohydrate chain is attached to amino acid residue 67. The N-terminus (=*N*-acetyl-methionine) is inside the cell. So this receptor has a somewhat unusual orientation: C-terminus *out* and N-terminus *in*. How many integral membrane proteins, other than those in red blood cells, have been sequenced? Are most of them oriented like the G protein of vesicular stomatitis virus, with the N-terminus *out* and the C-terminus *in*?

Bretscher: They are found both ways round. There are several examples.

Goldstein: What about the HLA antigen?

Bretscher: That has its N-terminus outside and its C-terminus inside (Robb et al 1978).

Rothman: Aminopeptidase of the intestinal brush border membrane has been found to have its C-terminus on the outside and its N-terminus inside.

Palade: Isomaltase is another example (Semenza 1981).

Helenius: So is neuraminidase in influenza virus.

Goldstein: But none of those proteins is localized in coated pits.

Sabatini: I believe that the sucrase–isomaltase of the intestinal brush border is anchored to the membrane by its terminal segment (Brunner et al 1979).

Bretscher: How large is this cytoplasmic N-terminal domain?

Goldstein: The cytoplasmic domain of the chicken asialoglycoprotein receptor constitutes about 12% of the mass of the protein (Drickamer 1981).

REFERENCES

Abercrombie M, Dunn GA 1975 Adhesions of fibroblasts to substratum during contact inhibition observed by interference reflection microscopy. Exp Cell Res 92:57-62

Abercrombie M, Heaysman JEM, Pegrum SM 1970 The locomotion of fibroblasts in culture. III: Movement of particles on the dorsal surace of the leading lamella. Exp Cell Res 62:389-398

Abercrombie M, Heaysmen JEM, Pegrum SM 1972 Locomotion of fibroblasts in culture. V: Surface marking with concanavalin A. Exp Cell Res 73:536-539

Anderson RGW, Vasile E, Mello RJ, Brown MS, Goldstein JL 1978 Immunocytochemical visualization of coated pits and vesicles in human fibroblasts: relation to low density lipoprotein receptor distribution. Cell 15:919-933

Bennett V, Branton D 1977 Selective association of spectrin with the cytoplasmic surface of human erythrocyte plasma membranes. J Biol Chem 252:2753-2763

Bleil JD, Bretscher MS 1982 Transferrin receptor and its recycling in HeLa cells. EMBO J 1:351-355

Blest AD, Maples J 1979 Exocytotic shedding and glial uptake of photoreceptor membrane by a salticid spider. Proc R Soc Lond B Biol Sci 204:105-112

Blest AD, Stowe S, Eddey W 1982a A labile, Ca^{2+}-dependent cytoskeleton in rhabdomeral microvilli of blowflies. Cell Tissue Res 223:553-574

Blest AD, Stowe S, Eddey W, Williams DS 1982b The local deletion of a microvillar cytoskeleton from photoreceptors of tipulid flies during membrane turnover. Proc R Soc Lond B Biol Sci 215:469-479

Bretscher MS, Thomson JN, Pearse BMF 1980 Coated pits act as molecular filters. Proc Natl Acad Sci 77:4156-4159

Brunner J, Hauser H, Braun H et al 1979 The mode of association of the enzyme complex sucrase-isomaltase with the intestinal brush border membrane. J Biol Chem 254:1821-1828

Campbell DG, Gagnon J, Reid KBM, Williams AF 1981 Rat brain Thy-1 glycoprotein. Biochem J 195:15-30

Cherry RJ 1979 Rotational and lateral diffusion of membrane proteins. Biochim Biophys Acta 559:289-327

Drickamer K 1981 Complete amino acid sequence of a membrane receptor for glycoproteins: sequence of the chicken hepatic lectin. J Biol Chem 256:5827-5839

Goldstein JL, Anderson RGW, Brown MS 1979 Coated pits, coated vesicles, and receptor-mediated endocytosis. Nature (Lond) 279:679-685

Harris A, Dunn G 1972 Centripetal transport of attached particles on both surfaces of moving fibroblasts. Exp Cell Res 73:519-523

Louvard D 1980 Apical membrane amino peptidase appears at the site of cell-cell contact in cultured epithelial cells. Proc Natl Acad Sci USA 77:4132-4136

Marcus PI 1962 Dynamics of surface modification in myxovirus-infected cells. Cold Spring Harbor Symp Quant Biol 27:351-365

Marsh M, Helenius A 1980 Adsorptive endocytosis of Semliki Forest virus. J Mol Biol 142:439-454

Poo M, Cone RA 1974 Lateral diffusion of rhodopsin in the photoreceptor membrane. Nature (Lond) 247:438-441

Poo M, Lam JW, Orida N 1979 Electrophoresis and diffusion in the plane of the cell membrane. Biophys J 26:1-22

Robb R J, Terhorst C, Strominger JL 1978 Sequence of the COOH-terminal hydrophilic region of histocompatibility antigens HLA-A2 and HLA-B7. J Biol Chem 253:5319-4324

Semenza G 1981 Intestinal oligo and disaccharidases. In: Randle PJ et al (eds) Carbohydrate metabolism and its disorders. Academic Press, London vol 3:425-479

Utermann G, Simons K 1974 Studies on the amphipathic nature of the membrane proteins in Semliki Forest virus. J Mol Biol 85:569-587

Wey CL, Cone RA, Edidin MA 1981 Lateral diffusion of rhodopsin in photoreceptor cells measured by fluorescence photobleaching and recovery. Biophys J 33:225-232

Final general discussion

Membrane recycling under experimental control

Baker: Membrane recycling must have some source of energy driving it. There must be switches and signals that determine where things move and when they move around the cell. But all these things happen inside the cell and few techniques can get at them. Dr Knight and I are interested in trying to gain control over the internal environment of cells in order to examine certain aspects of membrane turnover in conditions where we have control over the small-molecular-weight environment of the cell interior. We have studied nerve cells, in which some of the processes that have been discussed here are spatially separated: the synthetic machinery is in the cell body; transport of vesicular materials takes place along the axon; and exocytosis and membrane retrieval happen at the nerve terminal. To gain control over the internal environment, we have exposed cells to brief intense electric fields which cause membrane breakdown (Baker & Knight 1978, 1980, 1981). We simply charge up a capacitor and discharge it through a suspension of cells. For simple spherical cells placed between two plates this effectively means two holes per discharge, each with a diameter of about 4 nm. Depending on the cell type, these holes can remain patent for some hours (in adrenal medullary cells and hepatocytes), but in some cell types (e.g. red cells and macrophages) they seal up very quickly, and we don't really know why there is this marked difference. When the holes remain patent, one can control the internal environment of the cell.

Cells treated in this way remain morphologically like normal cells, provided that the medium contains anions such as glutamate and very little chloride, a low ionized calcium (10^{-8} M, i.e. in the physiological range), and some Mg-ATP buffered by Hepes or Pipes at about pH 7.0. This procedure allows control over small molecules inside the cell—something that is easily demonstrated by use of radioactive or fluorescent markers. One can wash out these molecules, add them back, and so on. We have applied this technique to suspensions of bovine adrenal medullary cells which normally release catecholamines by exocytosis. When these cells are made permeable, very little catecholamine comes out because it is packaged within vesicles. By

1982 Membrane recycling. Pitman Books Ltd, London (Ciba Foundation symposium 92) p 282-292

changing the Ca^{2+} concentration from 10^{-8} M into the micromolar range, we can evoke exocytosis. Mg-ATP in the millimolar range is essential for this and a pulse of calcium in the presence of ATP will evoke exocytosis, followed by membrane retrieval apparently by endocytosis. A cycle of membrane turnover is thus initiated. Various manipulations can be applied to the system to reveal what is needed for this cycle (see Table 1 and Baker & Knight 1981).

TABLE 1 (Baker) Properties of calcium-dependent catecholamine release from 'leaky' adrenal medullary cells

(1) Activation half-maximal at an ionized calcium concentration of 1 μM
(2) Requiremeńt for Mg-ATP is very specific. Half-maximal activation requires 1 mM
(3) Unaffected by
  (a) agonists and antagonists of acetylcholine receptors including acetylcholine, nicotine and hexamethonium (1 mM)
  (b) Ca-channel blocker D600 (100 μM)
  (c) agents that bind to tubulin (colchicine, vinblastine, 100 μM)
  (d) cytochalasin B (1 mM)
  (e) inhibitors of anion permeability (SITS, DIDS, 100 μM)
  (f) protease inhibitor TLCK (1 mM)
  (g) cyclic nucleotides (cyclic AMP, cyclic GMP, 1 mM)
  (h) *S*-adenosyl methionine (5 mM)
  (i) phalloidin (1 mM)
  (j) vanadate (1 mM)
  (k) Leu- and Met-enkephalins, substance P (100 μM)
  (l) somatostatin (1 μM)
  (m) NH_4Cl (30 mM)
(4) Inhibited by:
  (a) chaotropic anions: SCN > Br > Cl
  (b) detergents (complete inhibition after 10 min incubation with 10 μg/ml of digitonin, Brij 58 or saponin)
  (c) trifluoperazine (complete inhibition with 20 μg/ml)
  (d) high magnesium concentration: small increase in apparent K_m for calcium accompanies large reduction in V_{max}
  (e) high osmotic pressure: large reduction in V_{max} but no significant changes in the affinity for calcium
  (f) carbonylcyanid *p*-trifluormethoxyphenylhydrazone (FCCP) (45% inhibition by 10 μM)

(From Baker & Knight 1981)

Membrane turnover requires Mg-ATP, is activated by micromolar concentrations of ionized calcium and is unaffected by orthovanadate (1 mM), phalloidin (1 mM), cytochalasin B (1 mM) and NH_4Cl (30 mM).

For studying the movement of membranous material within the cell (Adams et al 1982), we have made use of the fact that particle movement by fast axoplasmic transport can be seen in axons isolated from the walking legs of crabs. One can make pairs of holes by discharging a capacitor between pairs of electrodes placed either side of the axon, and watching the movement of particles under phase-contrast microscopy. If one makes the axon permeable in a potassium–glutamate medium containing Mg-ATP, the particles continue to move in both orthograde and retrograde directions. If the ATP is removed, the particles stop moving and start again when ATP is added back (Fig. 1). This axoplasmic flow system, which is supposedly associated with microtubules, seems to be rather specific for Mg-ATP. It is not influenced by calcium over a wide range of concentrations and it can be inhibited by orthovanadate which perhaps suggests the involvement of a dynein-like molecule.

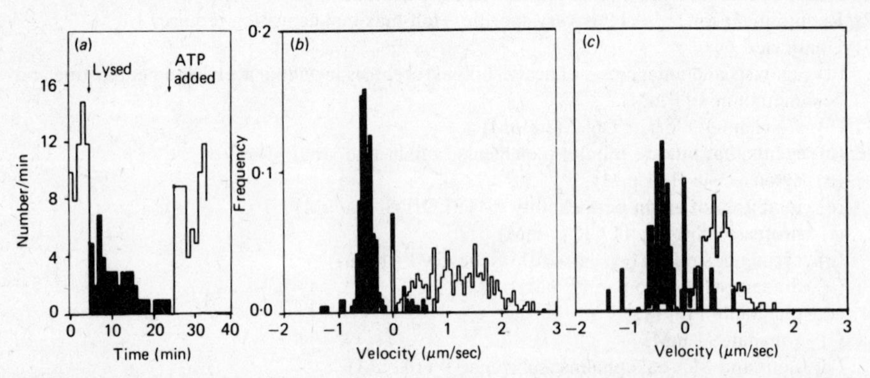

FIG. 1. (*Baker*) (a) Number of particles per minute passing a reference point in an axon before and after lysis (blocked bars) induced by exposure to brief intense electric fields, and after addition of 6 mM-ATP. (b), (c) Instantaneous velocities (over 1 μm) of particles in intact axon (b) and reactivated axon (c). Blocked bars show anterograde particles and open bars retrograde. (From Adams et al 1982, with permission of The Physiological Society.)

These two examples show that membrane movement can be subjected to experimental control and it is likely that this technique will have all sorts of other uses in the general field of membrane recycling. Our results are interesting because they already suggest that the process that moves membranous vesicles in axoplasmic transport is significantly different from the process that effects membrane turnover by exocytosis and endocytosis.

Rothman: You indicated that the holes heal more quickly in some cell types than in others. Have you applied this to any tissue culture cells?

Baker: It can be applied to any cell but the resealing seems to depend on cell type. For instance, red blood cells, HeLa cells and macrophages seem to reseal rather quickly and one can make them permeable at 0 °C but at 37 °C they tend to recover. Adrenal medullary cells and hepatocytes seem to remain permeable even at 37 °C. Perhaps cells that are normally free swimming in suspension culture or in the circulation are much more resistant and able to recover.

Sabatini: How do you determine the size of the holes?

Baker: By looking at the diffusion of markers of different molecular weights (see Baker & Knight 1982).

Sabatini: Is it possible to increase the size of those holes, or do you just simply make more of them if you try?

Baker: Once made, they cannot be increased in size in an easily controlled manner. But a more intense electric field or a longer time of exposure might make slightly bigger holes. We haven't systematically explored those possibilities (but see Zimmerman et al 1981). With holes of 4 nm, adrenal medullary cells have normal ultrastructure, and will respond for some hours.

Branton: What is the rationale for avoiding inorganic anions, e.g. Cl^-?

Baker: If glutamate is replaced by chloride (or Br^- or SCN^-) then both axon flow and endocytosis/exocytosis stop completely. Axon flow is highly dependent on the presence of an anion like glutamate. This is not a metabolic dependence because both D- and L- glutamate are equally effective. The exocytosis/endocytosis system will work with acetate and a number of other large anions. One explanation is that certain anions bind to proteins and in so doing dislodge or in some other way render ineffective proteins that are essential for these processes. In axons, neurofilaments are dissociated by anions. Axoplasm from, say, the squid axon is a stable gel in glutamate; but in chloride or thiocyanate it liquifies because the neurofilaments dissociate into subunits. This anion effect is largely reversible and should be distinguished from the irreversible proteolytic breakdown of neurfilaments that is activated by millimolar concentrations of calcium (see Rubinson & Baker 1979).

Sabatini: Must you avoid sodium?

Baker: No. The nature of the cation is relatively unimportant—one can have potassium, sodium or magnesium, or one can even replace the salt by sucrose and still have a normal response to calcium (see Baker & Knight 1981).

Hopkins: You measured the size of the holes using tracers. Have you used visible tracers and, once they get in, how accessible is the cell?

Baker: We have not done much with fluorescent tracers but mostly used radioactive markers. We measure the space occupied by tritiated water, which is a measure of the intracellular volume, and then we look at the marker in relation to that measurement (see Baker & Knight 1981). We find that about 70% of the cell water is accessible to the markers that get into the cells. The cell water that is not accessible is presumably within organelles which, because of their smaller size, are not affected by this technique. The discharge has to build up a field across the membrane in order to produce dielectric breakdown, and this is easier to do, the bigger the structure. Small intracellular structures are unaffected by fields that will render the surface membrane permeable.

Hopkins: Would you expect any holes in the internal membranes after this procedure?

Baker: Only if extremely high voltages are used. It is easy to find a voltage that affects the plasma membrane but leaves the intracellular organelles unaffected. Thus, catecholamine is not released from the storage granules, but can subsequently be released by a calcium pulse. Also, if these 'leaky' cells are exposed to radioactive Ca^{2+} buffers, the organelles inside the cells accumulate Ca^{2+} normally.

Rothman: Can you see these holes e.g. by using a fluorescent molecule on the outside and watching it enter at each hole?

Baker: If we use large cells such as those from sea urchin eggs, where the cells are about 60 μm in diameter, we need only a small voltage to make the holes. We can actually see the cytoplasm coming out like two ears on opposite sides of the cell (Baker et al 1980). When aequorin was injected into large amoebae and then an electric field was applied to them, fluorescence was visible at the two ends of the cell (D. L. Taylor, personal communication). But in the adrenal cells we do not normally see the holes and have to rely on the penetration of marker molecules.

Cohn: Can you separately block the movement of particles distally and proximally in the axon?

Baker: No. If ATP is removed, everything stops and if it is replaced, both movements re-start. The average velocity of particle movement away from the nerve terminal is slightly slower in the re-activated state than movement towards the terminal.

Cohn: What particles are moving?

Baker: We are using phase-contrast microscopy and therefore we are considering particles that we can see at the level of the light microscope. The particles moving retrogradely seem to be multivesicular bodies. The orthograde movement in crab axons requires further examination before we can establish what the particles are.

Sabatini: If, instead of using calcium to induce exocytosis, you were to use another stimulus, such as a secretagogue, or acetylcholine, would it work?

Baker: No. If the ionized calcium concentration in the 'leaky' cell is kept constant, one can apply the normal secretagogues, such as acetylcholine or carbamylcholine, and they are without effect. Presumably they normally work through a change in [Ca^{2+}] and we have kept the [Ca^{2+}] buffered and controlled.

Palade: You mentioned that vanadate inhibits movement by axoplasmic flow. Is it possible to dismantle the microtubules of the axon and see what happens to the movement?

Baker: Yes. Richard Adams is hoping to continue experiments in that direction. Invertebrate microtubules, including those in crab axons, do seem to be quite resistant to dismantling.

Sabatini: Are the strength of field and the duration of current sufficient to produce electrophoresis within the cell?

Baker: The voltage that we use decays with a time constant of about 200 μs, but there is quite a large current flowing through these small holes for a brief period of time. One possible explanation for why the holes do not reseal is that we have heated up and denatured a ring of protein around the hole and this has served to 'fix' the hole in the membrane.

Cohn: Does ATP have an effect only after you have put a hole in the membrane or does it affect the rate of movement in the absence of holes?

Baker: Extracellular ATP has no effect on adrenal medullary cells or on crab axons in the absence of holes. There is evidence, however, that in intact cells ATP is involved in both axoplasmic transport and membrane turnover; but all the experiments rely on the use of metabolic poisons which alter many parameters other than ATP levels. In our experiments we can switch axon flow or membrane turnover on and off just with ATP, whilst keeping ionized [Ca^{2+}] and other variables constant.

Blest: When the holes have been punched in the crab axon, how long does it take for extracellular solutes to enter and equilibrate?

Baker: It seems only to take a matter of minutes for the axon to equilibrate. This is most easily demonstrated after loading an axon with fluorescein diacetate which is hydrolysed to fluorescein once inside the axon. Subsequent exposure to brief intense electric fields permits the trapped fluorescein to emerge within minutes.

Crude export from ER to Golgi, and the role of coated vesicles

Sabatini: Dr Bretscher discussed the selectivity power of the coated vesicles and calculated that they could purify one molecule in 10 000 of those in the plasma membrane. Therefore one would expect that the vesicles could do exactly the same thing between the endoplasmic reticulum and the Golgi; the passage of a crude extract between both organelles would not be necessary in that case. This reminds me of Dr Rothman's previous work (Rothman et al 1980) in which he concluded that the coated vesicles that transport G protein contain this polypeptide in the ratio of about one molecule per three molecules of clathrin. This is an extraordinarily high concentration of G protein. If the vesicles between the endoplasmic reticulum and the Golgi are already filled up with G protein then there is no crude export from the endoplasmic reticulum to the Golgi and there is no need for the Golgi to act as a purifier.

Rothman: To address your second point first, in order to test directly the proposal that the coated vesicle filter is not quantitatively sufficient to prevent a crude export from ER, one needs to purify selectively those coated vesicles engaged in ER-to-Golgi transport, and then to determine the amount of 'ER' membrane proteins that they contain. This selective purification is not yet possible. The coated vesicle preparation from vesicular stomatitis virus (VSV)-infected Chinese hamster ovary (CHO) cells used in our work (Rothman & Fine 1980) indeed contains one molecule of G protein for every three of clathrin. Unfortunately, this preparation is a composite of coated vesicles in transit from ER to Golgi and from Golgi to plasma membrane. In fact, our pulse–chase experiments tell us that most of these are of the post-Golgi type. Therefore, the striking concentration of G protein in these vesicles does not reveal anything about the ER-to-Golgi vesicles (those of concern in the crude export from ER), but is mainly a measure of post-Golgi coated vesicles.

Then, to return to your first point about the selective power of coated vesicles: current numbers do suggest that coated vesicles are capable of enriching one membrane protein population relative to another by a factor of from 1000 to 10 000-fold. But unfortunately, really firm figures are lacking. To arrive at this figure, we need to know, first, the factor by which one set of proteins is concentrated in a pit relative to its bulk membrane concentration, and second, the factor by which the other set of proteins is excluded from the pits. The product of these two factors gives the selective power of a coated pit, which measures its efficiency as a sorting machine. The experiments of Bretscher et al 1980) suggest that the factor by which the concentration of bulk membrane protein is reduced in a pit is in the range of 10 to 100, depending on the protocol used. To measure the concentrative factor, we must examine the numbers for a receptor localized in the coated pit, e.g. the low density lipoprotein (LDL) receptor. For this receptor, about 30% is outside the pits and 70% is inside at any one time. The pits represent about 2% of the surface area. Therefore (and this is true for several receptors) we can make a reasonable estimate of the concentrative factor of the coated pit, which would be in this case $(70/30) \times (1/0.02)$, or roughly 100. Therefore, the selective power of a coated pit is in the range of 10^3–10^4, obtained by multiplying the two factors.

Now, is this selective power, used in a *single* step of selection, sufficient to purify plasma membrane protein precursors out of the ER membrane? To estimate this, we need to know what fraction of proteins in the ER membrane are plasma membrane protein precursors, and what fraction of proteins in the plasma membrane are ER contaminants. As discussed in my review (Rothman 1981) only about 10^{-4} of ER proteins are precursors of plasma membrane proteins. But what is the level of contamination of plasma membrane by ER? Is it 1%, or 0.01%? We really don't know. If, in fact, the level of ER contaminants in the plasma membrane were $\leqslant 1\%$, the actual selectivity factor needed to account for the overall process of purification would be $10^4 \times 10^2 = 10^6$. This could easily be greater than what now appears to be the selective power of an individual step of coated vesicle purification, a factor of 10^3–10^4 fold. But all these numbers are somewhat estimated! A coated vesicle therefore has the right qualitative properties for sorting out plasma membrane proteins but we cannot yet decide whether the selectivity is quantitatively adequate for the task.

Sabatini: There must be considerable purification in the coated vesicle.

Rothman: Yes. But is it so considerable that all the purification can be done in one go, or does it need five attempts? That is precisely what we don't know.

Sabatini: The presence of ER constituents in the Golgi can be interpreted in other ways: we may be dealing with an artifactual contaminant or with a permanent constituent that does not have to be sent back.

Rothman: Even if one accepts that the same proteins present in the ER membrane are also *bona fide* constituents of Golgi membranes, there is not yet any direct evidence about whether they are in transit. Indirect evidence (see Rothman 1981) suggests that the ER proteins in the Golgi are mainly on its cis side. If these ER proteins arrive from ER in continuous fashion, but do not leave the Golgi at the trans end, then they must be removed in between, and presumably be returned to the ER.

Sabatini: We have just accepted that the transitional element does exactly the same thing: it contains the ER components and lets go only the coated vesicles that already have a selected subset of polypeptides.

Rothman: But, quantitatively, *how* select is it, and *how well* can it exclude those endoplasmic reticulum proteins? We don't know.

Mellman: Perhaps there are different classes of highly specific coated vesicles in the cell, but we must remember that the proteins that are being concentrated (in theory) by Golgi-derived coated vesicles, and are destined for the plasma membrane, are the very proteins that Dr Bretscher (p 272-281) has said are preferentially excluded from the plasma membrane-derived coated vesicles.

Palade: Are you assuming, Dr Rothman, that membrane proteins coming from the ER to the Golgi are randomized within Golgi membranes? That is, do they lose their identity as distinct patches so that they have to be sorted out again?

Rothman: I am assuming the simplest picture: that the ER proteins are entering the Golgi via transport vesicles. They would be carried as contaminants along with the proteins whose export had been intended and selected for. Following the fusion of this vesicle with the Golgi (probably at the 'cis' face), I would naturally assume that lateral diffusion would rapidly mix the ER proteins together with the bulk of proteins that are already present in that cisterna.

Palade: Your basic assumption is, therefore, that randomization precedes resorting.

Can you explain if the countercurrent system described in your paper (p 120-137) works equally well on the *content* as well as on the molecules of the membrane containers?

Rothman: The countercurrent or, more accurately, distillation process that I had in mind would be needed for membrane proteins, but could also, through appropriate receptor systems, operate on luminal content as well. As the bulk of the soluble content in the ER is destined for secretion or export, in contrast to the bulk of the membrane, which is retained, the distillation process does not need to be invoked for the content, in general.

Sabatini: If the lumen of the Golgi cisternae can become loaded with markers such as the one that Dr Farquhar's paper described (see Farquhar & Palade 1981), which are not restricted to the trans cisternae but are also found in cis cisternae, we have to explain why they are never found within endoplasmic reticulum cisternae and why those vesicles that shuttle back from the Golgi to the ER never take any of the *content* molecules to the ER.

Rothman: I agree. That must be explained. Of course, we don't know whether these markers are sticking to membrane; in fact, it is likely that a particle bound electrostatically to the membrane is, in reality, a *membrane* marker and *not* a content marker. If the markers are really membrane-bound, your point is probably a moot one. There is not enough specificity in these experiments to permit any detailed interpretation.

Meldolesi: Since the presence of these ER-enriched protein markers in the Golgi fraction is becoming a controversial issue, I should correct the figures that you gave in your talk. In our hands, NADH-cytochrome b_5 reductase, which is one of those proteins, has the same or higher concentration in the heavy Golgi subfraction (GF1) than in the rough microsome fraction (Borgese & Meldolesi 1980). We got this result after we found conditions for preserving the activity of this enzyme, which is particularly labile in the Golgi. More recently Nica Borgese has obtained preliminary results by radioimmunoassay which confirm that in the Golgi subfraction (GF1) the concentration of the enzyme protein is as large as in the ER. I interpret this to mean that NADH-cytochrome b_5 reductase is a constitutive Golgi enzyme, and not an ER component that spills into the Golgi during intracellular transport and is then shuttled back.

Rothman: Is that the same protein that was also found in the outer mitochondrial membrane in the same work?

Meldolesi: Yes, but contamination by mitochondrial outer membranes cannot account for the high concentration of the enzyme in GF1.

Rothman: It may, nevertheless, not be the best protein on which to focus because of its unusual mode of biosynthesis.

Sabatini: This molecule is synthesized on free ribosomes and is distributed in many membranes of the cell. Its appearance in the Golgi or anywhere else may be a direct result of post-translational insertion without the need to pass through the ER. So I believe that this is the wrong protein to consider in this problem of recycling for the purpose of purification.

Palade: Is it necessary to introduce, at a certain moment, a selector into the system, to keep together the membrane proteins that you have concentrated by distillation?

Rothman: There would need to be selectivity in at least one of two possible ways. There could be a positive selection for the transport of the class of exported membrane proteins away from those of the endoplasmic reticulum. Alternatively, there could be selective retrieval of the endoplasmic reticulum proteins from the Golgi cisternae, in which case ER membrane proteins would presumably have to have their own address markers. But, one or the other (or both) would be needed for the multistage purification that I have proposed for the Golgi.

REFERENCES

Adams, RJ, Baker PF, Bray D 1982 Particle movement in crustacean axons that have been rendered permeable by exposure to brief intense electric fields. J Physiol (Lond) 326:7P

Baker PF, Knight DE 1978 Calcium-dependent exocytosis in bovine adrenal medullary cells with leaky plasma membranes. Nature (Lond) 276:620-622

Baker PF, Knight DE 1980 Gaining access to the site of exocytosis in bovine adrenal medullary cells. J Physiol (Paris) 76:497-504

Baker PF, Knight DE 1981 Calcium control of exocytosis and endocytosis in bovine adrenal medullary cells. Phil Trans R Soc Lond B Biol Sci 296:83-103

Baker PF, Knight DE 1982 Calcium-dependence of catecholamine release from bovine adrenal medullary cells after exposure to intense electric fields. J Membr Biol, in press

Baker PF, Knight DE, Whitaker MJ 1980 The relation between ionized calcium and cortical granule exocytosis in eggs of the sea urchin *Echinus esculentus*. Proc R Soc Lond B Biol Sci 207:149-161

Borgese N, Meldolesi J 1980 Localization and biosynthesis of NADH-cytochrome b_5 reductase, an integral membrane protein, in rat liver cells. I: Distribution of the enzyme activity in microsomes, mitochondria, and Golgi complex. J Cell Biol 85:501-515

Bretscher M, Thomson J, Pearse B 1980 Coated pits act as molecular filters. Proc Natl Acad Sci USA 77:4156-4159

Farquhar MG, Palade GE 1981 The Golgi apparatus (complex)—(1954–1981)—from artifact to center stage. J Cell Biol 91:77s-103s

Rothman J 1981 The Golgi apparatus: two organelles in tandem. Science (Wash DC) 213:1212-1219

Rothman JE, Fine RE 1980 Coated vesicles transport newly synthesized membrane glycoproteins from endoplasmic reticulum to plasma membrane in two successive stages. Proc Natl Acad Sci USA 77:780-784

Rothman J, Bursztyn-Pettegrew H, Fine RE 1980 Transport of the membrane glycoprotein of vesicular stomatitis virus to the cell surface in two stages by clathrin-coated vesicles. J Cell Biol 86:162-171

Rubinson KA, Baker PF 1979 The flow properties of axoplasm in a defined chemical environment: influence of anions and calcium. Proc R Soc Lond B Biol Sci 205:323-345

Zimmerman U, Scheurich P, Pilwat G, Genz R 1981 Cells with manipulated functions: new perspectives for cell biology, medicine and technology. Angew Chem Int Ed Engl 20:325-344

Chairman's closing remarks

GEORGE E. PALADE

Section of Cell Biology, Yale University School of Medicine, 333 Cedar Street, P.O. Box 3333, New Haven, Connecticut 06510, USA

This symposium has reviewed and discussed the substantial body of information that has been secured over the last five or six years on vesicular transport, especially on the recycling arcs of this general activity. We now know that cells operate a multiplicity of discontinuous circulatory systems, regulate the traffic of the corresponding vesicular carriers, and retain the membrane specificity of the compartments connected by these circuits.

This special field of research is promising because we understand the broad outlines of the overall process, while many important details remain to be investigated and fitted into their proper place in the general scheme, and also because the technology needed for further work is, at least in part, already available.

Questions for which we still need answers can be grouped under a few headings.

Extent of vesicular traffic

At present our interest is focused mainly on the endocytic and exocytic circuits, and on their connections and intracellular extensions. But present evidence does not rule out the use of vesicular traffic for membrane transport to other intracellular compartments, such as peroxisomes, mitochondria (outer membrane) and chloroplasts (envelope). There is, in fact, partial chemical overlap between endoplasmic reticulum (ER) membranes (microsomal) and outer mitochondrial membranes, and there is evidence on record, (open to question however) that suggests that some mitochondrial proteins

1982 Membrane recycling. Pitman Books Ltd, London (Ciba Foundation symposium 92) p 293-297

are synthesized on polysomes attached to the ER membrane (see Chua & Schmidt 1979).

Many features of the established circuits require further elucidation. For instance, the extent to which circuits overlap is unknown; yet this is an important question because some of them, e.g. the endocytic and the exocytic circuits, appear at first sight to duplicate each other. Circuits with more than two terminals apparently exist (see Farquhar, p 157-183, and Rodewald & Abrahamson, p 209-232, this volume), but we do not know how general is this formula, or what complications it introduces to the regulation of vesicular traffic, or how the discharge of content and the delivery of membranes is controlled at each terminus. Factors so far identified—e.g. compartmental pH, and electrostatic interactions of probe molecules with the membrane of the carrier—lack specificity, and may be only part of a larger repertoire of conditions and more specific agents.

Practically nothing is known about the nature and pathways of the vesicular carriers that transport newly synthesized membrane proteins from the Golgi complex to the plasmalemma, or to differentiated plasmalemmal domains. Louvard's experiments (1980) suggest, however, that biogenetic traffic may be coupled to the recycling of plasmalemmal vesicles that seem to have topographically defined ports of entry on the apical plasmalemma in polarized epithelial cells. It may be assumed that viral envelope proteins are transported to the cell surface by the same carriers that move, to the same location, the glycoproteins of the host cell plasmalemma. If so, viral glycoproteins may be used as tags for the identification and isolation of these vesicular carriers.

Volume and kinetics

Information about the amounts of different membranes involved in the process, and the rate at which recycling proceeds, is still fragmentary. Indirect evidence suggests that short-range recycling is extremely rapid (see Goldstein et al, p 77-95, this volume). The same may apply, according to Tartakoff (1980), to vesicles shuttling between the ER and the Golgi complex. He estimates that the transfer of secretory proteins from the ER to the Golgi complex requires 20 000 vesicles per second; this implies that each peripheral Golgi vesicle shuttles at least 10 times per second between the two termini. Tartakoff assumes that vesicle diffusion could not ensure such high rates of transport. To some extent the geometric bottle-neck represented by transfer from a relatively large ER to a comparatively small Golgi complex magnifies the problem of vesicular transport at this junction. Should we conclude that a motor is required to move transporting vesicles, just as a motor is required in axon transport (see Baker p 282-287, this volume)? Firm kinetic data on intracellular transport would be useful, but may be difficult to obtain.

Functional implications

The functional meaning of recycling raises other fascinating problems. At present, we proceed on the assumption that recycling occurs because endocytosis and exocytosis take place, and we assume that the two activities are essentially connected with the metabolic needs of the cell (import and export of matter) and with the regulation of surface receptors. But could the endocytic pathway be connected with the surveillance of the condition of the plasmalemma, as mentioned by Dr Cuatrecasas (p 000–000, this volume)? In constitutive endocytosis, cell surface membrane is continuously moved through a compartment of low pH, e.g. pre-lysosomes or lysosomes, before being returned to the surface. While in transit, only its originally external surface is exposed to the acid medium. Current evidence indicates that many experimental ligands of physiological significance dissociate from their receptors in these low pH compartments. It is therefore conceivable that fortuitous, opportunistic, non-specific ligands or adsorbed molecules are removed from the plasmalemma in the same fashion. The existence of such a process is suggested by the results of experiments with multivalent ligands (lectins, antibodies) that induce, in sequence, patching of antigens or unspecified receptors, capping and endocytosis followed by degradation of the incorporated ligands (Taylor et al 1971, Louvard 1980). Therefore, in addition to its function in the import of matter and the recycling of receptors, constitutive endocytosis may serve as a means for cleaning the plasmalemma by exposing it, in proper orientation, to an acid bath while in transit through pre-lysosomes or lysosomes. Moreover, as suggested by Dr Farquhar at this symposium (p 157-183), recycling membrane in transit through the Golgi complex may be modified or repaired by reglycosylation of partially degraded glycoproteins, for instance. If so, recycling may become an essential factor in a general maintenance operation that combines acid cleaning of membrane surface with repair or modifications of membrane glycoproteins.

Sorting of products and membranes during vesicular transport

Work on receptor-mediated endocytosis indicates that animal cells have the ability to sort out efficiently ligand molecules of physiological interest present in small concentrations in complex mixtures (plasma or interstitial fluid, for instance). Coated pits appear to play a major part in the sorting of both membrane components (receptors) and solutes (ligands). Coated pit equivalents exist—as already mentioned—along the secretory pathway at sites where they could play a similar role. Dr Rothman's hypothesis (1981), according to which the Golgi complex functions as the equivalent of a distillation tower, has the merit of providing the first possible explanation (but not necessarily the final one) for a specific yet still unexplained architectural feature of the Golgi complex—its stacked cisternae. But, here again, more evidence is needed before we can decide whether the mechanisms that cells use for sorting products and membrane components, at various termini, function in one step (as in coated pits) or in a succession of steps (as in Dr Rothman's hypothesis).

Key issues still unsolved

Further progress in understanding vesicular transport and membrane recycling depends on the solution of a few key issues, the foremost of which are the control of membrane specificity and the regulation of vesicular traffic.

As to the first problem, control of membrane specificity, we have obtained information about the molecules (clathrin and associates) of the most frequently encountered type of stabilizing infrastructure, and we partly understand the conditions that lead to the self-assembly of these molecules into geodetic cages. We are also building up a rapidly expanding inventory of receptors that appear to be the natural candidates for selection and stabilization by these infrastructures. Moreover, we know by now that coated pits and coated vesicles are involved in membrane traffic in many locations in which specific receptors have not yet been identified. As a result, we are facing a paradox: the cells use the equivalent of a quasi-universal device to select and stabilize specifically a wide variety of apparently unrelated membrane proteins. How can this paradox be explained? We may be dealing with successive levels of selectivity reflecting different degrees of affinity—low specificity, for example, for cationized ferritin, and high specificity, for instance, for specific physiological ligands. But even for specific ligands the paradox persists simply because a large variety of receptors is clustered, or induced to cluster, in apparently identical coated pits. The answer is probably connected with the structure of the endodomains of this variety of receptors. These endodomains are expected to recognize (and be recognized by) some molecular component(s) of the cage structure. For the moment we have only a number of analogies to guide our hypotheses. We know that cells use two different ways of channelling diverse inputs into common action. The first is to use macromolecules with variable and constant domain (as for immunoglobulins and histocompatibility antigens); the second is to rely on general details of tertiary structure of the pertinent molecular domains rather than on the fine details of their primary structure (as for the interactions between signal sequences and signal recognition proteins). Perhaps similar formulae, applied to either receptors or cage components, are used to solve the problem posed by this paradox. But we may expect interesting surprises, since solutions entirely novel (for us) cannot be excluded.

As far as the second key issue is concerned, the regulation of vesicular traffic, we are still at the very beginning of a long road. We recognize the problem, but have no concrete elements that can contribute to its solution at the level of molecular interactions.

Until now, our attention has been focused primarily on the ectodomains of membrane proteins, mostly because these domains carry the sites involved in co-translational and post-translocational modifications of proteins. Ectodomains will undoubtedly remain of considerable interest since they comprise the recognition sites for physiological ligands in the case of various receptors, including lysosomal enzyme receptors. But the endodomains of membrane proteins deserve more attention than they have so far received; they are now expected to carry the information required for the control of membrane specificity as well as for regulation of vesicular traffic.

Each key issue is too complex to analyse in the context of the intact cell. The hope is that the two or three partners involved in these operations are sturdy enough to emerge still active from the ordeal of cell fractionation so as to make possible their use in reconstituted systems *in vitro*. Evidence already published by Quinn & Judah (1978), Fries & Rothman (1980) and Rothman & Fries (1981) suggests that this may indeed be so.

REFERENCES

De Petris S, Raff MC 1973 Normal distribution, patching and capping of lymphocyte surface immunoglobulin studied by electron microscopy. Nature (New Biol) 241:257-259

Fries E, Rothman JE 1980 Transport of vesicular stomatitis virus glycoprotein in cell-free extract. Proc Natl Acad Sci USA 77:3870-3874

Louvard D 1980 Apical membrane aminopeptidase appears at sites of cell–cell contact in cultured kidney epithelial cells. Proc Natl Acad Sci USA 77:4132-4136

Rothman JE 1981 The Golgi apparatus: two organelles in tandem. Science (Wash DC) 213:1212-1219

Rothman JE, Fries E 1981 Transport of newly synthesized vesicular stomatitis virus glycoprotein to purified Golgi membranes. J Cell Biol 89:162-168

Quinn PS, Judah JD 1978 Calcium dependent Golgi-vesicle fusion and cathepsin B in the conversion of proalbumin to albumin in rat liver. Biochem J 172:301-309

Tartakoff A 1980 The Golgi complex: crossroad of vesicular traffic. Int Rev Exp Pathol 22:227-251

Taylor RB, Duffus WPH, Raff MC, dePetris S 1971 Redistribution and pinocytosis of lymphocyte surface immunoglobulin molecules induced by anti-immunoglobulin antibody. Nature (New Biol) 233:255-259

Index of contributors

Entries in **bold** *type indicate papers; other entries refer to discussion contributions*

Abrahamson, D. R. **209**
Anderson, R. G. W. **77**

Baker, P. F. 29, 31, 54, 71, 134, 135, 150, 272, 276, 282, 283, 285, 286, 287
Blest, A. D. 236, 237, 276, 287
Branton, D. 89, 135, 236, 237, 242, 243, 257, 258, 259, 261, 263, 264, 275, 278, 285
Bretscher, M. S. 28, 29, 55, 57, 73, 92, 115, 117, 136, 151, 177, 204, 237, 242, 258, 261, 262, **266,** 272, 273, 274, 275, 276, 277, 278, 279, 280
Brown, M. S. **77**

Cohn, Z. A. **15,** 28, 29, 30, 31, 32, 33, 53, 56, 70, 73, 90, 94, 113, 114, 152, 154, 155, 176, 177, 178, 202, 205, 206, 229, 274, 286, 287
Cuatrecasas, P. 70, 73, **96,** 104, 105, 106, 107, 108, 116, 117, 203, 204, 228, 237

Dean, R. T. 29, 55, 114, 135, 149, 237, 243

Farquhar, M. G. **157**

Gabel, C. A. **138**
Geisow, M. J. 33, 51, 52, 74, 91, 135, 152, 153, 175, 178, 179, 230, 273, 274
Goldberg, D. **138**
Goldstein, J. L. 28, 30, 31, 32, 52, 53, 55, 72, **77,** 89, 90, 91, 92, 93, 94, 104, 113, 117, 118, 134, 151, 228, 233, 236, 237, 241, 259, 263, 264, 272, 276, 277, 279, 280
Gowans, J. L. 89

Helenius, A. 30, 55, **59,** 69, 70, 71, 72, 73, 74, 75, 92, 106, 108, 203, 230, 231, 242, 278, 280
Hopkins, C. R. 91, 92, 175, 177, 180, 207, 229, 239, 241, 258, 260, 277, 285, 286
Hubbard, A. L. 30, 54, 69, 71, 73, 104, 105, 109, 112, 113, 114, 115, 133, 134, 153, 180

Ivanov, I. E. **184**

Jourdian, G. W. 55, 90, 150, 154, 260

Kornfeld, S. 92, 106, 118, 132, 133, **138,** 149, 150, 151, 152, 153, 154, 155, 156, 203, 275

Marsh, M. **59**
Meldolesi, J. 30, 31, 90, 112, 113, 133, 176, 178, 203, 228, 236, 259, 262, 263, 272, 278, 290, 291
Mellman, I. S. 33, **35,** 51, 52, 54, 56, 57, 72, 91, 93, 105, 106, 107, 180, 204, 229, 242, 259, 272, 278, 289

Palade, G. E. **1,** 29, 32, 56, 70, 74, 75, 93, 94, 107, 108, 114, 117, 118, 131, 133, 136, 150, 155, 174, 175, 176, 177, 178, 180, 181, 206, 207, 227, 231, 239, 241, 242, 243, 261, 262, 263, 264, 275, 278, 279, 280, 286, 289, 290, 291, **293**
Pearse, B. M. F. 134, 242, **246,** 257, 258, 259, 260, 261, 262, 263, 264

Raff, M. 30, 72, 73, 74, 91, 92, 105, 107, 204, 205, 226, 227, 230, 231, 260, 261
Reitman, M. L. **138**
Rindler, M. J. **184**
Rodewald, R. 56, **209,** 227, 228, 229, 230, 231, 260
Rodriguez-Boulan, E. J. **184**
Rothman, J. E. 31, 33, 51, 57, 71, 72, 75, 90, 93, 114, **120,** 130, 131, 132, 133, 134, 135, 136, 151, 154, 156, 174, 176, 179, 180, 205, 207, 231, 242, 243, 257, 258, 273, 278, 279, 280, 284, 286, 288, 289, 290, 291

Sabatini, D. D. 69, 72, 90, 130, 131, 132, 133, 136, 148, 149, 152, 153, 154, 155, 174, 176, 177, 179, 180, **184,** 202, 203, 204, 205, 206, 207, 236, 237, 242, 243, 261, 262, 263, 274, 275, 276, 280, 285, 286, 287, 289, 290, 291
Schneider, Y.-J. 114, 115, 228, 230
Sly, W. S. 30, 32, 56, 69, 73, 94, 105, 115, 149, 151, 152, 153, 154, 179, 227, 241, 260, 264, 265, 272, 273
Steinman, R. M. **15**

Varki, A. **138**

Widnell, C. C. 53, 92, 181, 206

Indexes compiled by John Rivers

Subject index

N-acetylglucosamine 122, 125, 139, 143, 146
N-acetylglucosamine-1-phosphodiesterase 146
  subcellular localization 146, 147
N-acetylglucosamine-1-phosphotransferase 143–146
  specificity for lysosomal enzymes 144–146
  subcellular localization 146, 147, 153
Acetyl-LDL, internalized, degradation of 52, 53, 55
  See also Low density lipoprotein
Acid hydrolases 139, 143, 146, 149, 152, 155
Acidification, kinetics of 64–68, 69, 74
Actin
  in formation of stabilizing infrastructures 3, 274, 276
  phagocytosis and 19
Acylation, fatty *See Fatty acylation*
Acyltransferase 130, 132
Adenosine triphosphate-mediated clathrin release 134, 135, 136, 182
Adrenal medullary cells, leaky 283, 284, 285, 286, 287
  ATP and 287
Adsorptive endocytosis 29, 98, 138, 139
  cell surface 246, 273
  of Semliki Forest virus 60, 61
  See also Endocytosis, Phagocytosis, Pinocytosis, Receptor-mediated endocytosis
Ammonium chloride, virus uptake inhibiting 61, 66, 69, 73
Ankyrin in formation of stabilizing infrastructures 3
Antigens, surface, capping and patching 269, 270, 295
Asialoglycoproteins
  binding of 87, 110, 114
  receptor-mediated endocytosis and 81
    in hepatocyte 109–115
Asialoglycoprotein receptors 79, 117, 280
  internal pool of 111–113, 115
  location 111–113, 115
  LDL receptors and 113
  recycling of 84, 85, 113
Asialo-orosomucoid 109, 110, 111
  sucrose-conjugated 114
Asialotransferrin 117
Axoplasmic flow system
  ATP and 287, 288
  membrane particle movement in 284–287
  vanadate inhibiting 284, 287

Baby hamster kidney (BHK) cells
  acidification in endocytic pathway 67
  virus uptake in 61, 63, 64
Budding
  of enveloped virions 4, 10, 121
  of viruses
    double infection experiments 188, 207
    polarity of 186, 187
    tunicamycin ineffective in 204, 205

Calcium dependency
  of catecholamine release 283, 284
  of exocytosis 284
  of ligand–receptor binding 87
  of LDL receptor binding 87, 113
Carbonyl cyanide *m*-chlorophenylhydrazone 154
Carboxylic ionophores 61, 66, 85
Carboxypeptidase Y 155
Cationic ferritin *See Ferritin*
Catecholamine release from leaky adrenal cells 283, 284
Cell compartments 3, 4, 6, 7, 12, 110, 111, 120
  See also Golgi apparatus, cis and trans compartments
Cell locomotion mechanisms 269, 270, 279, 280
Cell membranes *See Membranes, Plasma membranes*
Chemotactic peptide, receptor-mediated endocytosis and 81
Chinese hamster ovary (CHO) cells, mutant lines 122, 130
  fatty acylation in 156
  lysosomal enzymes in 143
  UT-1 cells, ER proliferation in 234–239

Chloroquine 66, 73, 85, 99, 105, 106
Cholesterol
ASOR receptor and 111
Crystalloid ER, in 236
deficiency 78, 83, 233, 234
familial hypercholesterolaemia and 78, 83
LDL transporting 82, 83
membrane exchange 17
placenta, human, in 259, 260
separation in Golgi stack 136, 271
virus entry into cell and 65, 66
Chorionic gonadotropin, receptor-mediated endocytosis in 80
Chylomicron remnants, receptor-mediated endocytosis in 80
Clathrin 78, 248, 256, 262, 296
ATP-released 134, 135, 182
amino acid sequence of 263, 264
antibodies 264
binding of 262, 263
-coated pits *See Coated pits*
-coated vesicles *See Coated vesicles*
cytosol-dependent release 134, 135
interaction with cell components 135, 260
packing 251, 255, 256, 258
phagocytes and 19
receptors, binding of 87, 94
recycling of 79, 82
stabilizing infrastructures and 3, 10, 11, 31
triskelion 250, 251, 252, 254, 258, 261, 262
Clathrin cages 134, 135, 249–252, 256, 261, 262, 296
binding to membrane 262
disruption by urea 252, 253, 257, 258
See also Geodetic cages
Coated pits 3, 4, 11, 12, 44
distribution of 272, 273
formation of 242, 243, 276, 277
intestinal absorptive cells, in 210, 211, 228
LDL receptors in 57, 79, 94, 266, 271, 277, 278, 289
membrane traffic and 296
molecular filter function 247, 255, 262, 267–269, 271, 277, 295
pinocytosis of immune complexes and 46, 49
receptors 94, 267, 278, 296
amino acid sequence of 280
See also LDL receptors
receptor recognition in 86, 87
receptor-mediated endocytosis, in 78, 79, 159, 241, 266
recycling of 246
secretory cells, in 179
selective power 86, 87, 271, 289
structure of 246–266
virus uptake by 60, 64, 72, 73
Coated vesicles 5, 10, 11, 12
antigen–antibody complexes, uptake of 56
core proteins of 254, 255, 256, 257, 258, 259, 261, 262
binding of 262
electrophoretograms of 264, 265
endocytosis and 267 *See also Endocytic vesicles*
in BHK-21 cells 63
ER–Golgi crude export and 287–291
fluid phase pinocytosis and 28, 29, 30, 31, 43, 44
fusion
with lysosomes 243
with membranes 260
Golgi cisternae, on 182, 292
heterogeneity 255–259, 263
IgG immune complex pinocytosis and 49
inner core 252–255, 256, 267, 258
internalization
of IgG complexes by 57
of plasma membrane by 56
of Semliki Forest virus particles by 60, 63, 64
intestinal absorptive cells, of 210, 211, 225
light chains and 262, 263
lipid content of 57
membrane, composition of 44
membrane binding site 258, 259
membrane traffic and 296
molecular filtration process and 247, 255
morphology and function 242, 243, 246–266
outer clathrin cage 249–252
placental, human 252–255
cholesterol in 259, 260
plasma membrane proteins in 44
preservation of 247, 252, 259
protein content 57, 248, 249, 260
purification 247, 248
receptors in 260, 261
receptor-mediated endocytosis and 78, 79
receptor-mediated pinocytosis and 46
selective power 288–290
selective recognition of membrane proteins 255, 256, 257, 258
size of 262, 263
sucrose affecting 247, 264
transport
of plasma membrane proteins by 57
of vesicular stomatitis virus by 186, 271
Triton treated 252, 253, 258
uncoated, protein:lipid ratio 279
uncoating of 79, 242, 243, 246, 260, 261, 267

virus uptake by 30, 72, 73
Coated vesicle-mediated uptake 28
Colchicine 179, 180
Colloidal gold
conjugates, immune complexes and 49, 52
immunolabelling 190–200, 207, 239
Compactin 233, 234, 235, 237
Concanavalin A
macrophages and 19
mitogenic effects 98
pinocytosis and 18, 31
Content markers 30, 42
Cycloheximide 85, 100, 105, 106, 207
Cytochalasin B 179
Cytochalasins inhibiting phagocytosis 19
Cytoplasmic free vesicle 12
Cytoskeletal drugs 179
Cytoskeleton
disassembly of 178
plasma membrane protein movement and 274, 275, 276

Dansylcadaverine 61, 99, 105
Dextrans as tracers 159
Diacytosis 6, 10, 112, 114, 117
Differential domains in eukaryotic cells 3, 4
Digitonin 111
Dihydrofolate reductase 234, 237
Diphosphatidylglycerol, specific to inner mitochondrial membrane 2
Diphtheria toxin, receptor-mediated endocytosis and 81, 106, 107, 108
Distillation in Golgi apparatus 128, 290, 291, 296
Dolichol phosphate 143
'Down regulation' 98, 100, 102, 105

Endocytosis 3, 4, 5
adsorptive *See Adsorptive endocytosis*
capping and, temperature correlations 276
enveloped animal viruses, of 59–76
fluid-phase 30, 31
IgG, of
constitutive 216, 217, 295
fluid-phase 216, 217, 227, 228
receptor-mediated 209, 217, 228
selective 210, 213, 225
ionic composition of medium and 31
LDL particles, of 268
membrane perturbants and 19
mitogenesis and 99
pH and 55
receptor-mediated 9, 11, 36
See also Receptor-mediated endocytosis
surface membrane, of 266
in secretory cells 157, 158, 159, 161, 163
temperature dependence for sucrose 29
Endocytic circuit 5, 6, 266, 267, 293, 294
Endocytic pits *See Coated pits*
Endocytic vesicles 15, 19, 21, 36, 49, 57, 79, 163
acidification 51, 52, 54, 64, 68, 69, 71
dissociation in 91
formation 71
fusion with lysosomes 64–67, 69
IgG uptake in rat intestine 216, 217
pH in 67, 70, 71, 74
receptors in 73, 74, 91
transport other than to lysosomes 79
virus particles in 64
See also Vesicles
Endoplasmic reticulum (ER) 5, 6, 9, 120, 122, 123, 124
crude export to Golgi 287–291
crystalloid
arthropod photoreceptor transductor membranes and 236
cholesterol content 236
guinea pig adrenal, in 236
proliferation 233–239
protein content 236, 237
lysosomal enzyme receptors, site of 153
membrane proteins of 289, 290
recycling of 135
sorting in Golgi compartments 125–128, 130, 131, 133, 134, 136
smooth 12, 126, 130
sorting of new proteins 270, 271
Endosomes *See Endocytic vesicles*
Entamoeba histolytica
cell compartments in 26
intermediate vesicles in 32
Enzymic iodination technique 21
Epidermal growth factor (EGF)
antibodies 101
binding and internalization 98
biotinylated 239, 241
degradation and 98
down regulation of 98
endocytosis and 98, 202
–ferritin conjugates 97
inhibition
of endocytosis 99, 104
of lysosomal degradation 104
intracellular accumulation 99
mitogenic activity 98, 99, 100, 101, 106
immediate and late effects 102
pH and 106, 107

Epidermal growth factor (EGF) (*cont.*)
receptor-mediated endocytosis and 80
early events in 239–242
Epidermal growth factor receptors 79, 91
affinity 100, 101
aggregation of 97, 98, 101, 102, 239, 241, 242
analogues, fluorescently labelled 97, 101
antibodies 98, 101, 105, 106, 107
CBrN analogue 101, 102
degradation 98
dimerization 101
entry into coated pits 105
internalization 97, 98, 99, 100, 104, 105, 107
lateral diffusion 97
non-recycling 99, 105
patching of 97, 98, 101
rotational diffusion 97
turnover 99, 105, 106
uptake and fate 96–108
Epidermal growth factor–receptor complexes
endocytosis and synthesis 98, 100, 102
intracellular accumulation 100
marker aggregation 99
proteolytic fragment of 100, 104
Epithelial cells
apical (luminal) region 185, 188, 194, 199, 203, 204, 206, 207
basolateral region 185, 188, 194, 199, 203, 204, 206, 207
membrane biogenesis 185
plasma membrane domains 185, 187, 192, 199
polarity 184–208
protein distribution in, specificity of 185, 186, 187
tight junctions 199, 202, 203, 204
transport of solutes 204
viral budding, polarity of 186, 187, 188
See also MDCK cells
Erythrocyte
IgG coated, phagocytosis of, internalization of Fc receptor and 44–46, 53
membrane protein interaction with infrastructure 275, 276
rosetting of 45, 53
Ethyleneglycolbis (aminoethylether) tetra-acetate (EGTA) 110, 111
Eukaryotic cell 1, 2, 3
coated membrane system 246
compartments of 3, 4, 120
stabilizing interactions 3, 4, 9, 10, 11
Exocrine pancreatic cell, secretory pathways in 171
Exocytosis 3, 4, 5, 157, 158, 159, 163, 283, 284
Exocytic circuit 5, 6, 293, 294

Fab fragments 37, 39
Familial hypercholesterolaemia 78, 83, 89, 93
Fatty acylation
Chinese hamster ovary cells, in 156
glycoproteins, of 123, 124, 156
membrane proteins and 117, 118
site of 123, 124, 132
Fc receptors
binding to IgG 107, 108
half-life 45, 56, 105, 106
immunoprecipitation 37, 39
internalization
fluid-phase pinocytosis and 39–43
phagocytosis and 44–46, 106
pinocytosis of immune complexes and 46–49, 51, 106
intestinal 213, 214, 215, 216, 227, 228
affinity 215
basolateral surface, on 231
disappearance in adult 229
IgG uptake 215, 216
pH dependence 215
recycling of 216, 228
lysosomal degradation 45, 52, 53
macrophages, of 21, 30, 33, 35–58
monoclonal antibodies to 38, 39, 40, 45, 91, 108
degradation of 45, 52, 53, 55
non-recycling 91
receptor-mediated uptake 18, 38
structure 38, 39
Ferritin
cationized 16, 44
LDL receptor studies, in 91, 92, 94
specific receptors to 181
tracer, as 158–167
transfer in interstitial cells 219–221, 228
double tracer studies 221–224
–EGF conjugates 97
immunolabelled 268
lactosaminated 74, 109, 110
native, as tracer 163, 165
placental coated vesicles, in 253, 254, 255
Fibroblasts, cultured, pinocytosis in 20, 29, 44
Fluid-phase pinocytosis 18, 20, 30, 31, 36
BHK-21 cells, in 63
iodination studies 40
pH and 55
phagocytosis and 18

temperature and 18, 23, 28, 29
virus uptake and 63
Fluorescence microscopy 97, 101, 239, 241
Fluorescence photobleaching 97, 275, 276
Forming granules 161, 163
Fusogens, 75

G protein *See Glycoproteins*
Galactose 125
β-galactosidase binding 150
Galactosyltransferases
Golgi apparatus, in 121, 124
monensin and 93
Geodetic cages 3, 10, 11, 94, 182, 296
See also Clathrin cages
Glycophorins, acylation of 118
Glycoproteins 121, 122, 123, 124
concentration in cis and trans Golgi compartments 133
fatty acylation 123, 124, 156
Golgi membranes, from, processing of 122, 123, 179
intracellular transport 121, 122, 123, 124, 130
pools of 123, 124
receptor-mediated endocytosis and 81
vesicular stomatitis virus-encoded 121, 190–200, 203, 204
viral 3, 10
polarized distribution of 185, 186, 187, 203, 204
See also Spike glycoproteins
Glycosylation, terminal
glycoproteins, of 120, 121, 122, 123, 295
lysosomal enzymes, of 152, 153
membrane recycling and 150
Glycosyltransferases 122, 123, 124, 125, 126, 131, 132, 133, 146, 207
β-glucuronidase 139, 140, 141, 142, 155
phosphorylation of 151
Gold
avidin-coated 239, 241
colloidal
conjugates 49, 52
immunolabelling 190–200, 207, 239
–macroglobulin-protein complexes 239–242
Golgi apparatus
cis and trans compartments 123, 124, 125, 171, 181, 182
coated vesicles of 246, 247, 259, 263
crude export to, from ER 287–291
distillation process in 128, 290, 291, 296
ER proteins
sorted in 125–128, 130, 131, 136
transport to 5, 6, 9
intestinal cell, cationic ferritin in 221
intracellular transport in 121, 122, 123, 124
membrane flow in 171, 174, 175, 176
membrane traffic in secretory cells 157–183
secretory pathways within 180, 181
structure and function 120–137, 172
subcompartments 181, 182
subfraction 1, ER enriched protein markers in 291
vesicular transport and 5, 6, 9, 12, 36
monensin affecting 93
viral glycoprotein transport and 186, 192, 194, 206, 207
Golgi cisternae 10, 12, 121, 124–128, 159, 161, 163, 171, 174–177, 182
Golgi complex *See Golgi apparatus*
Golgi–endoplasmic reticulum lysosome complex (GERL) 5, 6, 16, 98, 181
Golgi enzymic markers 12, 123, 124
Golgi–lysosome intermediate compartment 110
Golgi–plasmalemma vesicular transport 5, 6, 158, 294
Golgi stack 121, 122, 161, 163, 171, 296
compartments of 124–128, 174–177
ER protein sorting and 125–128, 136, 271

H 63 antigen, exclusion from coated pits 267–269
HA protein of influenza virus, distribution of 190–200
HeLa cells
Newcastle disease virus synthesis and transfer in 267
transferrin receptor in 115–118, 278
Hepatocyte
asialoglycoprotein receptor-mediated endocytosis in 109–115
intracellular sorting of ligands 230
High density lipoprotein 94
Horseradish peroxidase (HRP) 17, 18, 20, 25, 158, 159, 211, 216, 217, 219, 221, 222, 229
acidification of 67, 68
–anti-HRP IgG complex 211, 229
–ASOR conjugates 109, 110
–IgA conjugates 220
uptake
monensin and 92, 93
temperature and 29
Hydroxymethylglutaryl-CoA reductase 233–239
in crystalloid ER 236, 237

I-cell disease 143, 149, 151, 152, 155
Immune complexes
HRP–IgG 211
pinocytosis of 46–49
receptor-mediated endocytosis and 81
Immunoelectron microscopy 133
Immunofluorescence in immune complex binding and pinocytosis 46, 47, 51
See also Fluorescence microscopy, Fluorescence photobleaching
Immunoglobulin
maternal, receptor-mediated endocytosis and 81
myeloma cells, production in 165, 167
Immunoglobulin A, transport in hepatocytes 114, 115
Immunoglobulin E, activation, receptor aggregation and 102
Immunoglobulin G 36, 37, 38, 39
–coated erythrocyte ghosts 44–46, 52
–ferritin 211, 217
lysosomal uptake 227, 228, 231
–horseradish peroxidase 211, 227, 229
immune complexes 42, 46, 48, 51
degradation 52, 53
internalized in coated vesicles 57
pinocytosis 49
intestinal cells absorbing 210–214
iodination 40, 42
luminal uptake, intestinal 211, 213
monoclonal C7 antibody 84, 86
receptors *See Fc receptors*
receptor-ligand complex 216, 224, 227
selective transport 216–219, 226, 227
transport in neonatal rat intestine 209–232
γ-immunoglobulin in placental coated vesicles 253, 255
Influenza virus
asymmetric budding on epithelial cells 186, 187
double infection with vesicular stomatitis virus in MDCK cells 188
HA protein of, distribution 190–200
Influenza A virus 59
binding of 72
cleavage of spike glycoproteins 70
endocytosis of 70
fusion with liposomes 70
fusogenic protein 75
Insulin
activation, receptor aggregation and 102
human placental coated vesicles, in 255
mutagenic effects 98
receptors 79
receptor-mediated endocytosis and 80
transport to Golgi complex 172
Intermediate vesicles 32
Intestinal cells
absorption of IgG 210–214
apical vacuoles 210, 224, 225, 229
coated pits and vesicles of 210, 211
IgG uptake and 213, 219, 230
internal sorting site 216–219, 222, 224, 225, 226, 227, 230
secondary lysosome precursors 221, 229
Intestine, neonatal, IgG transport in 209–232
Intracellular membrane transport 1–14
control of 8
endocytosis and 4, 5
exocytosis and 5, 6
Iodination
endocytic vesicle membrane, of 36, 37
Fc receptor, of 39
lactoperoxidase-mediated 22, 24, 32, 37, 40, 169, 206
membrane proteins, of 54, 55
myeloma cells, of 169, 170, 177
pinocytic vesicles, of 40, 42
phagolysosomes, of 24, 32

Lacrimal glands, exocrine cells of, traffic to Golgi complex 159, 175
Lactoperoxidase conjugates
low density lipoproteins 51
Semliki Forest virus 51
Lactoperoxidase–glucose oxidase iodination 40, 169
Lactoperoxidase–latex iodination 24, 32
Lactoperoxidase-mediated iodination 22, 37, 206
Latex particles 17, 18, 19
cell surface interactions 32, 33
opsonized 33
Latex phagolysosomes, interiorization, clathrin and 31
Latex phagosomes 19
Fc receptors and 30
Lipid bilayers 2, 3
membrane proteins in 4, 279
Lipids, membrane 2
See also Low density lipoproteins
Low density lipoproteins (LDL)
acetylated, degradation of 52, 53
See also Acetyl-LDL
binding and internalization 90, 267
crystalloid ER and 236
endocytosis of 266, 267
lactoperoxidase conjugates 51
placental coated vesicles, in 255

receptor system 82–95, 113
See also LDL receptors
transport to lysosome 267
uptake 28, 246
receptor-mediated endocytosis and 77–95
temperature coefficient for 29, 30
Low density lipoprotein receptors 11, 79
ASGP receptors and 113
binding
calcium dependency 87, 113
specificity 94
cell surface concentration 273
characteristics 83
coated pits and 57, 79, 94, 266, 271, 277, 278, 289
deficiency of, in familial hypercholesterolaemia 78, 83, 89
dissociation of 86, 87, 92
endocytic pathway 275
internalization time course 113, 117
internalization-defective allele 83, 93, 94
monoclonal antibodies to 84, 85, 86, 89, 90, 91
pH-dependent dissociation 91
temperature-dependent dissociation 86
mutants of 82, 83
pathway 233
polyclonal antibodies to 85, 86, 91, 94
purification 83
receptor-negative allele 83, 93
recycling of 84–86, 90, 267
Lysosomal acid phosphatase 152
Lysosomal enzymes
mannose 6-phosphate containing 81
receptor recycling in 84, 85
mannose residues of 138, 139, 140
mouse P388D$_1$ 141, 146, 147, 149, 155
phosphorylation 138–158
receptors 153
substrates, as 148, 149
uncovered 146, 151, 152, 153
See also Lysosomal acid phosphatase, Lysosomal hydrolases
Lysosomal hydrolases 5, 6, 64, 79, 81, 82, 231
Lysosomes 5, 6, 64, 79
degradation in 32, 33, 52, 53, 55, 64, 80, 81, 82, 98, 213, 227, 228
degradative hydrolytic enzymes of *See Lysosomal hydrolases*
fusion
pH-dependence of 23, 63, 64, 65, 66
temperature-dependence of 69, 70
with coated vesicles 243
with endosomes 64–67
with phagosomes 23, 36, 70
lysosomal enzyme delivery to 139, 146, 149, 150, 152, 154, 155
primary 16, 25, 67, 71, 74, 79
role in endocytic cycle 267
secondary 16, 17, 18, 19, 25, 26, 31, 32, 33
virus uptake by 61, 64–67, 69
Lysosome–Golgi membrane traffic 165, 177, 178
Lysosomotropic weak bases 61, 66, 69, 85, 108, 230

α_2-macroglobulin, receptor-mediated endocytosis and 78, 81
calcium dependency of binding 87
receptor recycling in 85
α-macroglobulin–protease–gold complexes, receptor-mediated endocytosis in, early events in 239–242
α-macroglobulin–protease receptors, aggregation 239, 241
α-macroglobulin–trypsin complex 241
Macrophage
coated vesicles of 19
cultures 16, 36
fluid-phase vesicles in 28
mouse P388D$_1$ lysosomal enzyme in 141, 146, 147, 149, 155
phagocytosis by 18, 19
pinocytosis by 18, 20
plasma membrane of 16, 17, 18, 19
secretory pathways in 180
vacuolar apparatus 15–18
Magnesium-ATP 283, 284
Mammary tumour virus, mouse, budding on epithelial cells 187
Mammotroph, pituitary, secretory cell pathways in 159–165, 175, 177
Mannose 6-phosphate 141, 152, 155
receptor 141, 150, 153, 155
Mannose-conjugated proteins, receptor recycling 85
Mannose conjugates, lysosomal degradation and 55, 56
Mannose residues of lysosomal enzymes 138, 139, 140, 141
Mannoses of protein oligosaccharides 121, 124, 131, 132
See also Oligosaccharides
α-1,2-mannosidase 121, 124, 130, 131, 132, 133, 153
Marker enzymes 2, 12, 21, 22
MDCK cells 185
double infection with viruses of opposite polarity 187, 188, 205

MDCK cells (*cont.*)
mixed virions in 199
protein and enzyme polarity in 185, 186, 187
β-melanotropin, receptor-mediated endocytosis and 80
Membrane
binding 8, 37, 38, 39
cellular 2
chemical specificity 3, 4, 8–10, 13
chemistry, stabilization of 3, 4, 9
interiorization of 18, 19, 29
lipid content 2
stabilizing infrastructures 3, 4, 9, 10, 11, 296
flow 2, 3, 269, 270, 274, 275
membrane protein diffusion and 273
specificity of 278
fusion in virus endocytosis 60, 63, 65
fusion–fission 2, 8, 13
intracellular 1, 2
holes, experimentally induced 283–288
perturbants, endocytosis and 19
proteins
endodomains of 296, 297
integral 3, 4, 11
orientation of amino acid sequences 280
randomization of 3, 4, 269, 290
transport 5
turnover 36, 39, 41, 42, 54, 55
See also Glycoproteins, Proteins, Polypeptides
receptors 8
recovery 158, 163, 170
recycling 2, 3, 4, 5, 6, 8, 12, 23–26, 35–58, 267
BHK-21 cells, in 63
experimental control 282–287
functional implications 295
glycosylation and 150
myeloma cells, in 165, 177
protein synthesis and 150
secretory cells, in 157–183
site of reincorporation 267, 269, 274, 279, 280
volume and kinetics 294, 295
signals 8
specificity, control of 296
transport, intracellular 1–14
turnover 283, 284
ATP and 287, 288
See also Plasma membrane
Methotrexate 234
Methylamine 61, 85, 99, 105
Micropinocytic vesicles 5
Monensin 61, 73, 85, 90, 108
concentration affecting activity 92, 93
LDL receptor recycling and 90
Monoclonal antibodies
Fc receptors, to 38, 39, 40, 91, 108
LDL receptors, to 84, 85, 86, 89, 90, 91
plasma membrane proteins, to 36, 37
Mucolipidoses 143
Myeloma cells
membrane recycling in 165, 177
secretory pathways in 165, 167, 169, 170, 177
membrane protein in 180

NADH-cytochrome b_5 reductase 131, 133, 291
Nerve growth factor (NGF)
antibodies, extracellular NGF and 105
hormone–receptor accumulation in cell nucleus 99
receptor affinity 100
receptor-mediated endocytosis and 80
Neuraminidase 70, 84, 113
Newcastle disease virus
budding on epithelial cells 187
synthesis and transfer 267
Nigericin 61, 85
Nuclear polyhedrosis baculoviruses 187
5′-nucleotidase
activity in macrophages 17, 18, 21
asialoglycoprotein receptors and 111, 112
recycling in fibroblasts, monensin and 92

Oligosaccharides
binding to receptors 154
G protein of 122, 123, 124
lysosomal enzymes of
hybrid types 149, 150, 154, 155
phosphorylation of 139, 140, 141, 142, 143, 144, 145
uncovered 151, 153
trimming of 122, 123, 124, 153
Opsonized erythrocytes
monoclonal antibodies to 39
receptor-induced phagocytosis of 33
Opsonized latex particles, interaction with Fc receptor 33
Orthomyxovirus, pathways of entry into cell 60

Palmitate, attached to receptors 118
Pancreatic β-cell, secretory pathways in 171
Parotid gland, exocrine cells of, secretory pathways in 159
Peripheral intermediate compartment of hepatocytes 110, 111, 112, 113, 114
Phagocytic vacuoles 4

Phagocytic vesicles 20
  plasma membrane proteins in 44
Phagocytosis 18, 19, 36
  coat formation in 44
  IgG-coated erythrocyte ghosts, of 44–46
  pH and 55
Phagolysosome 21
  Fc receptor degradation in 45, 52, 53, 55, 56
  fusion mechanisms 23
  intralysosomal pH and 23
  lactoperoxidase–latex and 32
  latex, interiorization of
    clathrin and 31
    plasma membrane and 21
  membrane
    Fc receptor on 33
    nature of 21, 22, 24
Phagosomes 19, 20
  flow and direction 22
  formation 21
  fusion with lysosomes 21, 23
  peptides of 40
Phorbol myristate acetate 100
Phosphate uncovering 141, 146, 151, 152, 153, 154
Phosphorylating enzymes 138–156
  defective 143, 151
Pinocytic vacuoles 4
Pinocytic vesicles (pinosomes) 16, 17, 18, 19, 35
  flow and direction 22, 25, 26
  generation of 20, 21
  iodinated 40–43
  isolation of 39, 40
  membrane
    composition of 22, 40, 44
  turnover of proteins 41, 42
  selective labelling of 40
  uncoated 29
Pinocytosis, fluid-phase 18, 19, 20, 30, 31, 36, 73
  BHK-21 cells, in 63
  coated vesicles and 43, 44
  fibroblasts, in 29
  IgG immune complexes, of 49
  pH and 55
  phagocytosis and 18
  receptor-mediated
    internalization of Fc receptor in 46–49
    iodination studies 40, 42
  soluble immune complexes of Fc receptors and 46–49
  Semliki Forest virus particles in study of 61
  temperature and 18, 23, 28, 29
Pinosomes *See Pinocytic vesicles*
Pituitary cells, anterior, secretory pathways in 158, 159–165
Placenta, human, coated vesicles of 252–255
Plasma membrane 16, 17, 18, 19, 78, 79, 84
    See also plasmalemma
  centrifugal flow 32
  composition 36
  ER membrane proteins directed to 126
  fusion of 25, 36
  internalization 18, 19, 21, 29, 32, 35, 39
  phagolysosomes and 21, 22, 45
  polypeptides of 21
    bidirectional flow 23–25
    monoclonal antibodies to 36, 37, 40
  proteins of 36, 40–43
    stability 31
    turnover 36, 39, 41, 42, 54, 55
  protein:lipid ratio 41
  receptors 79
    in receptor-mediated endocytosis 78
  recycling process 23–26, 32, 267
  site of insertion of newly synthesized proteins 267
  vesicular flow 22, 25, 26, 36
  vesicular specificity 36
Plasmalemma 3, 5, 6, 12 *See also plasma membrane*
  cleansing by constitutive endocytosis 295
  microdomains in vascular endothelial cells 10
  receptors in 94
  stabilizing infrastructure 10
  vesicles 4, 5, 10
    recycling of 294
    uncoated 29
  vesicular transport and 5, 6, 9
Plasmalemma–Golgi membrane traffic 158, 165, 167, 169, 170, 177
  hormone transport and 172
  implications of 171, 172
  secretion stimulating 175
  surface membrane component 'repair' 172
Plasmalemma–lysosome membrane traffic 165
Polyclonal antibodies to LDL receptor 85, 86, 91, 94
Polypeptides
  of pinocytic vesicles 40
  plasma membrane 21
    bidirectional flow 23–25
    identification on Fc receptors 39
    monoclonal antibodies to 36, 37, 38, 39, 40
Polystyrene particles *See Latex particles*

Prelysosomal compartments of
hepatocytes 110, 111, 114
ASOR receptors in 112, 113
Prolactin 159, 161
receptor-mediated endocytosis and 81
transport into Golgi complex 172
Protein(s)
extracellular, receptor-mediated
endocytosis of 78, 80, 81
:lipid ratio in molecular exclusion 279
sorting of 120, 121, 125–128, 130, 131, 136
synthesis
intracellular 6, 7
membrane recycling and 150
transmembrane, acylation of 118
Proton pump 54, 64, 69, 74, 135
Pseudo-Hurler polydystrophy 143, 151
Pseudomonas toxin, receptor-mediated
endocytosis and 81
Pseudotype viruses 198, 199

Radioiodination *See Iodination*
Receptor(s)
cell-surface for viruses 60, 61, 72
coated pits, of 94, 267, 278, 296
dissociation of 86, 87
IgG *See Fc receptors*
lateral diffusion of 79, 87, 90, 97
location of 11
newly synthesized 274
non-recycling 91
palmitate and 118
recognition in coated pits 86, 87, 296
recycling of 79, 84–86, 87, 91, 274
See also ASGP receptors, EGF receptors, LDL receptors, Transferrin receptors, etc
Receptor–ligand complex, transfer to coated
pits 79
Receptor-mediated endocytosis 9, 11, 36,
40, 72, 209
asialoglycoprotein, of, in
hepatocytes 109–115
coated pits and 78, 79
EGF, of 98, 99
fluid *See Receptor-mediated pinocytosis*
functions of 77, 78
general characteristics 78
LDL receptor system and 82–95
pH controlling 55
proteins undergoing 80, 81
sequential steps 79, 82
toxins undergoing 81
viruses undergoing 81
See also Endocytosis, Pinocytosis
Receptor-mediated pinocytosis 39–44
immune complexes, of 46, 53, 54
intracellular iodination in 42
See also Endocytosis, Pinocytosis, Receptor-mediated endocytosis
Receptosomes 12, 79, 98
Rhabdovirus, pathways of entry into cell 60
Ribophorin glycoproteins, purified in ER or
Golgi 131
Ricin toxin, receptor-mediated endocytosis
and 81

Secretion granules 158, 161, 163, 177
membrane recycling and 163, 179
Secretion vacuoles, discharging 4, 5
Secretory cells
hybrid oligosaccharides in 154, 155
membrane recycling in 157–183
Secretory pathway within Golgi
apparatus 180, 181
Semliki Forest virus
binding of 60, 61, 72
degradation of 64, 65
endocytosis of 59–76
internalization of 61–64
–lactoperoxidase conjugate 51
lipid bilayer, protein content 279
membrane fusion
intracellular 66, 67, 69, 231
liposomes, with 70, 71
pH-dependent 65, 66, 69, 92
spike glycoproteins and 59, 60, 71
penetration into cell 64, 65, 66, 67, 70, 71,
74, 75
pinocytosis studied by 61
receptors for 72
receptor-mediated endocytosis and 81
spike glycoproteins of 59, 60, 71
extracellular occupancy 279
intracellular distribution 192
uptake 30, 44, 61, 63, 64, 65, 74, 75, 246
Sendai virus
asymmetric budding on epithelial cells 186
fusion
with liposomes 70, 71
with plasma membrane 69
Short-circuiting *See Diacytosis*
Sialic acid 125
Sialyltransferase 124, 131
Simian virus 5 (SV5)
asymmetric budding in epithelial cells
186
–vesicular stomatitis virus double infection
of MDCK cells 188
Sodium dodecyl sulphate (SDS) 21, 33, 55,
116, 248, 253, 254, 264
Sodium–potassium pumps 54

Spectrins in formation of stabilizing infrastructures 3
Spike glycoproteins
of influenza virus 70
of Semliki Forest virus, membrane fusion and 59, 60, 71
Substrate attachment sites 270, 274
Sucrose uptake, temperature and 29
Suramin 83

Thiamine pyrophosphatase 124, 181
Thy 1 antigen, exclusion from coated pits 267–269, 277, 278
Thyroid epithelial cell, secretory pathways in 170
Toga viruses, pathways of entry into cell 60
See also Semliki Forest virus
Toxins
receptor-mediated endocytosis and 81
uptake 72
Transcobalamin II, receptor-mediated endocytosis and 80
Transcytosis 6, 10
Transferrin
placental coated vesicles, in 253, 255
receptor-mediated endocytosis and 80
Transferrin receptors
concentration 273
endocytic pathway 275
fatty acids of 117, 118
placental coated vesicles, in 255
recycling of, in HeLa cells 115–118, 273
Trifluoperazine dihydrochloride inhibiting LDL binding and internalization 90
Triskelion, clathrin 250, 251, 252, 254
Tunicamycin
lysosomal enzyme secretion and 143, 152
viral budding and 204, 205
Tyrosine phosphorylation
EGF analogue producing 102
EGF receptor antibody producing 107

Ultra-thin frozen sections of MDCK cells 190–200

Vanadate inhibiting exoplasmic flow 284, 287
Vesicles
endocytic *See Endocytic vesicles*
epithelial, double viral infection and 205
granulosa cell, intracellular 241
hepatic, intermediate 114
intestinal 210, 211, 213, 216, 217, 219, 221, 222, 224, 225, 229
of macrophages 16, 20, 21
flow and directionality 22, 25, 36
fusion 23, 36
intermediate 32
Vesicular carriers 5, 6, 294
Vesicular stomatitis virus (VSV) 60, 121, 122, 130
acetylated proteins of 118
glycoprotein distribution 190–200, 203
influenza virus and, in double infection of MDCK cells 188
intracellular transport 186
maturation and secretion, cytoskeletal drugs and 179
receptors for 73
SV5 virus and, in double infection of MDCK cells 188
transfer after synthesis 271
tunicamycin and 153
Vesicular transport 3, 4–14, 293
compartmental operation of 6, 7
control of 7–12, 178, 296, 297
sorting of products and membranes and 295
viral glycoproteins, of 203, 204
volume and kinetics 294, 295
Viral-envelope glycoproteins
intracellular pathways of 190–200, 203–207
polarity of 185, 186, 187, 203, 207
synthesis and processing 185, 186, 187
tight junctions and 199, 202, 203, 204
Viral neuraminidase 70
Virion
budding 4, 10, 121
enveloped, stabilizing infrastructure and 3, 4, 10
Viropexis 60
Virus(es)
binding of 60, 61, 66, 72
budding
double infection experiments and 188, 207
polarity of 186, 187
tunicamycin ineffective in 204, 205
entry into cell 60
enveloped, endocytosis of 59–76
internalization of 60, 64
membrane fusion
permeability and 71
pH and 65, 66
proteins, transfer after synthesis
Newcastle disease virus 267
vesicular stomatitis virus 271
receptor-mediated endocytosis and 81
receptors 79
spike proteins 59, 60, 70, 71
extracellular surface occupancy 279

Virus(es) (*cont.*)
intracellular distribution 192
transport
lysosome, to 73
surface membrane 55, 69, 73
uptake
by cell membrane 55, 61, 63
fluid-phase pinocytosis and 63

Yeast cell mutants 155
Yolk protein, receptor-mediated endocytosis and 80